Porous Materials

Inorganic Materials Series

Editors:

Professor Duncan W. Bruce
Department of Chemistry, University of York, UK

Professor Dermot O'Hare
Chemistry Research Laboratory, University of Oxford, UK

Professor Richard I. Walton
Department of Chemistry, University of Warwick, UK

Series Titles

Functional Oxides
Molecular Materials
Low-Dimensional Solids
Porous Materials
Energy Materials

Porous Materials

Edited by

Duncan W. Bruce
University of York, UK

Dermot O'Hare
University of Oxford, UK

Richard I. Walton
University of Warwick, UK

WILEY

A John Wiley and Sons, Ltd, Publication

This edition first published 2011
© 2011 John Wiley & Sons, Ltd

Registered office
John Wiley & Sons Ltd, The Atrium, Southern Gate, Chichester, West Sussex, PO19 8SQ, United Kingdom

For details of our global editorial offices, for customer services and for information about how to apply for permission to reuse the copyright material in this book please see our website at www.wiley.com.

Library of Congress Cataloging-in-Publication Data

Porous materials / edited by Duncan W. Bruce, Dermot O'Hare, Richard I. Walton.
   p. cm. — (Inorganic materials series)
   Includes bibliographical references and index.
   ISBN 978-0-470-99749-9 (cloth)
   1. Porous materials. I. Bruce, Duncan W. II. O'Hare, Dermot. III. Walton, Richard I. TA418.9.P6P667 2010
   620.1′16—dc22

                                    2010026282

A catalogue record for this book is available from the British Library.

Print ISBN: 978-0-470-99749-9 (Cloth)
e-book ISBN: 978-0-470-71137-8
o-book ISBN: 978-0-470-71138-5

Set in 10.5/13 Sabon by Integra Software Services Pvt. Ltd, Pondicherry, India.

Contents

Inorganic Materials Series Preface

Back in 1992, two of us (DWB and DO'H) edited the first edition of *Inorganic Materials* in response to the growing emphasis and interest in materials chemistry. The second edition, which contained updated chapters, appeared in 1996 and was reprinted in paperback. The aim had always been to provide the reader with chapters that while not necessarily comprehensive, nonetheless gave a first-rate and well-referenced introduction to the subject for the first-time reader. As such, the target audience was from first-year postgraduate student upwards. Authors were carefully selected who were experts in their field and actively researching their topic, so were able to provide an up-to-date review of key aspects of a particular subject, whilst providing some historical perspective. In these two editions, we believe our authors achieved this admirably.

In the intervening years, materials chemistry has grown hugely and now finds itself central to many of the major challenges that face global society. We felt, therefore, that there was a need for more extensive coverage of the area and so Richard Walton joined the team and, with Wiley, we set about a new and larger project. The *Inorganic Materials Series* is the result and our aim is to provide chapters with a similar pedagogical flavour but now with much wider subject coverage. As such, the work will be contained in several themed volumes. Many of the early volumes concentrate on materials derived from continuous inorganic solids, but later volumes will also emphasise molecular and soft matter systems as we aim for a much more comprehensive coverage of the area than was possible with *Inorganic Materials*.

We approached a completely new set of authors for the new project with the same philosophy in choosing actively researching experts, but also with the aim of providing an international perspective, so to reflect the diversity and interdisciplinarity of the now very broad area of inorganic materials chemistry. We are delighted with the calibre of authors who have agreed to write for us and we thank them all for their efforts and cooperation. We believe they have done a splendid job and that their work will make these volumes a valuable reference and teaching resource.

DWB, York
DO'H, Oxford
RIW, Warwick
July 2010

Preface

Porosity in the solid-state is a topic of long-standing attention in materials science and the case of the zeolites exemplifies the importance of porous materials across many disciplines of science. Here, the study of naturally occurring silicate minerals led to the discovery of synthetic analogues in the laboratory that now have huge commercial value, ranging from large-scale industrial petroleum cracking catalysts to household applications in water-softening additives in detergents. This is an important example of how curiosity-driven, fundamental research in complex inorganic structures, and how they might be assembled in a controlled way, ultimately can lead to novel materials with societal benefit.

The field of porous materials has, however, undergone dramatic development in the past few decades, particularly with the increasingly routine use of advanced structural probes for studying the structure and dynamics of the solid state. Porosity in inorganic materials now extends from the nanoscale up to the macroscale and is a highly researched area, particularly since the idea of design in synthesis is being realised by control of solution chemistry in the crystallisation of complex extended structures. Properties are increasingly the goal in this field: novel solid hosts for confinement of matter on the nanoscale, highly specific shape selective catalysts for energy-efficient organic transformations, new media for pollutant removal, and gas storage materials for energy applications. The role of the synthetic chemist remains key to the discovery and classification of porous materials; the fact that novel porous materials are still reported at an increasing rate in the chemical literature demonstrates the vitality of the field.

The five chapters in this volume cover some of the key families of inorganic solids that are currently being studied for their porosity. The area of zeolites is still researched heavily since there remain long-standing questions in understanding crystallisation and the extent to which novel materials, with structure and chemical properties are tuned for particular applications, can be produced. The chapter on zeolite chemistry illustrates this and takes a novel angle, describing how the synthesis of porous materials can be inspired by nature. Various other important families are covered representing the scales of porosity from nanoporous through mesoporous, and also showing how various chemical classes of material can be rendered porous by elegant synthetic approaches.

We are very pleased that well-respected authors who are active in research in this important area agreed to prepare chapters for this volume and thank them for their excellent results. We hope that this collection will provide a useful and up-to-date introduction to an area of abiding interest in materials chemistry.

DWB, York
DO'H, Oxford
RIW, Warwick
July 2010

List of Contributors

Miguel Camblor   Instituto de Ciencia de Materiales de Madrid (CSIC), Madrid, Spain

Karen J. Edler   Department of Chemistry, University of Bath, UK

Suk Bong Hong   School of Environmental Science and Engineering, Pohang University of Science and Technology (POSTECH), Pohang, Korea

Cameron J. Kepert   School of Chemistry, University of Sydney, NSW, Australia

Robert Mokaya   School of Chemistry, University Park, University of Nottingham, Nottingham, UK

Masahiro Sadakane   Graduate School of Engineering, Hiroshima University, Higashi-Hiroshima, Japan

Wataru Ueda   Catalysis Research Center, Hokkaido University, Sapporo, Japan

Yongde Xia   School of Chemistry, University Park, University of Nottingham, Nottingham, UK

Zhuxian Yang   School of Chemistry, University Park, University of Nottingham, Nottingham, UK

1

Metal-Organic Framework Materials

Cameron J. Kepert

School of Chemistry, The University of Sydney, Sydney NSW, Australia

1.1 INTRODUCTION

In recent years there has been a rapid growth in the appreciation of molecular materials not just as arrangements of discrete molecular entities, but as infinite lattices capable of interesting cooperative effects. This development has arisen on many fronts and has seen the emergence of chemical and physical properties more commonly associated with non-molecular solids such as porosity, magnetism, and electrical conductivity. This chapter focuses on an area of molecular materials chemistry that has seen an extraordinarily rapid recent advance, namely, that of metal-organic frameworks (MOFs).[†] These materials consist of the linkage of metal ions or metal ion clusters through coordinative bridges to form

---

[†] Whilst certain qualifications on the use of the term 'metal-organic framework' have been put forward (*e.g.*, relating to formal bond valence and energy, ligand type, *etc.*),[3] the common usage of this term has spread well beyond these to become largely interchangeable with a number of more general terms such as 'coordination polymer', 'coordination framework', 'metallosupramolecular network' and 'hybrid material'. As such, this term is used here, with some reluctance, in its broadest general sense to encompass a very diverse range of material types in which metal atoms are linked by molecular or ionic ligands.

Porous Materials   Edited by Duncan W. Bruce, Dermot O'Hare and Richard I. Walton
© 2011 John Wiley & Sons, Ltd.

frameworks that may be one-dimensional (1D), two-dimensional (2D) or three-dimensional (3D) in their connectivity.[1–14]

In the broadest sense, the use of coordination chemistry to produce framework materials has been with us since the discovery of Prussian Blue more than 300 years ago, with developments throughout the last century providing an array of framework lattices spanning a range of different ligand types.[15, 16] The rapid expansion of this early work into more structurally sophisticated families of materials can be traced to two developments. First, the exploitation of the strong directionality of coordination bonding has allowed a degree of materials design (so-called 'crystal engineering') in the synthesis of framework phases. Here, the use of molecular chemistry has allowed both the rational assembly of certain framework topologies – many not otherwise accessible in the solid state – and the control over framework composition through the incorporation of specific building units in synthesis or through post-synthetic modification. Secondly, the capability to construct materials in a largely predictive fashion has led to the emergence of a range of new properties for these materials. This most notably includes porosity, as seen in the ability to support extensive void micropore volume, to display high degrees of selectivity and reversibility in adsorption/desorption and guest-exchange, and to possess heterogeneous catalytic activity. A range of other interesting functionalities have also emerged, many in combination with reversible host–guest capabilities. A particularly attractive feature of the metal-organic approach to framework formation is the versatility of the molecular 'tool-box', which allows intricate control over both structure and function through the engineering of building units prior to and following their assembly. The adoption of this approach has been inspired in part by Nature's sophisticated use of molecular architectures to achieve specific function, spanning host–guest (*e.g.* ion pumping, enzyme catalysis, oxygen transport), mechanical (*e.g.* muscle action), and electronic (*e.g.* photochemical, electron transport) processes. Following rapid recent developments the immensely rich potential of MOFs as functional solids is now well recognised.

At the time of writing this field is experiencing an unprecedented rate of both activity and expansion, with several papers published per day and a doubling in activity occurring every *ca* 5 years. Faced with this enormous breadth of research, much of which is in its very early stages, the aim of this chapter is not to provide an exhaustive account of any one aspect of the chemistry of MOFs, rather, to provide a perspective of recent developments through the description of specific representative examples, including from areas yet to achieve maturity. Following a broad overview of the host–guest chemistry of these materials in Section 1.2, particular focus is given to the incorporation of magnetic, electronic, optical, and mechanical properties in Section 1.3.

1.2 POROSITY

1.2.1 Framework Structures and Properties

1.2.1.1 Design Principles

1.2.1.1.1 Background
The investigation of host–guest chemistry in molecular lattices has a long history. Following early demonstrations of guest inclusion in various classes of molecular solids (*e.g.* the discovery of gas hydrates by Davy in 1810), major advances came in the mid twentieth century with the first structural rationalisations of host–guest properties against detailed crystallographic knowledge. Among early classes of molecular inclusion compounds to be investigated for their reversible guest-exchange properties were discrete systems such as the Werner clathrates and various organic clathrates (*e.g.* hydroquinone, urea, Dianin's compound, *etc.*), in which the host lattices are held together by intermolecular interactions such as hydrogen bonds, and a number of framework systems (*e.g.* Hofmann clathrates and the Prussian Blue family), in which the host lattices are constructed using coordination bonding.[15, 16] A notable outcome from this early work was that the host–guest chemistry of discrete systems is often highly variable due to the guest-induced rearrangement of host structure, and that the coordinatively linked systems – in particular those with higher framework dimensionalities – generally display superior host–guest properties with comparatively higher chemical and thermal stabilities on account of their higher lattice binding energies.

Whilst the excellent host–guest capabilities of coordinatively bonded frameworks have been appreciated for many decades, the extension of this strategy to a broad range of metals, metalloligands and organic ligands has been a relatively recent development. Concerted efforts in this area commenced in the 1990s following the delineation of broad design principles[1] and the demonstration of selective guest adsorption;[17] notably, these developments arose in parallel with the use of coordination bonds to form discrete metallosupramolecular host–guest systems.[18] A number of different families of coordinatively bridged material have since been developed, each exploiting the many attractive features conferred by the coordination bond approach. A consequence of this rapid expansion is that many inconsistencies have arisen in the terminology used to distinguish these various families. In this chapter, the broadest and arguably most fundamental distinction, *i.e.* the exploitation of coordination bonding to form frameworks consisting of metal ions and molecular or ionic ligands, is used to define this diverse class of materials.

1.2.1.1.2   MOF Synthesis

In comparing MOFs with other classes of porous solids many interesting similarities and points of distinction emerge. A comparison has already been made above with discrete inclusion compounds, for which it was noted that coordinative rather than intermolecular linkage confers a high degree of control over materials' structure and properties, whilst retaining the benefits associated with the versatility of molecular building units. At the other end of the spectrum, an equally useful comparison can be made with other porous framework materials, which notably include zeolites and their analogues (*e.g.* AlPOs). Here, some close parallels exist between the structural behaviours of the host lattices, but many important differences exist relating to synthesis, structure and properties. One principal point of distinction is that the building units of MOFs are commonly pre-synthesised to a high degree. This allows the achievement of specific chemical and physical properties through a highly strategic multi-step synthesis in which the comparatively complex structure and function of the molecular units are retained in the framework solid. This ability to retain the structural complexity of the covalent precursors is a direct result of the low temperature synthesis of MOFs (*i.e.* typically <100 °C, with the majority able to be performed at room temperature), which in turn may be attributed to the favourable kinetics of framework formation; whereas the synthesis of more conventional porous framework solids commonly requires high temperatures, the labile nature of the metal-ligand bond in solution means that MOF assembly with error-correction can occur at low temperatures and over nongeological time periods to produce highly crystalline, ordered structures. As such, whereas the achievement of structural metastability and complexity in zeolites is generally achieved through control of the kinetics of framework formation or through framework templation and subsequent calcination, for MOFs a high degree of complexity is intrinsic to the molecular building units and can thus be achieved to a large extent through thermodynamic control.

There are two important further consequences of the low temperature route to framework formation. First, the entropic penalty associated with the entrapment of solvent in channels and pores is less pronounced than for higher temperature synthetic routes. Secondly, and conversely, the enthalpic favourability of regular bond formation is a dominant driving force for framework formation. Through exploitation of the highly directional nature of coordination bonding, a reasonable degree of control over the structure of MOF lattices can thus be achieved. Extensive efforts in the use of well defined coordination geometries and suitably regular ligands have led to the development of relatively sophisticated 'crystal engineering' principles, albeit with absolute control over polymorphism in many cases being subject to the whims of crystal nucleation and subtle sensitivities to temperature, solvent, etc.

Among a range of useful design principles for MOFs are the 'node and spacer'[19, 20] and reticular 'secondary building unit' (SBU)[21, 22] approaches. Common to each of these is the concept of using multitopic ligands of specific geometry to link metal ions or metal ion clusters with specific coordination preferences. Using these approaches it is possible to distill framework formation to the generation of networks of varying topology‡[23–28] with the geometry of these being determined in large part by the geometry of the molecular building units (see Figure 1.1). In many cases the geometry of the building units defines a single possible network topology if fully bonded; for example, the use of octahedral nodes and equal-length linear linkers generates the cubic α-Po network [see Figure 1.1(i)]. In many cases, however, only the dimensionality of the resulting framework can be predicted with any reasonable degree of certainty, with very low energy differences arising due to torsional effects, intraframework interactions or subtle geometric distortions; for example, the use of tetrahedral nodes and linear linkers can generate a range of 3D 4-connected nets that include cristobalite [diamondoid; Figure 1.1(f)], tridymite (lonsdaleite), and quartz. In many further cases still, even the prediction of network dimensionality is not straightforward; for example, square nodes and linear linkers can produce a 2D square grid and a 3D NbO-type net [Figure 1.1(e) and (h)], and triangular nodes and linear linkers can produce a wide range of nets that vary only in their torsional angles through the linear linkers, e.g., 0° torsion produces the hexagonal (6,3)

---

‡A large number of different chemical classification systems exist for network topologies. These notably include those based on simple chemical compounds (e.g. diamondoid/cristobalite-type), an (n,p) system used by Wells related to that of Schläfli that classifies according to the number of links in a loop (n) and the node connectivity (p) [e.g. (6,4)],[23] a three-letter system derived from that used for zeolites (e.g. **dia,dia-a,dia-b**,etc.),[24] and a 2D hyperbolic approach (e.g. sqc6).[25] As an example, the chiral (10,3)-a network is known also as the $SrSi_2$ net, **srs**, Laves net, Y^*, 3/10/c1, K4 crystal, and labyrinth graph of the G surface.

net, 109.5° torsion produces the chiral (10,3)-a net [see Figure 1.1(a–c)], *etc*. A further point of considerable complication from a design perspective is the interpenetration of networks,[27] which has a profound influence over the pore structure and therefore host–guest properties.

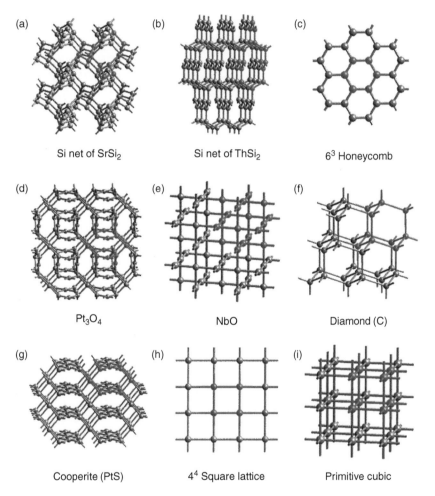

Figure 1.1 A selection of common network topologies for MOFs: (a) the 3-connected $SrSi_2$ [also (10,3)-a] net, shown distorted away from its highest symmetry; (b) the 3-connected $ThSi_2$ net; (c) the 2D hexagonal grid; (d) the Pt_3O_4 net, which contains square planar and trigonal nodes; (e) the NbO net, which contains square planar nodes; (f) the diamondoid net; (g) the PtS net, which contains tetrahedral and square planar nodes; (h) the 2D square grid; and (i) the α-Po net. Reprinted with permission from M. Eddaoudi, D.B. Moler, H.L. Li, B.L. Chen, T.M. Reineke, M. O'Keeffe and O.M. Yaghi, *Acc. Chem. Res.*, **34**, 319. Copyright (2001) American Chemical Society

An important consequence of both the versatility of the molecular build-ing units and the accessibility of novel framework topologies is that MOFs can readily be synthesised that are both chiral and porous. Efforts in this area have seen the emergence of the first homochiral crystalline porous materials through two primary routes (see also Sections 1.2.3.2 and 1.2.4.2): (1) the use of chiral ligands to bridge metal ions within network topologies that would otherwise be achiral,[29–39] as first seen in the use of a pyridine-functionalised tartrate-based ligand to form the porous homochiral 2D layered framework POST-1, which consists of honeycomb-type Zn^{II}-based layers;[29] and (2) the use of chiral co-ligands to direct the assembly of achiral building units into chiral framework topologies,[40–43] as first seen in the homochiral synthesis of an interpenetrated (10,3)-a network phase.[42]

In exploiting the favourable thermodynamics and kinetics of MOF crystal growth, very large pores of uniform dimension and surface chemistry are commonly achieved that would be inaccessible by other chemical routes.[44, 45] For example, whereas the synthesis of mesoporous silicates *(i.e.* those with pore dimensions in the range 20–500 Å) generally requires surfactant templation and calcination to leave behind amorphous hosts with regular mesopores,[46] crystalline MOFs with pores up to 47 Å in dimension[47] have been synthesised by the assembly of molecular building units from solution. In addition to favouring the formation of complex mesoscale architectures, the strength and directionality of the coordination bond also imparts a relatively high degree of stability to these. This is seen, for example, in their reasonably high thermal (up to $\sim$500 °C in some cases) and chemical stabilities (albeit with susceptibility to strongly coordinating guests such as water being common), extremely low solubilities, and robustness to guest desorption (see Section 1.2.1.2). Achievement of the latter feature, which is most common in higher dimensionality (*i.e.* 2D and 3D) framework sys-tems, has led to this field providing the most porous crystalline compounds known, with void volumes occupying as much as $\sim$90 % of the crystal volume. The achievement of such low volumetric atom densities through the use of moderately light elements means that the gravimetric measures of porosity and surface area are also extremely high. Among a number of notable families of highly porous MOFs are members of the MOF/IRMOF family (see Figure 1.2),[22, 48–51] MIL-*nnn* (in particular *nnn* = 100, 101),[52, 53] ZIF-*nnn* (in particular *nnn* = 95, 100)[54, 55] and NOTT-*nnn* series (in particular *nnn* = 100–109),[56, 57] which provide some of the most extreme measures of porosity and surface area yet achieved: *e.g.* among these ZIF-100 (see Section 1.2.3.1.1 and Figure 1.12) and MIL-101 have the largest pores, of dimension 35.6 and 34 Å, respectively; and MOF-177 and MIL-101 have Langmuir surface areas of 5640[58] and

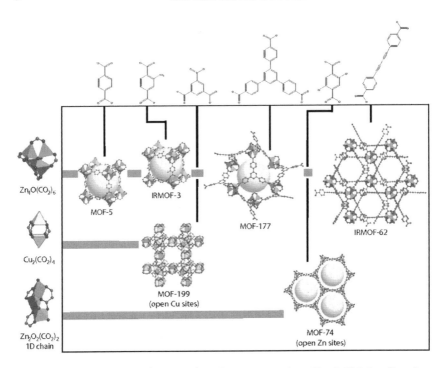

Figure 1.2 A selection of MOFs based on tetranuclear $Zn_4O(CO_2)_6$, dinuclear $Cu_2(CO_2)_4$ and 1D $Zn_2O_2(CO_2)_2$ secondary building units (left) and a range of multitopic carboxylate ligands (top). Reprinted with permission from D. Britt, D. Tranchemontagne and O.M. Yaghi, *Proc. Natl. Acad. Sci. U.S.A*, **105**, 11623. Copyright (2008) National Academy of Sciences

$5\,500\,m^2\,g^{-1}$,[53] each more than double that of porous carbon, and gravimetric pore volumes of 1.69[58] and 1.9 $cm^3\,g^{-1}$,[53] respectively.

A further distinguishing feature of MOFs over other classes of porous materials is the extreme diversity of their surface chemistry, which can range from aromatic to highly ionic depending on the chemical nature of the building units used. This notably includes the achievement of multiple pore environments within individual materials.[31] An important consequence of this versatility is that the surface chemistry can be tuned for highly specific molecular recognition and catalytic processes (see Sections 1.2.2, 1.2.3 and 1.2.4).

1.2.1.1.3 Post-Synthetic Modification of MOFs
In addition to the high degree of control over framework structure that can be achieved prior to and during MOF synthesis, considerable control can be exercised following framework assembly by exploiting the porosity of MOFs.[1] Developments here have seen the emergence of a range of

post-synthetic approaches in which framework structure and pore chemistry are modified *via* low energy chemical pathways that involve the internal migration of guest species. These processes occur topotactically, *i.e.* with some retention of the parent structure, to generate metastable phases that are commonly inaccessible through 'one-pot' syntheses.[59]

The simplest and most common form of post-synthetic modification is the desorption of guest molecules. This process, which in some cases is achieved most optimally at low temperature in multiple low-energy steps (*e.g.* through activation by volatile solvents[60] or supercriticial CO_2[61]), commonly leads to apohost phases that are structurally stable despite having very high surface energies. This is particularly so in cases where guest desorption leaves behind bare metal sites (see Sections 1.2.2 and 1.2.4), an example being the desorption of bound water molecules from the $Cu_2(CO_2)_4(H_2O)_2$ 'paddlewheel' nodes within $[Cu_3(btc)_2(H_2O)_3]$ (HKUST-1,[62] also MOF-199; btc = 1,3,5-benzenetricarboxylate) (see Figure 1.3). Guest desorption influences the host–guest properties of the framework in two ways. First, in generating a large unbound surface it allows the subsequent adsorption and surface interaction of guest molecules that would not otherwise have displaced those present at the surface following MOF synthesis (*e.g.* gases, aromatics into polar frameworks). Secondly, the modification of pore contents can have a pronounced influence on framework and pore geometry, thereby greatly modifying the adsorption properties of the host (see Sections 1.2.1.2 and 1.2.3.1.2).

The exchange of guest species can also dramatically influence host framework properties. This is particularly the case for the exchange of ions within charged frameworks – a process that can change both the

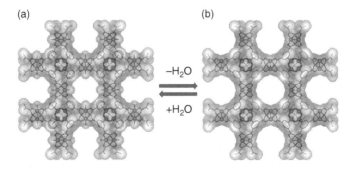

(a)   (b)

$-H_2O$
$+H_2O$

Figure 1.3 Reversible desorption of bound water molecules from the $Cu_2(CO_2)_4(H_2O)_2$ nodes within $[Cu_3(btc)_2(H_2O)_3]$ (a) to produce $[Cu_3(btc)_2]$ (b). This process occurs following the desorption of unbound guests (not shown). Cu atoms are drawn as spheres and a transparent van der Waals surface is shown

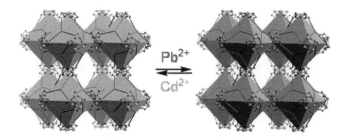

Figure 1.4 Reversible exchange of framework metal ions within $Cd_{1.5}(H_3O)_3$ $[(Cd_4O)_3(hett)_8]$ *via* a single-crystal-to-single-crystal process. Reprinted with permission from S. Das, H. Kim and K. Kim, *J. Am. Chem. Soc.*, **131**, 3814. Copyright (2009) American Chemical Society

relative polarity of the framework surface and the framework geometry. In contrast to zeolites, which in consisting of anionic frameworks generally only display cation exchange, MOFs can undergo both cation[63–65] and anion[1, 66, 67] exchange depending on their framework charges. Whilst such processes commonly involve the exchange of labile ions within the pores, the former notably also includes the reversible exchange of metal nodes from within the framework itself, as has been seen with the replacement of Cd^{II} within $Cd_{1.5}(H_3O)_3[(Cd_4O)_3(hett)_8]$ (where hett is an ethyl-substituted truxene tricarboxylate) by Pb^{II} (see Figure 1.4);[68] in contrast to the analogous dealuminisation process in zeolites, which requires multiple steps under extreme thermal and chemical conditions, this exchange process occurs at ambient temperature. Notably, the development of ion-exchange capabilities in MOFs has numerous other points of significance, for example in the development of proton conducting frameworks.[69, 70]

The incorporation of metal sites and other charged species into the pores of MOFs is in many cases driven by the energetics of complexation at the framework surface. Such a process may occur either through cation/anion exchange or salt inclusion. The former has been achieved, for example, with the exchange of protons with titanium(IV) di-isopropoxide at chiral BINOL units (BINOL = 1,1′-di-2-naphthol) to generate materials that display enantioselective catalytic activity.[35, 71] The latter may involve either the complexation of metal ions at binding sites on the framework surface with concomitant inclusion of charge-balancing anions, or cation/anion complexation at bare surface metal sites with concomitant inclusion of metal complex anions/cations into the pores.[72] The complexation of neutral metal species has also been used to modify pore chemistry, as seen with the reaction of MOF-5 with $Cr(CO)_6$ to form $[Zn_4O((\eta_6$-1,4-benzenedicarboxylate)$Cr(CO)_3)_3]$, in which the aromatic linkers now take the form of the organometallic $Cr(benzene)(CO)_3$ piano-stool complex.[73]

A further strategy for framework modification involves electron transfer between host and guest, a process that in principle provides amongst the strongest of enthalpic driving forces for the inclusion (or removal) of cations or anions and for the modification of framework properties. Redox activity at both the metal and ligand sites within the framework has been achieved. An example of the former is the oxidation of $[Ni^{II}_6(C_{26}H_{52}N_{10})_3(btc)_4] \cdot n($-guest) (BOF-1; btc = 1,3,5-benzenetricarboxylate) by I_2, in which oxidation of some of the Ni^{II} sites to Ni^{III} results in the inclusion of triiodide ions into the pores.[74] Examples of the latter include a number of dicarboxylate framework systems in which post-synthetic framework reduction leads to the inclusion of alkali metal ions and to dramatic changes in hydrogen gas adsorption properties of the modified framework.[75, 76]

An equally powerful although less studied form of post-synthetic modification treats MOF crystals as chemical substrates at which covalent grafting can occur. The first use of this approach was the alkylation of unbound pyridyl units within the homochiral framework POST-1 (described in Section 1.2.1.1.2), a process that deactivates these sites catalytically.[29] More recently, this approach has been used to confer a range of desirable host–guest properties to MOFs, with particular success seen with the grafting of a range of functional groups to the unbound amine group on the NH_2-bdc (bdc = 1,4-benzenedicarboxylate) linker within IRMOF-3.[59] A notable consequence of this process is the modification of chemical surface properties and the fine tuning of the dimensions of the pores and pore windows, with the systematic increase in organic chain length leading to a corresponding decrease in surface area of the framework due to pore occlusion.[77] Another notable example is the two-step attachment of a catalytically active vanadium complex through ligand grafting (with $\sim$13 % conversion of the amine groups) followed by metal complexation to yield a material that exhibits heterogeneous catalytic activity at the vanadium centres (see Figure 1.5).[78]

1.2.1.2   Structural Response to Guest Exchange

A common synthetic goal in MOF synthesis is the generation of frameworks that display zeolite-like rigidity to guest desorption and exchange[31, 50, 51, 79–90] (so-called '2nd generation materials') rather than collapse irreversibly upon guest removal ('1st generation materials').[5] The host–guest chemistry of such systems is readily interpretable

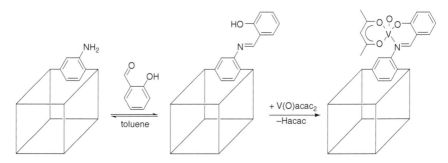

Figure 1.5 Schematic for the functionalisation of IRMOF-3 (see Figure 1.2) with salicylaldehyde and subsequent binding of a vanadyl complex (acac = acetylacetonate). Reprinted with permission from M.J. Ingleson, J.P. Barrio, J.B. Guilbaud, Y.Z. Khimyak and M.J. Rosseinsky, *Chem. Commun.*, **28**, 2680–2682. Copyright (2008) Royal Society of Chemistry

using standard models, with rapid guest transport commonly occurring within the pores and Type I adsorption isotherms displayed. Importantly, these features lead to a high degree of predictability in the host–guest chemistry, with the framework structure able to be simulated as a rigid host within which dynamic guest molecules migrate and bind,[91, 92] and with guest selectivity depending principally on the size and shape of the guest molecules and the strength of the host–guest surface interactions. Such properties are highly desirable for a wide range of host–guest applications.

In addition to the considerable interest in rigid frameworks, a very interesting feature of many MOFs is their high degree of framework flexibility. Materials of this type, which have been classified as '3rd generation materials',[5] display flexing of their framework lattices in response to various stimuli; this most commonly involves response to the desorption and exchange of guest molecules, but may also arise due to changes in temperature, pressure, irradiation, *etc*. The adsorption iso-therms of materials that display guest-induced flexing typically exhibit hysteretic behaviour due to the fact that the apohost phase has a different pore structure from that of the adsorbed phase, with transformation between the two being an activated process. Structurally, the adsorption properties can range from intercalative behaviours in which staged adsorption occurs through the gradual guest-induced opening of pores (*cf.* clays) to more cooperative behaviours in which guest adsorption influences the structure of the entire MOF crystal (*i.e.* crystal and pore homogeneity are retained throughout the adsorption process). In materi-als of this general type the guest-selectivity is considerably more complex than that of the zeolitic phases, with adsorption commonly depending on the strength of the host–guest interaction (which needs to be sufficient to drive the framework deformation), as well as guest size and shape considerations. This is particularly the case for mixtures of guests, where cooperative effects are commonly seen; *e.g.* the adsorption of one guest can have a 'gate-opening' function to allow the inclusion of a second guest that would not otherwise be adsorbed. Despite being generally less predictable than rigid frameworks, such materials have potential use in a range of applications that make use of their chemically selective adsorption and/or hysteretic behaviour (*e.g.* for guest storage). A further point of interest here is that structural modification upon guest loading provides a mechanism for molecular sensing.

At the present time it is not straightforward in all cases to predict in advance whether MOFs will survive guest desorption, or the extent to which their frameworks might distort upon desorption and subsequent adsorption. Some clear guiding principles exist, however. First, the rigidity of the building units has a clear influence on framework flexibility, with the strength of coordination bonding providing a useful initial guide as to the energetics of bond bending as well as thermal stability. Secondly, the extent of connectivity and topological underconstraint within the framework lattice has a key influence over whether low energy deformations might occur; *e.g. cf.* rigidity of triangular network *vs* scissor action of square grid. In considering whether host–guest interaction energies are sufficient to drive framework deformation or decomposition, a particularly important consideration is whether guests may bind at the metal nodes and thereby favour pronounced structural flexing, framework interconversion or even dissolution; a relatively common limitation of MOFs is their sensitivity to water vapour, with the metal nodes in some systems being susceptible to water binding and ligand displacement. More subtle effects such as hydrogen bonding interactions, or even weak intermolecular forces involving small gaseous guests, can frequently be sufficient to cause pronounced framework flexing.

1.2.1.2.1 Flexible Frameworks

Two different types of guest-induced flexibilities exist in MOF host lattices. The first can be considered as essentially static in nature, involving bulk framework deformations that may be readily characterised using diffraction-based techniques and which are frequently observable at the macroscale through changes in crystal dimensions. The second are dynamic and arise due to molecular vibrations or local guest-induced framework deformations away from the 'parent' structure. The latter are not so readily detectable by diffraction methods and are commonly inferred based on geometric considerations; for example, local distortions away from the bulk crystallographic structure have been shown to be necessary in certain cases to allow migration of guests through narrow pore windows.[93] Given these complexities, considerations of the guest selectivity of flexible systems need necessarily extend beyond simple 'size and shape' arguments towards the more complex consideration of guest-driven host lattice modification.

A broad array of interesting flexing behaviours have been seen in MOF systems, spanning intercalative-type behaviour in 2D layered systems to the deformation of individual frameworks and the translation of interpenetrated frameworks.[74, 79, 85, 94–99] The interdigitated 2D layer compound [Cu(dhbc)$_2$(4,4'-bpy)]·n(guest) (dhbc = dihydroxybenzoate; 4,4'-bpy = 4,4'-bipyridine) displays pronounced interlayer contraction upon guest desorption, with a 30 % decrease in the c-axis length.[100] This process occurs without loss of polycrystallinity and involves the gliding of aromatic units with respect to each other. Subsequent adsorption of guest molecules leads to regeneration of the more open structure, with the corresponding adsorption isotherms displaying activated, hysteretic behaviour in which a 'gate-opening' pressure is required before adsorption can occur.

The interdigitated bilayer phase [M^{II}_2(4,4'-bpy)$_3$(NO$_3$)$_4$]·n(guest) (M = Ni, Co, Zn)[79, 89, 93, 101] displays zeolite-like robustness upon desorption of ethanol guests from the parent phase[82] and two types of framework flexibility upon adsorption of other guests.[79] In situ single crystal diffraction characterisation during guest adsorption showed that molecular guests with dimensions too large for the pores of the apohost can be adsorbed due to a progressive widening of the 1D pores with increasing guest size associated with low energy scissor-type flexing of the bilayers. Even larger guests are adsorbed into this phase through a different pore expansion mechanism in which translation of the interpenetrated bilayer nets with respect to each other leads to an increase in the height of the 1D channels.

The MIL-53 family of 3D frameworks, with formula [M^{III}(OH,F)(bdc)]·n(guest) (M = Al, Cr, Fe; bdc = 1,4-benzenedicarboxylate), also display scissor-type flexing as a function of temperature and guest adsorption with considerable variation in the dimensions of the 1D channels.[102, 103] A comprehensive in situ powder X-ray diffraction examination of guest adsorption into the Fe analogue, MIL-53(Fe), has demonstrated that the guest-induced breathing effect depends principally on the strength of the interaction between host and guest rather than being particularly dependent on guest size (see Figure 1.6).[102] The principal

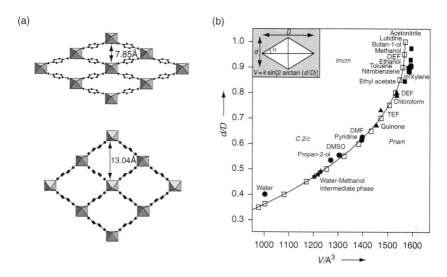

(a)

(b)

Figure 1.6 (a) Diagrammatic representation of the structural breathing in the MIL-53 family of materials. (b) Variation of the height to width ratio (d/D) of the diamond-shaped 1D channels within MIL-53(Fe) as a function of guest desorption and adsorption. Reprinted with permission from F. Millange, C. Serre, N. Guillou, G. Férey and R.I. Walton, *Angew. Chem. Int. Ed.*, **47**, 4100. Copyright (2008) WILEY-VCH Verlag GmbH & Co. KGaA

mechanism for this effect appears to be the interaction of included guests with the framework OH groups, leading to variation in the geometry of the 1D μ_2-OH bridged chains. This transformation occurs cooperatively throughout the lattice, such that only small amounts of guest adsorption are sufficient to cause long-range bulk structural flexing. The desorbed Al analogue, MIL-53(Al), displays a similar breathing effect induced purely by changes in temperature.[103] The structural transformation here occurs with hysteresis about room temperature, with the open high temperature (HT) form collapsing to the low temperature (LT) form with cooling below *ca* 200 K and the LT form converting back to the HT form with warming above *ca* 350 K. The consequences of this breathing action have been seen clearly in adsorption isotherm measurements on this phase: whereas the CH_4 adsorption causes little framework flexing, as evidenced by a Type I isotherm and invariant physisorption enthalpy, CO_2 adsorption occurs *via* a stepped isotherm in which pressures above *ca* 6 bar yield the more open framework phase, which has a lower CO_2 adsorption enthalpy.[104]

1.2.1.2.2 Framework Interconversions

A number of more extreme forms of structural response exist in which guest adsorption/desorption or variation of other parameters (*e.g.*

temperature and light irradiation) leads to a modification of framework connectivity. These may be classified into cases in which structural interconversion requires the breakage and formation of coordination bonds, and those where changes to the covalent connectivity arises.

Coordinative interconversion   Whilst the majority of porous MOFs retain their structural connectivity during guest-exchange processes, in an increasing number of systems the dynamic nature of the metal-ligand bond in solution has been mirrored in the solid state, yielding highly pronounced structural interconversions. Lability at the metal nodes in these systems can arise due either to a dissociative or associative mechanism, with each of these being influenced by neighbouring coordinating guests within the pores and/or by unbound donor sites on the framework ligands. Confirmation that the structural interconversions are topotactic processes within the solid state rather than solvent-assisted recrystallisations has been provided by *in situ* diffraction measurements in which the interconversions are followed in real time. This coordinatively dynamic nature of some MOF lattices is evidenced also by the demonstration that MOF synthesis can be achieved under essentially solvent free conditions at ambient temperature following initiation of the solid state reaction by ball milling,[105] and by the single-crystal-to-single-crystal exchange of metal nodes by immersion of MOF crystals in the solution of other metal ions.[68]

The simplest and most common form of framework interconversion involves disassociation of terminal ligands followed by intra-/interframework complexation. In $[Fe^{II}(pmd)(H_2O)(M^I(CN)_2)_2]\cdot H_2O$ (pmd = pyrimidine; M^I = Ag, Au), for example, thermal desorption of the bound water molecules leads to the coordination of a pmd ligand from an interpenetrated network, thereby linking the networks together.[106] This results in a topochemical conversion in which there is a change in the framework topology from the interpenetration of three separate 3D networks to a single 3D network. Interesting changes to the spin-switching properties result from this transformation (see Section 1.3.2.1). Another form of thermally induced structural interconversion is seen in $[Cu(CF_3 COCHCOC(OCH_3)(CH_3)_2)_2]$, for which temperature pulsing causes a conversion from a porous phase containing exclusively the *trans*-isomer of the Cu^{II} complex to a dense phase containing a mixture of *cis*- and *trans*-isomers.[107] Subsequent exposure of the dense phase to adsorptive vapour reverts the material to its porous form.

Exposure to solvent vapour can also drive pronounced framework interconversion in which coordination bond breakage and formation occurs.[41, 42, 93, 108] An example here is a family of frameworks incorporating the 1,3,5-benzenetricarboxylate (btc) linker, for which exchange of bound solvent at the metal nodes is accompanied by structural conversions between a range of different network topologies.[41, 42] Guest desorption from the homochiral framework [Ni$_3$(btc)$_2$(py)$_6$(1,2-pd)$_3$]·n(guest) (btc = 1,3,5-benzenetricarboxylate; py = pyridine; 1,2-pd = 1,2-propanediol), which consists of the interpenetration of two (10,3)-a networks, leads to an amorphous phase in which the long-range structural order is lost. Subsequent exposure to ethanol vapour leads to the regeneration of the ordered double-network structure, which upon prolonged exposure converts to a more dense quadruply interpenetrated (10,3)-a network phase (see Figure 1.7).[43] The latter is the same phase as crystallises from ethanolic solution, indicating that ethanol adsorption into the doubly interpenetrated network leads to a metastable topotactic phase that gradually converts to a more stable phase over time. In contrast, exposure to pyridine leads to the formation of a 2D hexagonal sheet structure, also requiring breakage and formation of coordination bonds, whereas 3-picoline adsorption leads to stabilisation of the doubly interpenetrated phase above that of the parent material; this phase is homochiral with 47 % permanent porosity and displays enantioselective guest-exchange.[41]

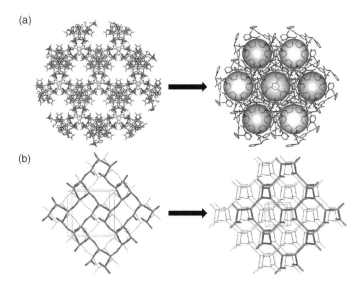

Figure 1.7 Interconversion of the highly porous, homochiral 3D MOF [Ni$_3$(btc)$_2$(py)$_6$(1,2-pd)$_3$]·n(guest) upon guest desorption and exposure to ethanol vapour (a), involving transformation from a doubly interpenetrated, distorted (10,3)-a network phase to one in which four regular (10,3)-a nets interpenetrate (b). Reprinted with permission from C.J. Kepert, T.J. Prior and M.J. Rosseinsky, *J. Am. Chem. Soc.*, **122**, 5158. Copyright (2000) American Chemical Society

Covalent interconversion The modification of MOF structure through topochemical reactions between organic linkers rather than through coordinative exchange is comparatively rare due to the less labile nature of covalent bonds. Some noteworthy examples exist, however, in which the organic units are favourably aligned within the framework to react with each other if induced to do so thermally or by photoexcitation. A particular point of interest here is the stereoselectivity of this process, with the regularity of the framework structure often imparting isomeric purity in the covalent product.[109]

A highly strategic example of this approach is seen in the [2 +2] cycloaddition of adjacent *trans*-1,2-bis(4-pyridyl)ethene (tvp) linkers within a number of 1D ladder-type frameworks in which dinuclear metal complex 'rungs' align the tvp molecules side-to-side.[110–112] In the ladder framework $[((CF_3CO_2)(\mu-O_2CCH_3)Zn)_2(\mu-tvp)_2]$ this dimerisation process occurs *via* a single-crystal-to-single-crystal transformation, allowing detailed structural characterisation of the product (see Figure 1.8).[112] Ligand polymerisation can also occur, as seen for example with the irradiation of $[Ca(C_4H_5O_2)_2(H_2O)]$ with ^{60}Co γ-rays to yield high molecular weight calcium poly(3-butenoate) (average $400\,000$ g mol^{-1}) in 97 % yield.[113] While not a form of framework conversion it warrants mention here that the polymerisation of adsorbed organic guests within MOFs has also been achieved with a high degree of stereoselectivity.[114]

1.2.2 Storage and Release

The very large pores and high surface areas of MOFs, combined with their potential low cost, low toxicities and industrial scalability, makes them outstanding candidates for the storage and release of a range of different guest molecules.[115] In the development of guest storage technologies, considerable current focus is on industrially important gases

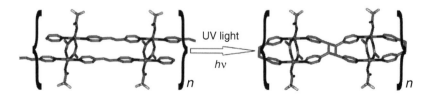

Figure 1.8 Structural conversion of a 1D ladder framework induced by photochemical excitation. Reprinted with permission from N.L. Toh, M. Nagarathinam and J.J. Vittal, *Angew. Chem. Int. Ed.*, **44**, 2237. Copyright (2005) WILEY-VCH Verlag GmbH & Co. KGaA

such as hydrogen,[116–119] methane[51, 88, 120] and acetylene,[121, 122] and extends also to the controlled release of larger guests such as pharmaceuticals.[123–125] Given that each of these areas is currently undergoing very rapid advance, a brief description is given here to the development of hydrogen storage materials, for which reasonably detailed structural–property relationships have emerged.

1.2.2.1   Hydrogen Storage

The efficient storage of hydrogen gas is a critical challenge that needs to be met if hydrogen-based energy cycles are to displace those based on fossil fuels. The US Department of Energy 2010 targets for vehicular hydrogen storage systems are a capacity of 6 wt% H_2, 45 g H_2 L^{-1}, an ability to operate in the temperature range -30 to 50 °C to a maximum pressure of 100 bar, reversibility over 1000 cycles, and a refuelling rate of at least 1.5 kg H_2 min^{-1}; the 2015 targets stipulate a *ca* 50 % improvement to these numbers. Among a wide range of physisorption phases (*i.e.* those where dihydrogen interacts with the surface through an intermolecular interaction), MOFs have shown the highest H_2 uptakes, being greater than those for porous carbons and zeolites and exceeding both the gravimetric and volumetric targets for hydrogen adsorption.[116–119] In comparing the potential of MOFs with that of chemisorption phases (*i.e.* those where hydrogen reacts to form a covalent or ionic compound; *e.g.* metal hydrides) a number of advantages exist relating to the much lower enthalpy of adsorption/desorption, resulting in much reduced heat flow requirements during refilling and to the highly reversible nature of the process, both for hydrogen and various contaminants.

Among a large number of different MOFs investigated for their hydrogen storage capabilities, the IRMOF (isoreticular metal-organic framework) series with formula [Zn_4OL_3] (where L = a range of dicarboxylates)[60, 126, 127] and NOTT-10n (n = 1–7) series with formula [Cu_2L] (where L = a range of tetracarboxylates)[56, 57] display adsorption properties that are both impressive and highly informative. A notable trend to emerge from the study of these and related systems is a reasonably strong correlation between hydrogen adsorption and BET and Langmuir N_2 surface area (see Figure 1.9), a relationship that provides a useful predictive tool for surface saturation loading of H_2.[117] Of these materials, the highest excess uptake (*i.e.* the uptake beyond that which would be contained within a free volume equivalent to that of the sample) is seen in MOF-5, which adsorbs 7.1 wt% at 77 K and 40 bar, with a

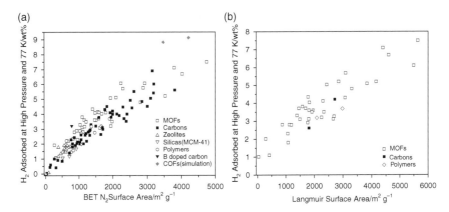

Figure 1.9   The variation of H_2 adsorbed (wt%) at saturation at 77 K with BET (a) and Langmuir surface area (b) for porous MOFs, carbons, zeolites, silicas, polymers, and covalent organic frameworks (COFs). Reprinted with permission from K.M. Thomas, *Dalton Trans.*, **9**, 1487–1505. Copyright (2009) Royal Society of Chemistry

total uptake of 10.0 wt% at 100 bar and a volumetric storage density of 66 g l^{-1}.[127] The larger pore material MOF-177, [Zn$_4$O(BTB)$_2$] (where BTB = 1,3,5-benzenetribenzoate), achieves an excess uptake of 7.1 wt% at 77 K and 66 bar, but with a lower volumetric density of 49 g l^{-1} on account of its lower surface density.[128] A comparable volumetric uptake is seen in NOTT-103.[56]

Whilst a number of important benefits exist for the physisorption approach to hydrogen storage over that of chemisorption materials, the weakness of the hydrogen-surface interaction, which is typically in the order of 4–6 kJ mol^{-1}, represents a considerable technological limitation. For such interaction enthalpies surface saturation can only be achieved at low temperatures or prohibitively high pressures; as a representative example, the hydrogen capacity of MOF-5 drops to 0.57 wt% and 9.1 g l^{-1} at 298 K and 100 bar.[127] Theoretical calculations indicate that enthalpies of *ca* 20 kJ mol^{-1} are required for optimal performance at ambient temperature with pressures up to 30 bar.[116] Recent efforts to develop hydrogen storage capabilities at ambient conditions have focused on the generation of highly charged framework surfaces, in particular those having bare metal sites where dihydrogen can interact directly;[65, 75, 76, 129–132] notably, this approach has also proven useful for the optimal storage of methane[120] and acetylene.[122] Currently, the maximum isosteric enthalpies of adsorption in MOFs are ~12–13 kJ mol^{-1}, achieved through dihydrogen binding at bare Ni^{II} and Cu^{II} sites.[133, 134] The highest ambient temperature hydrogen uptake is seen, however, in Mn$_3$[(Mn$_4$Cl)$_3$(BTT)$_8$]$_2$ (where

(a)   (b)

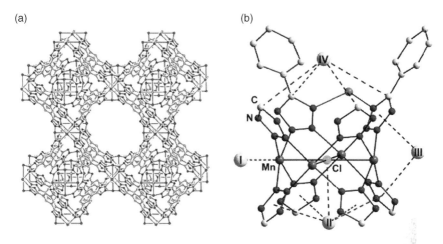

Figure 1.10 (a) The anionic framework structure of $[Mn(DMF)_6]_3[(Mn_4Cl)_3(BTT)_8(H_2O)_{12}]_2 \cdot n(guest)$ shown with $[Mn(DMF)_6]^{2+}$ units and guests removed. (b) Principal surface dihydrogen binding sites as determined by powder neutron diffraction. Reprinted with permission from M. Dincă, A. Dailly, Y. Liu, C.M. Brown, D.A. Neumann and J.R. Long, *J. Am. Chem. Soc.*, **128**, 16876. Copyright (2006) American Chemical Society

BTT = 1,3,5-benzenetristetrazolate), which combines a high surface area $(2100 \, m^2 \, g^{-1})$ with a high surface density of bare Mn^{II} sites (physisorption enthalpy = 10.1 kJ mol^{-1} at zero loading) to adsorb 12.1 g l^{-1} (7.9 g l^{-1} excess) of hydrogen at 90 bar and 298 K (see Figure 1.10);[129] this is 77 % greater than the density of compressed hydrogen gas under these conditions.

A further potentially beneficial feature of MOFs is their hysteretic adsorption behaviour, which in principle can be exploited to allow hydrogen loading at very high pressures/low temperatures and release at lower pressures/higher temperatures. Such a property is seen in $[Ni_2(4,4'-bpy)_3(NO_3)_4]$, a framework with narrow windows that displays scissor-type flexibility upon adsorption and desorption of guests (see Section 1.2.1.2.1); adsorption isobars reveal that H_2 desorption commences only upon warming above ~110 K, a property that is attributed to the kinetic trapping of dihydrogen in this phase.[135]

1.2.3 Selective Guest Adsorption and Separation

Many of the properties that make MOFs excellent candidates for molecular storage, such as their very high surface areas, adjustable pore sizes and tunable surface properties, also make them particularly well suited for application in the separation of molecular and ionic mixtures. Of particular technological interest here is the development of materials that are able to perform highly selective separations efficiently, rapidly, and on the bulk scale. Strategically important species range from gases (*e.g.* H_2, He, O_2, N_2, CO, CO_2, CH_4 and other alkanes, H_2S, NO_x, NH_3, *etc.*, as present in air, flue gases, natural gas, syn-gas, *etc.*)[136] to ions (*e.g.* $[NO_3]^-$, $[SO_4]^{2-}$, $[TcO_4]^-$, Cs^+, Sr^{2+}, *etc.*)[137] to large molecules such as pharmaceuticals and their precursors (many requiring enantioseparation).[138, 139] Large-scale target technologies for the former include H_2 and CH_4 purifications, CO_2 capture, CO removal for fuel cell technology, and desulfurisation of transportation fuels.

Adsorptive separations by porous materials are commonly achieved by one or more of the following mechanisms: (1) size/shape exclusion; (2) selective adsorbate–surface interactions; (3) different guest diffusion rates; and (4) the quantum sieving effect, in which small guests are adsorbed faster than larger ones due to their more rapid diffusion through narrow pore windows. Manipulation of these effects requires control over both pore structure and surface chemistry, each of which may be achieved strategically to a high level of sophistication in the synthesis and post-synthetic modification of MOFs. Moreover, in contrast to more conventional porous materials such as zeolites, the structural flexibility of many MOFs gives them a vastly more complex behaviour in which selectivity may depend, uniquely, on the ability of the guest molecule to distort the host framework.

1.2.3.1 Gas Adsorption

Considerable recent efforts have been devoted to the investigation of gas separation in both rigid and porous MOF phases. Whilst only a relatively small number of systems have been investigated by selective gas adsorption measurement (as opposed to multiple measurement with various pure gases, which due to cooperative effects provides only a guide to the separation capabilities), a number of distinct separation mechanisms have been evidenced.[136]

1.2.3.1.1 Rigid MOFs
Gas separation based on size/shape exclusion has been achieved in a relatively large number of small pore frameworks. The 3D diamond-type framework $[Mn^{II}_3(HCOO)_6]$, for example, selectively adsorbs H_2 over N_2 and Ar at 78 K and CO_2 over CH_4 at 195 K. Uptakes of the excluded gases N_2, Ar and CH_4 are almost zero due to their inability

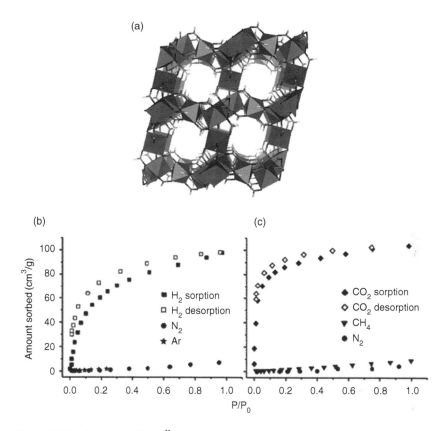

Figure 1.11 Structure of $[Mn^{II}_3(HCOO)_6]$ (a) and gas adsorption isotherms at 78 K (b) and 195 K (c). Reprinted with permission from D.N. Dybtsev, H. Chun, S.H. Yoon, D. Kim and K. Kim, *J. Am. Chem. Soc.*, **126**, 32. Copyright (2004) American Chemical Society.

to migrate through the very narrow pore windows of this phase (see Figure 1.11).[140] The size-selective separation of N_2 and O_2, which have extremely similar molecular dimensions and are separated industrially by ion-exchanged zeolites (*e.g.* CaX/NaX) according largely to the difference in their quadrupolar moments, has reportedly been achieved by $[Mg_3(ndc)_3]$ (ndc = naphthalenedicarboxylate),[141] $[Zn(dtp)]$ (dtp = 2,3-di-1*H*-tetrazol-5-ylpyrazine),[142] and $[Zn_4O(H_2O)_3(adc)_3]$ (PCN-13; adc = 9,10-anthracenedicarboxylate).[143]

In addition to molecular sieving effects, which make use of different diffusivities of the different guests, the fine control over MOF surface chemistry has seen the emergence of materials that discriminate gases according to the strength of their adsorption interaction. An example here is the 3D pillared layer phase $[Cu_2(pzdc)_2(pz)]$ (where pzdc = pyrazine-2,3-dicarboxylate and pz = pyrazine), which selectively adsorbs acetylene over CO_2 due the preferential docking of the former between two basic oxygen atom sites within the framework's highly constrained 1D pores (dimensions 4 × 6 Å) rather than to a size-selective molecular sieving effect.[121] The much larger pores of zeolitic imidizolate frameworks (ZIFs) also display a high degree of chemical selectivity, with capture of CO_2 from CO_2/CO mixtures attributed to the different binding affinities of these gases.[144] Breakthrough experiments, in which a continuous flow of 1:1 CO/CO_2 was passed through packed columns of ZIF-68, 69 and 70, further demonstrated the complete retention of CO_2 and passage of CO. In the related phases ZIF-95 and 100, which contain narrow pore windows (3.65 and 3.35 Å, respectively) between colossal pores, a similarly high selectivity for CO_2 adsorption over other gases was attributed to a combination of molecular sieving and surface selectivity effects, the latter arising due to quadrupolar interactions of CO_2 with the nitrogen atoms on the pore surface (see Figure 1.12).[54]

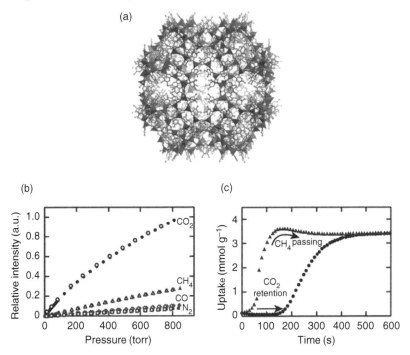

Figure 1.12 (a) The 35.6 Å cage within ZIF-100, which is accessed through narrow pore windows. (b) CO_2, CH_4, CO and N_2 gas adsorption isotherms for ZIF-100 at 298 K. (c) CO_2/CH_4 breakthrough curves, showing the retention of CO_2 within a packed column. Reprinted with permission from B. Wang, A.P. Cote, H. Furukawa, M. O'Keeffe and O.M. Yaghi, *Nature*, **453**, 207 (2008).

Whilst the above separations require cycling through adsorption and desorption processes, typically through swings in pressure or temperature, excellent potential exists for the achievement of continuous-flow separation through the development of MOF membranes. This has been achieved through oxidative electrodeposition of thin films of $[Cu_3(btc)_2]$ (HKUST-1) at a Cu metal mesh.[145] These membranes display permeabilities and selectivities that are superior to those of conventional zeolite membranes, with the high H_2 permeation flux ($0.107\,mol\ m^{-2}\ s^{-1}$) leading to ambient temperature separation factors of 7.04, 6.84, and 5.92 from 1:1 mixtures of H_2/N_2, H_2/CO_2, and H_2/CH_4, respectively. The recyclability and high chemical and thermal stability of these membranes makes them strong candidates for bulk-scale hydrogen separation and purification.

1.2.3.1.2 Flexible MOFs

Selective gas adsorption and separation in flexible MOFs is considerably more complicated than that in their rigid counterparts. Due to the high degree of cooperativity in these systems (e.g. in inducing framework deformation, the uptake of one guest can dramatically alter the uptake of another), comparison of adsorption isotherms of pure gases is of limited use and competitive measurements are essential if separation capabilities are to be determined. Due to the fact that such measurements remain very rare, and that structural information is often unavailable for the mixed-sorbed phases, only limited understandings of gas separations in flexible MOFs currently exist.

The simplest form of selective adsorption in flexible MOFs involves a molecular sieving effect where, for example, small guests deform the host lattice to allow their inclusion while larger guests are excluded. This property is seen in [Cd(pzdc)(bpee)] (where pzdc = pyrazine-2,3-dicarboxylate and bpee = trans-1,2-bis(4-pyridyl)ethene, also tvp), with water and methanol adsorption occurring and inducing expansion of the 1D channels whilst larger guests such as ethanol, tetrahydrofuran and acetone are excluded under all conditions.[146] More complicated are cases where differing degrees of pore expansion occur for different guests, as seen in materials such as $[Cu^{II}(dhbc)_2(4,4'\text{-}bpy)] \cdot n(guest)$ (see Section 1.2.1.2.1).[100] The separation capabilities of such phases are yet to be determined through competitive measurements and seem likely to depend principally on host–guest interaction enthalpies.

For weakly interacting guests, the exploitation of temperature rather than guest-induced gate-opening as a means to vary pore window dimensions – an effect that is seen in a number of zeolite and related materials – has led to the development of size-selective adsorption properties in MOFs that may be varied thermally. The flexible 3D framework phase [Ni^{II}_8(5-bbdc)$_6$(μ_3-OH)$_4$] (MAMS-1; 5-bbdc = 5-tert-butyl-1,3-benzenedicarboxylate) contains narrow pore windows that range in dimension from 2.9 to 5.0 Å depending on temperature,[147] an effect that is considerably larger than that seen in zeolites. Adsorption isotherms for a range of small gases into this phase show strong temperature dependencies; for example, no appreciable N_2 is adsorbed at 77 and 87 K whereas considerable adsorption is achieved at 113 K. These effects were attributed principally to thermally induced gate-opening rather than increased thermal excitation of the guests through the narrow hydrophilic pore windows.

1.2.3.2   Liquid Phase Adsorption

The separation of larger molecules, which requires adsorption from liquids/solutions rather than the gas phase, has also received some attention. Indeed, arguably the first demonstrations of guest selective adsorption in this field involved the separation of solvent mixtures using Hofmann-type clathrate phases of formula [M(1)^{II}L$_2$M(2)^{II}(CN)$_4$]·n(guest) (where M(1) = divalent octahedral transition metal; L = unidentate ligand such as NH_3; M(2) = Ni, Pd, Pt).[148, 149] These materials can be used as liquid chromatographic stationary phases for the separation and purification of a range of small molecules, with high degrees of guest selectivity and moderately rapid rates of guest exchange (albeit too slow to be competitive industrially) occurring despite their 2D collapsible nature. Guest-specific docking in these phases occurs in rigid elongated pores that are bound by the M(2) and L groups, with the docking of aromatics such as benzene being particularly favoured at these sites.

In targeting large-scale separation processes from the liquid phase, for which well-established technologies exist, MOFs offer arguably the greatest promise in specialist areas that make use of their unique guest selectivities. One particularly notable opportunity is in the field of enantioseparations, for which classical porous materials such as zeolites have made little impression due to the difficulty of preparing enantiomerically pure examples with free void volume; current enantioseparation technologies are based largely on nonporous surface-modified formulations that have limited chemical stability. Following the successful generation of the first homochiral porous phases (see Section 1.2.1.1.2), subsequent investigations of enantioselective guest-exchange have led to some of the first demonstrations of chiral separation using crystalline porous materials.[34, 42, 150] The robust 3D framework [Ni$_3$(btc)$_2$(3-pic)$_6$(1,2-pd)$_3$]·n(guest) (3-pic = 3-picoline), which is obtained through post-synthetic modification of the pyridine analogue (see Section 1.2.1.2.2 and Figure 1.7) and contains a helical pore structure that occupies *ca* 50 % of the crystal volume, enantioselectively adsorbs 1,1'-bi-2-napthol with an enantiomeric excess (*ee*) value of 8.3 %. Among a diverse range of frameworks constructed using chiral BINOL-based ligands,[35, 151–154] [Gd(*R*-L-H$_2$)(*R*-L-H$_3$)(H$_2$O)$_4$] (where *L* = 2,2'-diethoxy-1,1'-binaphthalene-6,6'-bisphosphonate) achieves a 13.6 % enantio-enrichment of racemic *trans*-1,2-diaminocyclohexane. The highest degrees of enantioselectivity have been seen using smaller pore frameworks, in which a more effective multi-point interaction between guest and host surface typically occurs. The 3D pillared layer framework [Ni$_2$(L-asp)$_2$(4,4'-bpy)]·n(guests) (L-asp = L-aspartate), which has 23 % pore volume and pore windows of dimension 3.8 × 4.7 Å, enantioselectively adsorbs a range of racemic mixtures, with the highest *ee* value of 53.8 % achieved for the separation of racemic 2-methyl-2,4-pentanedione.[30] Most impressively, almost complete enantioseparation of 2-butanol (*ee* = 98.2 %) is achieved with the immersion of the robust diamond-type framework [Cd(QA)$_2$] (QA = 6'-methoxyl-(8*S*,9*R*)-cinchonan-9-ol-3-carboxylate) in the racemic liquid.[155]

Enantioselective exchange of chiral cations has also been achieved in homochiral MOF systems. The 2D framework POST-1, described in Section 1.2.1.1.2, contains 1D homochiral pores within which both enantioselective guest exchange and catalysis occurs.[29] Upon suspension of L-POST-1 in a methanol solution of racemic [Ru(2,2'-bpy)$_3$]Cl$_2$ (2,2'-bpy = 2,2'-bipyridine), 80 % of the exchangeable protons are exchanged with the propeller-type cation, with *ee* = 66 % in favour of the D form.

1.2.4   Heterogeneous Catalysis

1.2.4.1   Overview

Heterogeneous catalytic activity was one of the first proposed[1] and demonstrated[156] host–guest properties of MOFs, with subsequent research providing a range of different catalytic activities across a diverse array of framework systems.[157, 158] Most notable among these systems are cases in which catalysis arises due to chemical activation at specific surface binding sites.

In conceptualising the catalytic capabilities of MOFs a useful first point of comparison is with zeolites, with which they share a number of common attributes: considerable pore sizes and volumes, allowing inclusion and reaction of large precursor molecules; robustness to guest desorption and adsorption, allowing rapid molecular transport; regularity of pore structure, imparting a degree of size selectivity to the catalytic process based on the shape/size of the reactant, intermediate, or product; and, in some cases, high thermal stability (up to *ca* 500 °C), albeit lower than that of most zeolites. At the other end of the spectrum, MOFs notably share many attributes with enzymes, in that sophisticated catalytic sites can, in principle, be incorporated into the framework structure to yield specific types of activity.[159] A notable point here is that highly active surfaces can be achieved through two means: the intrinsic strain of MOF lattices, in which the high lattice binding energies in many cases favour the stabilisation of unusual molecular and coordination geometries; and through post-synthetic modification to generate metastable high energy sites not otherwise accessible through 'one-pot' reactions. Given this positioning, the pursuit of catalytic applications for MOFs has to date centred principally on high-end reactions (*e.g.* to produce enantiomers and fine chemicals) rather than those requiring forcing conditions.

1.2.4.2 Synthetic and Post-Synthetic Approaches

1.2.4.2.1 Metal Sites
Several different approaches have been used to incorporate active metal sites onto the interior surfaces of MOFs. Foremost among these is the often adventitious generation of reactive metal nodes by framework formation, with the first example of this type being the 2D layered structure $[Cd(4,4'-bpy)_2(NO_3)_2]$, which catalyses the cyanosilation of aldehydes.[156] The shape-selective activity of this system is attributable to the Lewis acidity of the labile Cd^{II} nodes and their geometric constraint within the square grid layers. The subsequent generation of framework phases that are stable to metal site activation through the desorption of bound solvent molecules has generated a number of more advanced catalytic systems.[160–163] These include the well-known $[Cu_3(btc)_2]$ (HKUST-1, MOF-199; btc = 1,3,5-benzenetricarboxylate) (see Figures 1.2 and 1.3),[160, 161] and $[Cr_3F(H_2O)_2O(bdc)_3]$ (MIL-101; bdc = 1,4-benzenedicarboxylate),[162] each of which catalyse cyanosilylation reactions, with the latter acting as a stronger Lewis acid than the former due to the greater relative acidity of Cr^{III} over Cu^{II}.

More strategic efforts to incorporate specific metal site function into MOFs have seen the construction of lattices using dedicated metallo-ligands. An excellent example here is the use of chiral Mn-salen units to pillar square grid layers of form $Zn_2(bpdc)_2$ (bpdc = biphenyldicarboxylate) (see Figure 1.13).[164] This material functions as an enantioselective catalyst for olefin epoxidation, yielding *ee* values >80 %. Framework confinement of the manganese salen entity enhances catalyst stability, imparts substrate size selectivity, and permits catalyst separation and reuse, whilst retaining the

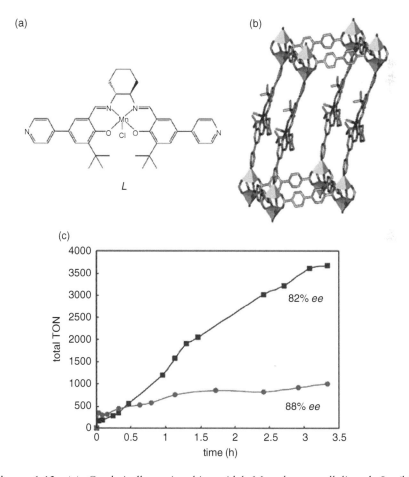

Figure 1.13 (a) Catalytically active bis-pyridyl Mn-salen metalloligand *L*. (b) Framework structure of $[Zn_2(bpdc)_2L]\cdot n$(guest). (c) Plots of total turnover number (TON) *vs* time for the enantioselective epoxidation of 2,2-dimethyl-2*H*-chromene catalysed by $[Zn_2(bpdc)_2L]\cdot n$(guest) (squares) and the free ligand *L* (circles). Reprinted with S.H. Cho, B.Q. Ma, S.T. Nguyen, J.T. Hupp and T.E. Albrecht-Schmitt, *Chem. Commun.*, 2563–2565. Copyright (2006) Royal Society of Chemistry

enantioselective performance of the free molecular analogue. Catalytic metal sites have also been incorporated as extra-framework species, an example being the encapsulation and stabilisation of free base metalloporphyrins into rho-ZMOF.[165] Among these encapsulated phases, the Mn-metallated 5,10,15,20-tetrakis(1-methyl-4-pyridinio)porphyrin analogue shows catalytic activity toward the oxidation of cyclohexane with very high turnover numbers and cyclability.

Post-synthetic incorporation of metal sites into MOFs has proven to be a particularly powerful technique for generating reactive surfaces that would otherwise be inaccessible.[35, 59, 71, 163] Of particular note here is the generation of a chiral framework that displays enantioselective catalytic activity,[35] achieved in two synthetic steps: first, the synthesis of a chiral nanoporous phase $[Cd_3Cl_6L_3]\cdot n$(guest) ($L = (R)$-6,6'-dichloro-2,2'-dihydroxy-1,1'-binaphthyl-4,4'-bipyridine); and, second, the chemisorption of titanium(IV) isopropoxide sites onto the hydroxyl units of the chiral bridging ligands of the apohost. The resulting solid was found to catalyse $ZnEt_2$ additions to aromatic aldehydes with efficiencies and enantioselectivities comparable with those for the free $Ti(O^iPr)_2$-functionalised BINOL-type ligand.

Finally, in an approach analogous to that used for porous carbons and zeolites, highly robust MOFs have been used as surface supports for metal atoms and clusters.[158] An example here is the chemical vapour deposition of various metals into MOF-5, yielding materials classified as metal@MOF-5 for which the nature of metal inclusion and the extent of exogenous loading is currently unknown. Of these, Cu@MOF-5 is active in the synthesis of methanol from syngas and Pd@MOF-5 catalyses the reduction of cyclooctene by hydrogen.[166]

1.2.4.2.2 Other Surface Sites

Whilst metal centres have provided the majority of known catalytic sites in MOFs, organic units have also provided a number of compelling examples. The size-selective catalytic activity of POST-1, described in Sections 1.2.1.1.2 and 1.2.3.2, in the transesterification of alcohols is attributed to the presence of unprotonated pyridyl groups that project into the channels and which likely assist in the deprotonation of the alcohol reactants. Catalytic yields in excess of 77 % were achieved with an $ee = \sim 8$ % using this homochiral system.[29]

The post-synthetic generation of Brønsted acid surface sites – a structural feature that is largely in conflict with the coordination conditions for framework synthesis – has also led recently to catalytic activity in MOFs. Protonation of the bound carboxylato groups within the framework [Cu(L-asp)(bpee)$_{0.5}$] (L-asp = L-aspartate; bpee = 1,2-bis(4-pyridyl)ethylene) leads to a material that catalyses epoxide methanolysis with up to a 65 % yield and *ee* = 17 %.[39] The strongly acidic nature of this material arises from the binding of the protonated carboxylato unit to Cu^{II}, which increases the proton acid strength *via* the stabilisation of the conjugate base. Notably, such an arrangement is inaccessible in the homogeneous phase, where protonation of the amino acid at either the COO or NH$_2$ site leads to its dissociation from the metal centre. Brønsted acid sites have also been achieved through grafting of protonated ethylenediamine units onto the bare Cr^{III} sites of MIL-101, yielding a material that is active in the condensation of benzaldehyde and ethyl cyanoacetate. Notably, the inclusion and subsequent gentle reduction of charge-balancing anionic metal complexes leads to the inclusion of catalytically active Pd nanoparticles within this framework.[72]

1.3   INCORPORATION OF OTHER PROPERTIES

Whereas conventional porous solids act largely as selective scaffolds within which reversible guest-exchange and catalysis can occur, the synthetic control over the structure and composition of MOFs, in addition to providing the impressive array of host–guest properties outlined above, has notably allowed the incorporation of many other interesting chemical and physical functions into these lattices. Many of these functions have been achieved for the first time in porous media, yielding host lattices that are able to respond and interact with guest molecules in entirely new and 'intelligent' ways. These notably include a range of magnetic, electronic, optical and mechanical phenomena, with the achievement of these commonly requiring the development of specific materials design principles relating both to the individual molecular building units and to their arrangement within the framework lattices. Whilst many such phenomena have been known and investigated for some time, it has only been recently that their design principles have been extended to the formation of open porous frameworks, allowing the combination of these properties with reversible host–guest chemistry for the first time. A strong motivation for these efforts has been the derivation of detailed structure–property relationships, with the exchange of guests providing both a powerful and very convenient means through which framework structure and, therefore, the property of interest can be perturbed systematically. For many of these properties the influence of exchangeable guest on host lattice function has been found to be highly pronounced, offering potential scope in areas such as molecular sensing. Moreover, the recognition that unique synergies may exist between framework function and guest-exchange has led to the generation of materials with exotic new materials properties, including direct interplay between the various chemical and physical functions.

1.3.1 Magnetic Ordering

1.3.1.1 Overview

The accomplishment of magnetic exchange-coupling in coordinately linked frameworks has been investigated in detail since the early 1900s[167–170] and, as such, has been both a pioneering and enduring motivation for the study of this class of materials. There are two principal approaches for the achievement of magnetic ordering in these systems: (1) the linkage of transition or lanthanoid metal ions through diamagnetic bridging ligands to achieve coupling between the metal spins; and (2) the linkage of these ions through radical ligands in which coupling between metal and ligand spins occurs. In each case the dimensionality of the framework lattice has important implications on the magnetic properties, as do issues such as the extent and nature of the coupling and the magneto-anisotropy of the metal ion. In each approach, the extent of exchange coupling and, therefore, magnetic ordering temperature commonly decreases rapidly as the number of bridging atoms between spin centres increases. Whereas many pure metals order magnetically at temperatures over 1000 K and metal oxide systems up to 900 K, those for two-atom bridged phases are below 350 K, whilst those for three-atom bridged frameworks[171] do not exceed 50 K. At atomic separations beyond this the ordering temperatures typically drop to below 2 K.[172] The investigation of coordinatively linked systems has therefore focused on a range of short bridges, which notably include hydroxo, cyanido, carboxylato, azido, dicyanamido, oxalato, and oxamato ligands. In contrast, the incorporation of radical ligands into framework lattices, which has received considerably less attention than the diamagnetic ligand approach due in part to their more difficult syntheses, allows the greater separation of metal ions to yield comparable magnetic ordering temperatures.

Given the very considerable literature on molecule based magnets (MBMs), which includes a number of detailed books and reviews,[13, 167–169, 172] principal attention here is given to describing the recent emergence of porous magnets.

1.3.1.2 Porous Magnets

A considerable challenge in the formation of porous magnets lies in their dual requirement of magnetic exchange and interconnected pore volume. Whereas porosity in MOFs commonly requires the use of relatively long molecular connecting units, long-range magnetic ordering above milli-kelvin temperatures requires relatively short exchange pathways between nearest neighbour spin sites. Efforts to combine these seemingly inimical requirements within the one material have therefore focused on 2D and 3D framework compounds constructed through the bridging of certain transition metal ions with one- or two-atom bridges (*e.g.* hydroxo, carboxylato and cyanido, phosphonato, halido) or through the use of radical bridging ligands.[13] An important requirement in such syntheses is the achievement of neutral framework lattices, with many early examples of open magnetic frameworks being nonporous on account of their pores being filled with counterions; examples here include systems with oxalato[173–175] and formato,[176, 177] bridges. Following early reports of 'magnetic sponges' that change their magnetic ordering properties upon irreversible guest desorption,[178] these approaches have generated a number of novel porous magnetic phases. Whilst most of these have arisen through dedicated syntheses, some have notably derived from materials that have been known for many years, for which the guest-exchange capabilities were either unappreciated or regarded as an experimental inconvenience rather than a property worthy of exploitation.

In the absence of any success to date in the formation of metal oxides that are both magnetic and porous to molecular (as opposed to ion) inclusion, the main success in the use of one-atom bridges has been in the formation of hybrid materials containing the μ_3-hydroxide and formate bridges. Principal among two-atom bridged materials are a range of metal cyanides, with Prussian Blue phases providing the majority of these. The emergence of these porous magnets[13, 172, 179–188] has been important in allowing the investigation of both guest-induced perturbation of magnetic properties[180–183, 185, 188] and magnetic-exchange interactions between host and guest.[184] Further, the discovery of solvatomagnetic effects in such materials has been of particular interest, both for the systematic elucidation of magnetostructural relationships and for possible applications in areas such as molecular sensing.

1.3.1.2.1 Hybrid Metal Hydroxide Frameworks
In utilising hydroxo bridges between metal ions, approaches have focused principally on the use and modification of the brucite $M(OH)_2$ structure,

which consists of edge-shared layers of octahedral transition metal ions (CdI_2-type), and the rutile-type structure, in which both edge- and vertex-sharing of metal octahedra occurs.[172] Among a range of different modifications to the brucite parent structure are the formation of 1D edge-shared chains and the interruption of the layer through removal of some of the metal sites. Bridging or pillaring of these low dimensionality magnetic chains and layers with a range of multitopic organic ligands has then led in some cases to the formation of porous 3D framework lattices. Notable also has been the report of a 3D fully hydroxo-bridged lattice, $[Co_5(OH)_2(OAc)_8]\cdot2H_2O$, which displays canted antiferromagnetic behaviour below $30\,K$.[189]

Among a large number of 1D hydroxo-bridged hybrid magnets[172] is the squarate-bridged 3D framework $[Co^{II}_3(OH)_2\ (C_4O_4)_2]\cdot3H_2O$,[190] which consists of 1D $[Co^{II}_3(\mu_3\text{-}OH)_2]^{4+}$ brucite-type ribbons linked by squarate anions to form a porous 3D network that houses 1D water-filled channels of dimensions 4.0×6.7 Å. Upon reversible dehydration/ rehydration, single crystal diffraction measurements indicate that the framework experiences only minimal changes in its unit cell parameters and bond distances and angles, with a remarkable accompanying interconversion from antiferromagnetic to ferromagnetic ordering at $8\,K$ (Figure 1.14).[191] It is not currently known whether this pronounced change in magnetic properties results from the steric perturbation of the framework lattice, in which dehydration leads to a *ca* 2° change in some of the squarate binding angles, or whether magnetic-exchange coupling

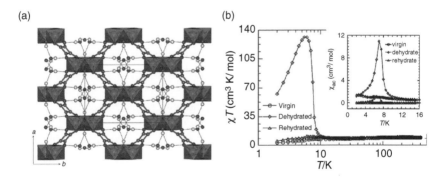

Figure 1.14 Structure of $[Co^{II}_3(OH)_2(C_4O_4)_2]\cdot3H_2O$ (a) and guest-dependent magnetic behaviour (b), showing the DC susceptibility χT product and AC susceptibility (inset) for the virgin hydrated, dehydrated, and rehydrated frameworks. Reprinted with permission from S.O.H. Gutschke, D.J. Price, A.K. Powell and P.T. Wood, *Angew. Chem. Int. Ed.*, **38**, 1088. Copyright (1999) Wiley-VCH Verlag GmbH & Co. KGaA.

through the hydrogen-bonding pore water molecules in the hydrated phase has some influence. Notable other 1D systems include a number of hybrid magnets based on the edge- and vertex-shared [110] ribbon within the rutile structure, having general formula $[M_3(OH)_2(dicarboxylate)_2(H_2O)_4] \cdot n(guest)$ (M = Co, Ni, Mn). These materials similarly order magnetically in the vicinity of 10 K, with the Co and Ni chains commonly forming ferrimagnets. A particularly notable example among these is the 3D framework $[Ni_3(OH)_2(cis-1,4-cyclohexanedicarboxylate)_2(H_2O)_4] \cdot 2H_2O$, which contains 1D water-filled pores and converts from ferrimagnetic (2.1 K) to ferromagnetic ordering (<4.4 K) upon partial dehydration and rehydration.[192]

More open porous frameworks containing 1D chains have been achieved through the bridging of metal centres by both oxide/hydroxide and carboxylate linkers. Two examples of such materials are the pseudo-isostructural $[V^{IV}O(bdc)] \cdot n(guest)$ (MIL-47)[193] and $[Cr^{III}(OH)(bdc)] \cdot n$ (guest) (MIL-53(Cr)),[194] in which 1D channels of dimensions 7.9 × 12 Å run parallel to the 1D metal chains. Whilst strong magnetic coupling is achieved in each material, the ordering is antiferromagnetic, with Néel temperatures (T_Ns) for the as-synthesised phases of 95 K (MIL-47) and 65 K (MIL-53(Cr)). Guest desorption leads to pronounced flexibility, with the pores of V analogue opening to have dimensions 10.5 × 11.0 Å, and a shift in the T_Ns to 75 K (MIL-47) and 55 K (MIL-53(Cr)·H_2O).

In the search for higher dimensionality magnetic pathways in the formation of porous phases some success has been achieved also in use of 2D hydroxo-bridged layers. Whereas the metal sites within the parent brucite-type $Co(OH)_2$ structure (which is metamagnetic with T_N = 10 K) are fully coordinated with an octahedral environment, variation in the synthesis conditions has allowed the replacement of some fraction of these sites with tetrahedral sites that lie out of the 2D layer and, in some cases, the replacement of hydroxo units with other μ_3-bridging anions.[172] The tetrahedral Co^{II} sites provide tethering points above and below the layer through which pillaring with bis-unidentate ligands has been achieved to produce materials with interlayer solvent-filled galleries. An example is the $[Co_8(OH)_{12}(SO_4)_2(diamine)] \cdot nH_2O$ family (diamine = 1,2-ethylenediamine (en), n = 3; diazabicyclooctane (dabco), n = 1), in which layer neutrality is achieved through replacement of 1 in 7 of the octahedral Co^{II} sites with two tetrahedral sites and 1 in 7 of the hydroxide sites with sulfate.[179, 180] Diamine bridges then link the layers, with the ethylenediamine analogue displaying intercalative properties with interlayer collapse upon guest desorption, whereas the dabco analogue displays robust porosity. Both display metamagnetic properties, with the higher ordering temperature of the dabco phase ($T_N = 21$ K, *cf*. 14 K for the en phase) attributed to the greater exchange coupling through the triple pathway of the dabco pillar. A distinct but related layer is seen in $[Co_5(OH)_8(chdc)]\cdot 4H_2O$ (chdc = *trans*-1,4-cyclohexanedicarboxylate), in which 1 in 5 of the octahedral Co^{II} sites are replaced with tetrahedral sites and charge neutrality of the framework is achieved through the use of a dicarboxylate pillar (see Figure 1.15).[195] This material is ferrimagnetic with critical temperature, $T_c = 61$ K, implying ferromagnetic coupling between the layers, and has a very high coercive field of 22 kOe at 2 K.

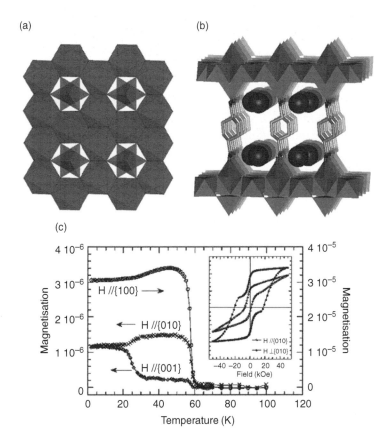

Figure 1.15 (a) Projection of the $Co_5(OH)_8L_2$ layer, consisting of edge-shared octahedral Co^{II} and vertex-shared tetrahedral Co^{II}. (b) Pillared layer structure of $[Co_5(OH)_8(chdc)]\cdot 4H_2O$. (c) Temperature and field (inset; measured at 2 K) dependent magnetisation of $[Co_5(OH)_8(chdc)]\cdot 4H_2O$, measured with respect to the crystallographic axes. Reprinted with permission from M. Kurmoo, H. Kumagai, S.M. Hughes and C.J. Kepert, *Inorg. Chem.*, **42**, 6709. Copyright (2003) American Chemical Society

The higher magnetic ordering temperature of this phase over that of related systems is attributable to the higher proportion of tetrahedral Co^{II} sites.[172] Upon dehydration the material undergoes a reversible single-crystal-to-single-crystal transformation in which the pillars rotate and the interlayer separation decreases slightly. This pronounced transformation has little observable influence on the magnetic properties.

1.3.1.2.2 Metal Formates

Whereas carboxylate units most commonly link metal ions through a three-atom bridge, it is not uncommon for these units to bridge two ions through a single oxygen atom. Whilst considerable magnetic exchange coupling may be achieved in the former case, particularly if multiple carboxylate bridges are present,[196] the latter binding mode has been exploited to great effect in a family of formate-bridged frameworks of formula $[M_3(HCOO)_6]\cdot$ n(guest) (where M = Fe, Mn, Co and Ni; and guest = a wide range of solvents) (see also Section 1.2.3.1.1 and Figure 1.11).[140, 197–203] The framework topology of this family is that of a distorted diamond-type, in which each formate coordinates to three metal ions and each metal is linked to its nearest neighbours by one single oxygen atom of the formate and one three-atom carboxylate bridge. Guest desorption from this phase can be achieved with retention of the framework structure, with subsequent adsorption with other guests leading to pronounced breathing effects in which the lattice has been seen to expand in volume by up to 12 %. Despite the extended network of M-O-M linkages throughout the structures of these phases, the magnetic ordering temperatures are rather modest, with the maximum being 22 K for the ferromagnetic Fe analogue. As expected given the considerable structural perturbation seen with guest exchange, the magnetic properties of these systems are highly variable, with the ordering temperature of the Fe analogue lying in the range 15–22 K depending on the identity of the adsorbed guest.

1.3.1.2.3 Metal Cyanides

The Prussian Blue family of materials, in addition to providing the first coordination compound back in 1704, has provided and continues to provide a wide range of interesting magnetic behaviours.[204–206] The family consists of a diverse array of frequently misassigned structures of general formula $C_m A_x[B(CN)_6]_y\cdot nH_2O$ (where C = cation, e.g. K^+, Cs^+; A and B = octahedral transition metal ions). Examples include Prussian Blue itself, $Fe^{III}_4[Fe^{II}(CN)_6]_3\cdot 14H_2O$, in which vacancies at the ferrocyanide sites within the cubic network rather than cation inclusion lead to charge balance, and a range of other vacancy $(A_x[B(CN)_6]_y\cdot nH_2O; x \neq y)$

and nonvacancy ($C_mA[B(CN)_6] \cdot nH_2O$; $0 \leq m \leq 2$) phases. Following the report of ferromagnetic ordering at 5.6 K in Prussian Blue in 1928,[207] the magnetic ordering temperatures of these frameworks have been increased through the variation of metal ions and framework composition. Following early work in which the diamagnetic low spin Fe^{II} sites within Prussian Blue were replaced with paramagnetic metal ions, more strategic efforts have been directed towards optimising the sign and magnitude of the magnetic exchange coupling through variation of the orbital occupancies (*e.g.* making use of the ferromagnetic $t_{2g}(B)-e_g(A)$ pathway) and relative energies. Notable successes from this strategy include $CsNi^{II}[Cr^{III}(CN)_6] \cdot 2H_2O$ (90 K ferromagnet),[208] $Cs_{0.75}Cr^{II}_{1.125}$ $[Cr^{III}(CN)_6] \cdot 5H_2O$ and $Cr^{II}_3[Cr^{III}(CN)_6]_2 \cdot 10H_2O$ (190 K and 240 K ferrimagnets, respectively),[209] $V^{II}_{0.42}V^{III}_{0.58}[Cr^{III}(CN)_6]_{0.86} \cdot 2.8H_2O$ (315 K ferrimagnet),[210] $K_{0.058}V^{II/III}[Cr(CN)_6]_{0.79} \cdot (SO_4)_{0.058} \cdot 0.93H_2O$ (372 K ferrimagnet),[211] and $KV^{II}[Cr^{III}(CN)_6] \cdot 2H_2O$ (376 K ferrimagnet).[212] The achievement of room temperature ordering in the latter V/ Cr systems, albeit with very small coercive fields (*e.g.* 25 Oe at 10 K for $V^{II}_{0.42}V^{III}_{0.58}[Cr^{III}(CN)_6]_{0.86} \cdot 2.8H_2O^{[210]}$), represents a major advance. A second broad family of cyanide-bridged magnets are bimetallic systems in which the hexacyanidometallate $[B(CN)_6]^{n-}$ metalloligands are linked through coordinatively unsaturated $[A(L)_x]^{m-}$ units (where L = polyamine ligands for example).[213] Early examples from this family include $[Ni(en)_2]_3[Fe(CN)_6]_2 \cdot 2H_2O$ (en = 1,2-ethylenediamine), which contains a ladder-type Ni^{II}-Fe^{III} network that orders ferromagnetically at 18.6 K,[214] and $[Mn(en)]_3[Cr(CN)_6]_2 \cdot 4H_2O$, which consists of a 3D Mn^{II}-Cr^{III} network that orders ferrimagnetically at 69 K.[215]

Of the relatively small number of reports of reversible guest-exchange in cyanide-based magnets, the Prussian Blue family provides many interesting examples. Following the demonstration of robust porosity in this family of materials,[216–219] it has been found that reversible water desorption from $CsNi^{II}[Cr^{III}(CN)_6] \cdot 2H_2O$ and $Cr^{II}_3[Cr^{III}(CN)_6]_2 \cdot 10H_2O$ leads to apohost phases with BET surface areas of 360 and 400 m^2 g^{-1} and magnetic ordering temperatures of T_c = 75 K and T_N = 219 K, respectively;[184] these are only slightly decreased from those of the hydrated phases (see above), with the latter being the highest ordering temperature yet observed for a porous magnet. Upon adsorption of the paramagnetic O_2 guest molecule, opposite magnetic couplings between host and guest are seen for each material; for $CsNi^{II}[Cr^{III}(CN)_6]$, there is an increase in the magnetic moment, indicating ferromagnetic exchange, whereas for $Cr^{II}_3[Cr^{III}(CN)_6]_2$ there is a reduction of the coercivity from 110 to 10 G and of the remnant magnetisation from 1200 to 400 emu

$G\ mol^{-1}$, indicating antiferromagnetic exchange. Notably, through examination of the O_2 adsorption energetics it was concluded that the magnetic interaction has at most a negligible influence on the adsorption energetics, suggesting that the proposed exploitation of internal magnetic field for the separation of O_2 from air is unrealisable.

In related systems, solvatomagnetic effects have been reported in the vacancy phase $Co^{II}_{1.5}[Cr^{III}(CN)_6]\cdot 7.5H_2O$, which converts from a peach-coloured and ferromagnetically coupled ($T_c = 25$ K) solid to a blue and antiferromagnetically coupled phase ($T_N = 18$ K) of formula $Co^{II}_{1.5}[Cr^{III}(CN)_6]\cdot 2.5H_2O\cdot 2EtOH$ on exposure to ethanol, an effect attributed to a change from six- to (average) four-coordination at the Co^{II} centre.[220] A similar effect is seen with the systematic variation of water occupancy in $Co[Cr(CN)_6]_{2/3}\cdot nH_2O$, which upon reversible desorption of bound and unbound water guests converts from pink with octahedral Co^{II} to blue with tetrahedral Co^{II}; accompanying this humidity-dependent transformation is a change from ferromagnetic to antiferromagnetic coupling and a decrease in the magnetic ordering temperature from 28 to 22 K (see Figure 1.16).[221] A more pronounced change in ordering temperature is seen in $K_{0.2}Mn_{1.4}Cr(CN)_6\cdot 6H_2O$, where T_N increases from 66 to 99 K upon water desorption.[222]

A number of more structurally diverse cyanide-bridged materials have also been shown to display reversible solvatomagnetic effects. These include the flexible host lattice $[Mn(NNdmenH)(H_2O)][Cr(CN)_6]\cdot H_2O$ (NNdmen = N,N-dimethylethylenediamine), which undergoes a reversiblesingle-crystal-to-single-crystal transformation from the 2D layer stru-cture of the parent phase to a 3D pillared-layer framework, $[Mn(NNdmenH)][Cr(CN)_6]$, a transformation that involves the generation and cleavage of Mn-N bonds. This structural change leads to an increase in the ferrimagnetic ordering temperature from 35.2 to 60.4 K.[223] Among a range of interesting porous phases based on the $S=1/2$ octacyanidotungstate(V) unit, pronounced solvatomagnetism is seen in the 2D framework $[Ni(cyclam)]_3[W(CN)_8]_2\cdot 16H_2O$ (cyclam = 1,4,8,11-tetraazacyclotetradecane), which converts from ferromagnetic behaviour to canted ferromagnetic upon reversible dehydration; this effect is attributed to a large change in the Ni-NC-W angles.[224] Similarly, exchange of water for n-propanol in the 3D framework $Cu_3[W(CN)_8]_2(pmd)_2\cdot 8H_2O$ (pmd = pyrimidine) to form $Cu_3[W(CN)_8]_2(pmd)_2\cdot 1.5PrOH\cdot 2.25H_2O$ results in an increase in magnetic ordering temperature from 9.5 to 12.0 K and a dramatic increase in coercive field; these changes are attributed to a decrease in antiferromagnetic coupling to a Cu site that converts from 6- to 5-coordinate.[225]

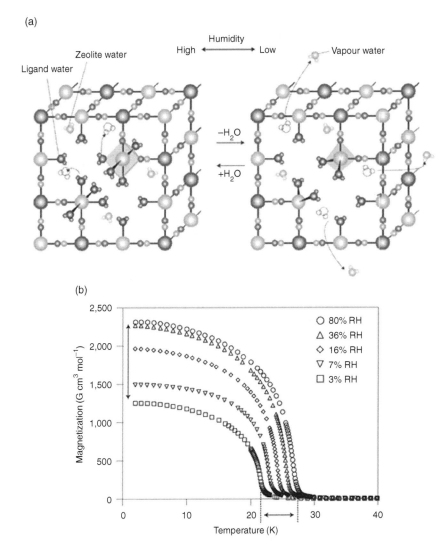

Figure 1.16 Adsorption and desorption of bound and unbound water from the vacancy Prussian Blue $Co[Cr(CN)_6]_{2/3} \cdot nH_2O$ (a) and the influence of relative humidity (RH) on magnetic ordering temperature, as seen in the low temperature magnetisation (b). Reprinted with permission from S.I. Ohkoshi, K.I. Arai, Y. Sato and K. Hashimoto, *Nat. Mat.*, 3, 857. Copyright (2004) Nature Publishing Group

1.3.1.2.4 Radical ligands

Principal attention in the incorporation of radical multitopic ligands into MOF phases has focused on the well known π-acceptors TCNE (tetracyanoethylene) and TCNQ (7,7,8,8-tetracyanoquinodimethane), which in their mononegative forms have an unpaired spin that can couple with spins on metal ions to which they are coordinated.[226] Among a range of framework materials constructed with these ligands is the amorphous room temperature magnet $V[TCNE]_x{\cdot}nCH_2Cl_2$ ($x \sim 2$; $n \sim 0.5$), which is proposed to have a glass-like 3D framework structure of the form $V^{II}[TCNE]^-_z$ $[TCNE]^{2-}_{1-z/2}$ ($1 < z < 2$). This material orders magnetically at 125 °C, which is marginally higher than the Prussian Blue phases. The same radical 4-connecting linker is seen in the phases $[Fe(TCNE^-)(C_4(CN)_8)_{1/2}]{\cdot}$ nCH_2Cl_2[227] and $[Fe^{II}(TCNE^-)(NCMe)_2]^+[Fe^{III}Cl_4]^-$,[228] the structures of which were determined by synchrotron powder diffraction measurement. These materials order magnetically in the vicinity of 100 K. Among a number of chemically functionalised TCNQ-based framework magnets are the 2D layered framework $[(Ru_2(O_2CCF_3)_4)_2(TCNQF_4)]{\cdot}3$ (p-xylene), which orders magnetically at 95 K,[229] and the 3D Ru framework $[(Ru_2(O_2CPh\text{-}m\text{-}F)_4)_2(BTDA\text{-}TCNQ)]$ (where BTDA-TCNQ= bis(1,2,5-thiadiazolo)tetracyanoquinodimethane), which is a ferromagnet with $T_c = 107$ K.[230] Host–guest properties have yet to be reported for a porous TCNE or TCNQ based magnet, with the closest example being the demonstration of reversible guest-exchange in a diamagnetic pillared layer phase $[Zn^{II}(TCNQ^{2-})(4,4'\text{-bpy})]{\cdot}6MeOH$.[231] The radical ligand approach has, however, led to the successful generation of porous 2D frameworks constructed using the highly stable polychlorinated tris(4-carboxyphenyl)methyl (PTMTC) radical.[182, 232, 233] Most notable among these is the highly flexible porous phase $[Cu^{II}_3(PTMTC)_2$ $(py)_6(EtOH)_2(H_2O)]$ (MOROF-1; see Figure 1.17), which shrinks and expands by up to 30 vol% with ethanol desorption/adsorption and displays subtle solvatomagnetic effects associated with framework collapse and the removal of coordinated guests.[182]

1.3.2 Electronic and Optical Properties

Among a wide range of electronic and optical phenomena known in molecular systems, many have been achieved and investigated in MOFs. As with the magnetic systems described above, recent efforts to incorporate such properties into porous systems have led to the first investigations of guest-induced perturbations of these phenomena, leading to a sensing function. Unlike porous magnets, for which magnetostriction effects are generally negligible, a further particular interest here lies in the often highly pronounced coupling of electronic excitation with lattice energetics, leading to direct interplay between electronic/optical and host–guest function.

(a)

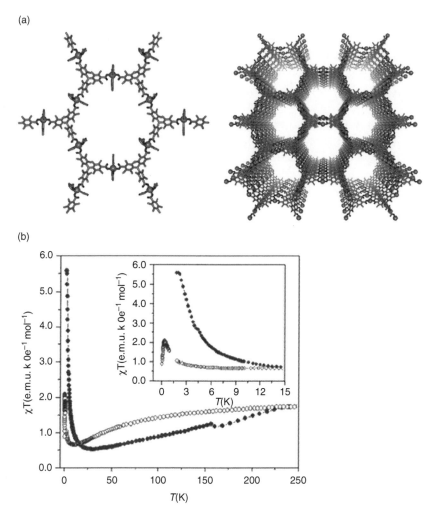

(b)

Figure 1.17 (a) Structure of MOROF-1, which consists of 2D hexagonal layers in which Cu^{II} ions are bridged by the radical $S = 1/2$ ligand PTMTC. (b) χT product for the solvated (filled circles) and desorbed (open circles) framework. Reprinted with permission from D. Maspoch, D. Ruiz-Molina, K. Wurst, N. Domingo, M. Cavallini, F. Biscarini, J. Tejada, C. Rovira and J. Veciana, *Nat. Mater.*, **2**, 190. Copyright (2003) Nature Publishing Group

1.3.2.1 Spin crossover

Spin crossover (SCO) in transition metal compounds is a well known form of molecular bistability in which the transition between a low-spin (LS) and a high-spin (HS) state can be induced by a variation of temperature, pressure, magnetic field or light irradiation. Physical consequences of this transition include pronounced changes in colour, magnetic moment, and coordination bond distances and strengths. The observation of abrupt transitions and memory effects in SCO materials, which arise due to electron–phonon coupling between the SCO sites and long-range elastic interactions within the lattice,[234] has led to suggestions that these materials may be candidates for application in information storage and retrieval, temperature sensing and visual displays.[235]

Early efforts to incorporate this electronic molecular switch into framework materials were driven primarily by an interest in elucidating the nature of cooperativity in SCO lattices, with the ultimate goal of controlling the switching properties to deliver bistable systems at ambient temperature.[236–248] Classical examples of such systems are members of a family of 1,2,4-triazole bridged 1D chain compounds of type $[Fe^{II}(R\text{-}trz)_3](anion)_2$, which undergo abrupt SCO transitions and wide thermal hysteresis loops ($\Delta T = 35$ K) spanning room temperature.[235–237] Elaboration of the ligand design to include alkane-linked bis-triazoles (btr) and bis-tetrazoles (btzb) subsequently yielded compounds of types $[Fe(btr)_2(NCS)_2] \cdot H_2O$[243] and $[Fe(btzb)_3](ClO_4)_2$,[248] which have 2D and 3D framework structures, respectively, and quite diverse SCO behaviour with hysteresis present in some cases. Cyanidometallate bridges between SCO centres have also led to pronounced hysteresis near ambient temperature, most notably in the 2D layered Hofmann type materials $[Fe(py)_2M(CN)_4]$ (M = Ni, Pd, Pt; py = pyridine)[249] and the 3D pillared Hofmann frameworks $[Fe(pz)M(CN)_4] \cdot 2(H_2O)$ (pz = pyrazine), which have ΔT up to 33 K.[240] Among other rare examples of 3D SCO framework phases is the Prussian Blue analogue, $CsFe^{II}[Cr^{III}(CN)_6]$, which undergoes SCO both thermally[250] and upon irradiation with X-rays.[251]

The recent achievement of porosity in SCO frameworks provides a new approach for investigating features such as the ligand field, electronic and steric communication between SCO centres, and lattice dynamics, as well as providing materials with completely new host–guest properties.[252] Among a range of porous spin crossover frameworks (SCOFs) are an extensive isotopological family of the form $[Fe(L)_2(NCX)_2] \cdot n(guest)$ (L= *trans*-1,2-bis(4-pyridyl)ethene (tvp, also bpee),[239] 4,4'-azopyridine (azpy),[97] 1,2-bis(4-pyridyl)ethane (bpe),[253] 1,2-bis(4'-pyridyl)-1,2-ethanediol (bped),[254] and 2,3-bis(4'-pyridyl)-2,3-butanediol (bpbd);[255, 256] and X = S, Se), which consist of interpenetrated rhombic grids between which 1D channels lie.[254] Investigation of the guest-exchange chemistry of the azpy, bpe, bpbd and bped analogues has uncovered a range of subtle guest-dependent structural and electronic behaviours. The azpy analogue, $[Fe^{II}_2(azpy)_4(NCS)_4] \cdot n(guest)$, displays a broad half SCO transition that depends on the nature of guest inclusion. Desorption of the unbound ethanol guests of the parent phase leads to a transformation in which the 1D pore channels collapse partially, with the open framework geometry being returned with the adsorption of a range of different alcohol guests. The guest-loaded phases display subtly different SCO properties, whereas the apohost is HS to low temperature; removal of the SCO function is attributed to the weakening of the Fe^{II} ligand field caused by nonideal coordination geometries following guest removal.[97] The bped analogue was the first porous material in which pore environment can be varied by excitation by light. The SCO in this material, which may also be induced thermally and/or influenced by the desorption/sorption of guest ethanol molecules, leads to a subtle breathing of the framework structure and modification of the pore chemistry.[253] The bpe analogue, $[Fe(bpe)_2(NCS)_2] \cdot n(guest)$, displays a guest- and spin-state dependence of considerable complexity. Through variable temperature synchrotron powder and single crystal X-ray diffraction measurement, coupled with characterisations of the host–guest, magnetic and photomagnetic properties, it was shown that this material can exist in at least nine subtly distinct structural forms as a function of guest loading, temperature and light irradiation. These uniquely include a half-spin state in which there is a chequerboard arrangement of HS and LS Fe^{II} sites at the two-step plateau (see Figure 1.18).[253] The most structurally robust of these phases incorporates the bpbd linker, with almost perfect framework rigidity resulting from a network of hydrogen-bonding interactions between the interpenetrated $[Fe(bpbd)_2(NCS)_2]$ grids. The switching temperature of this phase can be controlled in a predictable fashion by the incorporation of guests with differing polarities, an effect that emerges because the influence of steric interactions between host and guest are minimised. This material is unique among the SCOF family in displaying bistability, with thermal hysteresis in the SCO transition being attributed to the high degree of lattice cooperativity. Intriguingly, this can be turned on and off by the inclusion of different guests, indicating that host–guest rather than solely intraframework effects can influence the extend of lattice cooperativity and resulting memory effects in SCO systems.[255]

The use of cyanidometallate linkers between SCO metal centres has also generated a range of interesting porous phases. One 3D example, $[Fe^{II}(pmd)(H_2O)(M^I(CN)_2)_2] \cdot H_2O$ (pmd = pyrimidine; M^I = Ag, Au) displays the multifunctional properties of SCO, hysteresis ($\Delta T = 8$ K) and a reversible dehydration/rehydration structural interconversion in

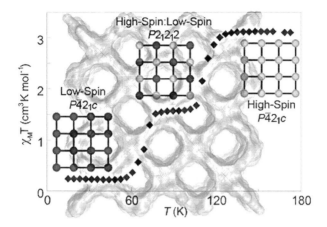

Figure 1.18 Two-step spin crossover in the interpenetrated square grid framework [Fe^{II}(bpe)$_2$(NCS)$_2$]·n(guest) (SCOF-4; shown in the background viewed down the 1D pores). The switching in this porous system proceeds from fully high-spin (right) to fully low-spin (left) via a chequerboard-type arrangement of high-spin and low-spin Fe^{II} nodes (centre). Reprinted with permission from G.J. Halder, K.W. Chapman, S.M. Neville, B. Moubaraki, K.S. Murray, J.F. Létard and C.J. Kepert, *J. Am. Chem. Soc.*, **130**, 17552. Copyright (2008) American Chemical Society

the crystal phase, the latter yielding the substitution of H$_2$O for pmd at the Fe^{II} centres.[106] This conversion results in large changes to the switching properties: for the Ag phase the SCO transition moves to lower temperature and has a larger hysteresis, whereas for the Au phase the transition is eliminated completely. In contrast, a very high degree of structural robustness has been found in the SCO Hofmann phases [Fe^{II}(pz)M^{II}(CN)$_4$]·2(H$_2$O) (M^{II} = Ni, Pd, Pt; pz = pyrazine), which consist of square grid [FeM(CN)$_4$] layers pillared by pyrazine (see Figure 1.19).[240] Dehydration of the Pt analogue leads to an increase in both the temperature and width of the SCO hysteresis loop.[257] Subsequent guest-dependent measurements on this family have uncovered a range of unprecedented materials properties, which include both guest-induced switching (providing a selective molecular sensing mechanism) and switch-induced changes to host–guest function (enabling manipulation of pore chemistry and therefore guest uptake/release through external stimuli).[258, 259] Further, exploitation of the electronic bistability of this system allows these processes to occur with a degree of molecular memory; for example, the framework can be switched to its alternate state by adsorption then desorption of one guest, then switched back by use of a different guest. Adsorption measurements in the bistable

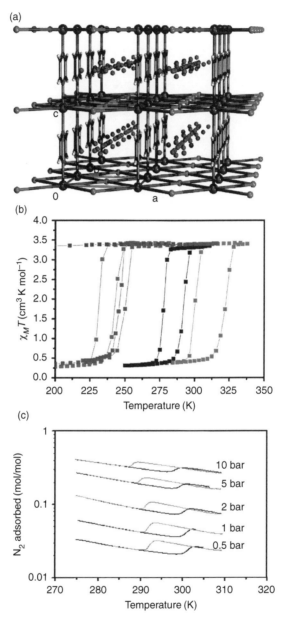

Figure 1.19 (a) Structure of $[Fe^{II}(pz)Ni^{II}(CN)_4]\cdot2(H_2O)$. (b) Influence of the adsorption of (from left to right) toluene, acetone, ethanol, methanol, and acetonitrile on the thermal SCO behaviour. (c) Dinitrogen isotherms collected on cooling (grey) and warming (black), showing the influence of the hysteretic spin transition on the gas adsorption properties. Reprinted with permission from P.D. Southon, L. Liu, E.A. Fellows, D.J. Price, G.J. Halder, K.W. Chapman, B. Moubaraki, K.S. Murray, J.F. Léard and C.J. Kepert, *J. Am. Chem. Soc.*, **131**, 10998. Copyright (2009) American Chemical Society

temperature region yield a range of unique behaviours. The HS and LS framework states display different guest affinities due to a *ca* 0.25 Å difference in pore dimension associated with the *ca* 0.2 Å difference in Fe-N bond lengths. Direct interplay between the host–guest and switching properties is also seen in adsorption isobar measurements, in which the hysteretic nature of the SCO is mirrored in the guest adsorption and desorption (see Figure 1.19).

1.3.2.2 Electron Transfer

The observation and investigation of electron transfer processes in MOFs pre-dates the exploration of porosity in these systems by some decades and, in the case of Prussian Blue, by more than two centuries. Among various types of electron transfer are a range of inner sphere processes, many corresponding to class II mixed valency[260] in which thermal energies or photoexcitation are sufficient to excite electrons between the different centres within the framework. In principle, the combination of porosity and electron transfer represents one of the great current challenges in the field, with strong coupling between these effects expected to lead to interesting synergies. The achievement of electrically conducting porous phases, in particular, is of interest for possible applications in molecular sensing and selective electrode materials as well as a number of more advanced functions. With electron transfer being largely unexplored in porous framework phases, only brief attention to this property is given here.

Prussian Blue phases provide a range of examples where electron transfer occurs from metal to metal (intervalence charge-transfer, IVCT). Of particular interest is the influence of photo-induced transfer on magnetic properties. In $K_{0.2}Co_{1.4}[Fe(CN)_6]\cdot 6.9H_2O$,[261] IVCT can be tuned by irradiation with photons of different frequencies, with red light enhancing the magnetisation and increasing the ferrimagnetic ordering temperature from 16 to 19 K through electron transfer from Fe to Co. Blue light, or heating to 150 K, reverses this effect. Similarly, photo-excitation of paramagnetic $Rb_{0.66}Co_{1.25}[Fe(CN)_6]\cdot 4.3H_2O$[262] (with charge distribution $Rb_{0.66}Co^{III}_{0.84}Co^{II}_{0.41}[Fe^{II}(CN)_6]$) yields a defect pair of Fe^{III} (LS) and Co^{II} (HS) that cause ferrimagnetic ordering at 15 K. Reversible photomagnetism has also been observed in rubidium manganese hexacyanidoferrates,[263] with photo-demagnetisation in $Rb_{0.91}Mn_{1.05}[Fe(CN)_6]\cdot 0.6H_2O$ occurring due to conversion from Fe^{II}-CN-Mn^{III} to Fe^{III}-CN-Mn^{II}.[264] A novel further property is that of photo-induced magnetic pole inversion, seen in $(Fe_{0.40}Mn_{0.60})_{1.5}[Cr(CN)_6]\cdot 7.5H_2O$.[265] Octacyanidometallates ($M = Mo, W$) have also produced a number of interesting photo-active phases. Photo-excitation of $Cu_2[Mo(CN)_8]\cdot 8H_2O$,[266] leads to conversion from a paramagnet to a ferromagnet with $T_c = 25$ K. Both temperature- and irradiation-induced IVCT are seen in $Cs[Co^{II}(3\text{-cyanopyridine})_2][W^V(CN)_8]\cdot H_2O$[267] and $Co^{II}_3[W^V(CN)_8]_2(pmd)_4\cdot 6H_2O$ (pmd = pyrimidine) (see Figure 1.20),[268] for which conversion from $Co^{II}(HS, S = 3/2)\text{-}W^V(S = 1/2)$ to $Co^{III}(LS, S = 0)\text{-}W^{IV}(S = 0)$ occurs with broad thermal hysteresis (167–216 K and 208–298 K, respectively). Reversal of this charge transfer with irradiation at low temperature yields metastable ferromagnets with ordering temperatures of 30 and 40 K, respectively.

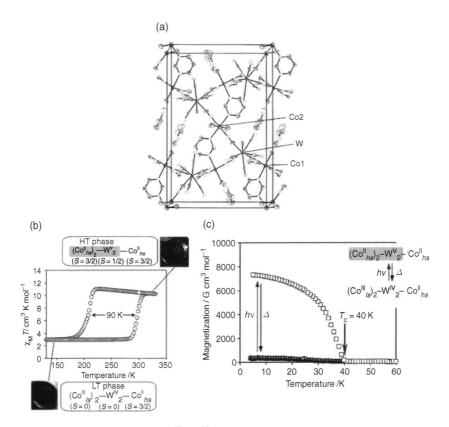

Figure 1.20 (a) Structure of $Co^{II}_3[W^V(CN)_8]_2(pmd)_4\cdot 6H_2O$. (b) Thermal hysteresis in the thermal interconversion between $Co^{II}\text{-}W^V$ (high temperature) and $Co^{III}\text{-}W^{IV}$ (low temperature) forms. (c) Influence of photo-excitation at low temperatures on the magnetisation. Reprinted with permission from S. Ohkoshi, S. Ikeda, T. Hozumi, T. Kashiwagi and K. Hashimoto, *J. Am. Chem. Soc.*, **128**, 5320. Copyright (2006) American Chemical Society

In the parallel investigation of systems where metal-ligand charge-transfer (MLCT) occurs, studies into transition metal complexes of the redox-active quinone ligand have unravelled crossover behaviour that accompanies electron transfer.[269] This reversible process, known as valence tautomerism, has been observed with bistability in the solid state.[270] Exotic photomechanical behaviours, such as the bending of crystals of 1D chain materials with IR irradiation,[271, 272] have been attributed to the unique structural consequences of electron transfer within the solid.

Whereas metal chalcogenides have made some important inroads into the challenge of generating electrically conducting porous phases,[273] little to no progress has been made to date on the merging of these two properties in MOF systems, with only weakly conducting materials achieved. For example, measurements on the fully dehydrated Prussian Blue, $Fe_4[Fe(CN)_6]_3$, indicate that the very modest semiconducting nature of this phase arises due to electron hopping between the Fe^{II} and Fe^{III} sites, a process that is also responsible for its intense blue colour. Electron delocalisation to give metallic conductivity is, however, well known in a number of nonporous phases. These include the Cu-DCNQI system (DCNQI = a range of N,N-dicyanoquinonediimines),[274] in which electron delocalisation and metallic conductivity occur due to a close matching of donor–acceptor electronic energy levels and strong orbital overlap. Also of note are a range of layered organic/inorganic materials in which electrical conduction occurs within electron-delocalised organic layers, such as those containing the bis(ethylenedithio)tetrathiafulvalene (BEDT-TTF) donor molecule;[275] examples here are the paramagnetic supercon-ductor $(BEDT-TF)_4A[Fe(C_2O_4)_3]\cdot C_6H_5CN$ (A = $[H_3O]^+$, K^+, $[NH_4]^+$)[276] and the ferromagnetic metal $(BEDT-TTF)_3[MnCr(C_2O_4)_3]$,[277] each of which contain magnetic oxalate based layers between the conducting organic layers.

1.3.2.3 Photoluminescence

The high level of control over chemical structure conferred by metal-organic synthesis makes MOF materials a fertile area for the achievement of novel optical properties. Such properties include nonlinear optics (NLO), achieved through the generation of noncentrosymmetric frame-works, and luminescence, achieved through the strategic arrangement of photo-active metal ions and organic ligands. A motivation for the latter is that MOF formation commonly leads to local geometric constraints that can lead to increased quantum efficiencies and fluorescence lifetimes.

Most notably, particular recent attention has been given to the synthesis of materials that are both porous and luminescent due to an interest in solvato-optical properties.[10, 13, 278, 279]

In addition to the processes described in Section 1.3.2.2 involving electron transfer between neighbouring metals (IVCT) and metal/ligand (MLCT and LMCT), luminescence can arise due to metal-based emission (e.g. for lanthanoid and d^{10} transition metal ions), ligand based emission (e.g. for conjugated organic linkers), guest molecule emission, and exciplex formation between host and guest.[278] Each of these processes can be influenced strongly by the presence of adsorbed guest molecules, providing a sensitive mechanism for molecular sensing that in principle promises detection levels approaching the single-molecule limit.[278]

The incorporation of luminescent lanthanoid nodes into framework lattices has led to a number of materials in which guest-dependent luminescence is seen. In the highly porous [Tb$_2$(tatb)$_2$(dma)$_3$] (tatb = triazine-1,3,5-(4,4′,4″-trisbenzoate); dma = N,N-dimethylacetamide), which is remarkable in containing 47 and 39 Å pores within a cubic lattice with cell parameter a = 123.901(1) Å, adsorption of ferrocene molecules leads to a quenching of the Tb^{III} emission that is attributable to a nonradiative energy-transfer pathway between host and guest.[47] Further, it was found in this system that emission from the included ferrocene guests was higher than that expected, suggesting that the framework acts as an antenna in harvesting photons for the guests. Similar reversible guest-induced quenching is seen with the adsorption of I$_2$ into [Eu$_2L_3$(DMSO)$_2$(MeOH)$_2$] (L = 4,4′-ethyne-1,2-diyldibenzoate)[280] and aromatic molecules into [Cu$_6L_6$] (where L is 5,6-diphenyl-1,2,4-triazine-3-thiol)[281] and [(ZnCl$_2$)$_3$(tpdpb)] (where tpdpb = 1,3,5-tris(p-(2,2′-dipyridylamino)phenyl)benzene).[282] Among a number of systems in which guest molecules coordinate to bare sites on lanthanoid ions and thereby change their luminescent properties,[283–286] the desorption of bound water from [Ln_2(fum)$_2$(ox) (H$_2$O)$_4$] (Ln = Eu, Tb; fum = fumarate; ox = oxalate) leads to the almost complete quenching of luminescence, a process that is reversible.[283] The desorption and subsequent adsorption of ammonia onto the bare Tb^{III} sites within [Tb$_2$(1,4-bdc)$_3$(H$_2$O)$_4$] (MOF-76) leads to a change in the fluorescence decay constants from 1.13 ms^{-1} (H$_2$O) to 0.74 ms^{-1} (apohost) to 1.00 (NH$_3$).[284] The luminescence of its methanol-exchanged analogue, MOF-76b, is enhanced upon exposure to solutions of anions, with fluoride exchange leading to a fourfold increase due to the formation of hydrogen bonding interactions between bound methanol and included anion.[287] Similar anion sensing capabilities are dis- played by [Tb^{III}(mucicate)$_{1.5}$(H$_2$O)$_2$][288] and [Zn$_2$(4,4′-bpy)(H$_2$O)$_8$(ClO$_4$)$_2$ (4-aminobenzoate)$_2$]·2(4,4′-bpy),[289] with the latter arising due to replacement of the bound $[ClO_4]^-$ ions. Cation sensing capabilities have also been exhibited.[290, 291]

In addition to quenching and enhancement effects, guest-induced shifts in the luminescent emission of the host have been reported. In principle, spectral changes of this type provide a more versatile, albeit potentially less sensitive, approach for molecular sensing than those given above. In $[Zn_4O(ntb)_2]$ (ntb = 4,4',4''-nitrilotrisbenzoate) the presence of host–guest π–π interactions leads to a shift in λ_{max} from 435 nm (pyridine) to 456 nm (methanol) to 466 nm (benzene).[292] The luminescence of this phase likely originates from the ntb linker, although may also result from LMCT within the Zn_4O cluster. A similar behaviour is seen with the adsorption of a range of different guest molecules into $[Zn_4O(sdc)_3]$ (sdc = trans-4,4'-stilbenedicarboxylate).[293] The absence of any clear relationship between spectral shift and guest polarity for this system suggests that the luminescence is sensitive to the specific nature of the host–guest interaction rather than being determined purely electrostatically.

1.3.3 Structural and Mechanical Properties

In the same way that the subtle energetics associated with guest adsorption and desorption can be sufficient to drive pronounced structural deformations in underconstrained MOF lattices (as described in Section 2.1.2), it has been found that variations in temperature and pressure can also lead to significant structural variation in these systems, both dynamic and static in nature, to yield novel mechanical properties.

1.3.3.1 Anomalous Thermal Expansivities

The expansion of chemical bonds with increasing temperature leads the vast majority of known solids to expand with heating (positive thermal expansion, PTE), a property once thought to be an immutable law of nature. A relatively small number of materials are known that defy this expectation and contract upon heating (i.e. display negative thermal expansion, NTE) or are temperature-invariant (i.e. display zero thermal expansion, ZTE). These novel behaviours arise due to a range of physical mechanisms that include IVCT,[294–296] magnetostriction[297] and, most commonly, transverse lattice vibrations.[298–300] Examples in the latter class include a family of oxide based materials, the most prominent being ZrW_2O_8,[301] which has a coefficient of thermal expansion $\alpha = d\ell/\ell dT = -9.1 \times 10^{-6}$ K^{-1}.

Investigation of the thermal expansivities of MOFs has recently led to the discovery of ZTE[302] and NTE[303–314] in both cyanide-[302–311] and benzene(di/tri)carboxylate-bridged frameworks.[312–314] Structural and theoretical analyses have shown that the multiply hinged molecular linkages of each class confer unprecedented vibrational flexibility to their framework lattices – a feature that is in contrast to all other NTE systems known. In the cyanide phases the double-hinged M-CN-M bridge uniquely allows each metal centre to achieve rotational and translational freedom from its neighbours,[303] whereas the mechanism for NTE in the polycarboxylato systems is considerably more complex, arising from both local and long-range vibrations.[312–314]

A direct consequence of the existence of numerous low energy transverse vibrational modes in MOFs is that these materials exhibit extreme NTE behaviours. Among a range of cubic metal cyanide systems that display isotropic NTE,[303–307] the interpenetrated diamondoid phases $Zn(CN)_2$ and $Cd(CN)_2$ have $\alpha = -16.9 \times 10^{-6}$ K^{-1} and -20.4×10^{-6} K^{-1}, respectively.[303] The desorption of volatile guests from single diamondoid network $Cd(CN)_2$ to achieve a 64 % porous apohost phase leads to the largest isotropic NTE yet reported for any material, with $\alpha = -33.5 \times 10^{-6}$ K^{-1} (see Figure 1.21).[304] The thermal expansivity of this phase can be tuned by adsorbing guest molecules into the porous framework, as seen for example with N_2 adsorption below 150 K to yield PTE behaviour.

More recently, the generation of noncubic metal cyanide frameworks has led to the discovery of colossal uniaxial NTE in these systems.[309, 315] In $Ag_3[Co(CN)_6]$, which consists of a 3D lattice of hexagonal symmetry, variation in temperature leads to a highly pronounced temperature-dependent hinging of the structure (see Figure 1.22), resulting in colossal thermal contraction along the c-axis ($\alpha_c \cong -125 \times 10^{-6}$ K^{-1}) and colossal expansion in the ab-plane ($\alpha_a \cong +140 \times 10^{-6}$ K^{-1}).[309] This property arises due to the very fine balance between the energetics of framework distortion and argentophilic interactions, with the latter favouring increased deformation away from the more regular α-Po (cubic) network geometry with decreasing temperature. Confirmation that the argentophillic interactions play a critical role in this property was provided by analysis of an isostructural Ag-free analogue, $H_3[Co(CN)_6]$, which exhibits conventional expansivities.

1.3.3.2 Compressibilities

Compressibility is an important materials property, both from a fundamental viewpoint in providing information on the energetics of structural deformations, and technologically, with many proposed

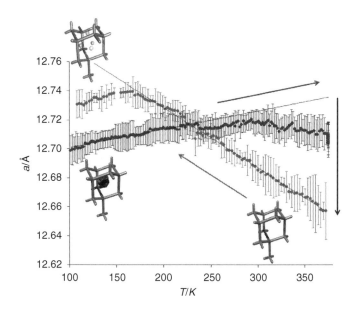

Figure 1.21 Variation in the cubic lattice parameter of Cd(CN)$_2$·n(guest) with temperature. The parent phase Cd(CN)$_2$·CCl$_4$ displays positive thermal expansion, whereas the apohost Cd(CN)$_2$ displays the most pronounced isotropic negative thermal expansion behaviour known. Reprinted with permission from A.E. Phillips, A.L. Goodwin, G.J. Halder, P.D. Southon and C.J. Kepert, *Angew. Chem. Int. Ed.*, **47**, 1396. Copyright (2008) Wiley-VCH Verlag GmbH & Co

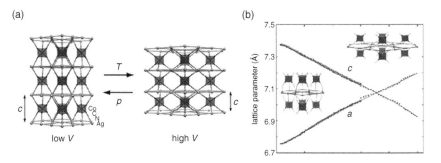

Figure 1.22 (a) Diagrammatic representation of the effect of changes in temperature and isotropic pressure on Ag$_3$[Co(CN)$_6$]. (b) Variation in the a and c parameters of the hexagonal unit cell with temperature, showing the colossal uniaxial NTE along the c-direction. (a) Reprinted with permission from A.L. Goodwin, D.A. Keen and M.G. Tucker, *Proc. Natl. Acad. Sci. USA*, **105**, 18708. Copyright (2008) National Academy of Sciences. (b) Reprinted with permission from A. L. Goodwin, M. Calleja, M. J. Conterio, M. T. Dove, J. S. O. Evans, D. A. Keen, L. Peters, M. G. Tucker, *Science* 2008, **319**, 794–797. Copyright (2008) AAAS

applications for MOFs (*e.g.* in gas separation and storage) requiring sample pelletisation and with pressure potentially representing a useful mechanism for post-synthetic modification of framework structure and adsorption properties. Whereas the compressibilities of materials such as metal oxides have been the subject of considerable investigation, little is currently known about the response of MOFs to external pressures.

Very high compressibilities are expected in MOF lattices due to their relative softness, topological underconstraint and stuctural openness. A high pressure synchrotron powder X-ray diffraction investigation of $[Cu_3(btc)_2]$ has confirmed this to be the case, with a bulk modulus $[K = 1/\beta = -V(\partial P/\partial V)_T$ where β is the compressibility] of $+30$ GPa determined at ambient temperature through the use of nonadsorbing pressure media.[317] The application of pressure using small-molecule liquids led, in contrast, to interesting behaviours in which the framework was found to be comparatively incompressible at low pressures due to the pressure induced adsorption of the liquids into the pores of the material.

High pressure measurements have also been performed on nonporous MOF phases and novel behaviours observed. The interpenetrated NTE phase $Zn(CN)_2$ (see Section 1.3.3.1) has $K_0 = 34.2(2)$ GPa and becomes more compressible at higher pressures.[318] The NTE behaviour of this phase increases at a rate of $-1 \times 10^{-6}\,K^{-1}$ per 0.2 GPa due to a pressure-induced softening of the low energy transverse vibrations. Application of isotropic pressure to the colossal uniaxial NTE phase $Ag_3[Co(CN)_6]$ (see Section 1.3.3.1 and Figure 1.22) yields the largest negative linear compressibility (NLC) yet seen for an inorganic material, with $\beta_\ell = -(\partial ln\ell/\partial P)_T = -76(9)$ TPa^{-1} along the *c*-axis.[316] Positive compressibility is seen in the *ab*-plane, with $\beta_a = 115(8)$ TPa^{-1}. The bulk modulus for this material is very small $[K = +6.5(3)$ GPa], reflective of very high compressibility.

Lastly, it has been predicted that auxetic properties (*i.e.* negative Poisson ratio; orthogonal contraction upon axial compression, and *vice versa*), which are closely related to NTE, may also arise in MOFs,[319–321] but this highly sought after behaviour is yet to be reported.

1.4 CONCLUDING REMARKS

The synthesis of MOFs offers enormous scope for the realisation of highly impressive and very useful materials properties. In combining the versatility and diversity of coordination chemistry, organic chemistry and supramolecular assembly, an unprecedented degree of structural complexity can be incorporated through multiple synthetic steps. The rational design of these materials covers both the identity of the individual building units, with fine chemical control over these being possible prior to, during, and after MOF synthesis, and the way in which they are arranged in space, with an appreciable degree of control over framework structure arising due to the strong directionality of the coordination linkages and to the capacity for post-synthetic modification. Exploitation of the many novel synthetic and structural aspects of these systems has led to the achievement of a diverse range of remarkable chemical and physical properties, many of which are superior to those of all other known classes of material and some of which are unprecedented. Foremost among these has been the achievement of the highest known surface areas for porous materials, leading to unprecedented gravimetric and volumetric uptakes of technologically important gases such as hydrogen and methane, and the generation and stabilisation of the largest known pores within crystalline materials. The fine control over pore structure and surface chemistry has in turn seen the achievement of very high selectivities for guest adsorption, leading to current scale-up efforts for industrially important separation and purification processes. Moreover, the unique ability to generate chiral frameworks through homochiral syntheses rather than chiral surface modification has led to some of the first demonstrations of enantioselective adsorption and heterogeneous catalysis within porous materials. In targeting other advanced forms of physical function, exploitation of the unique magnetic, electronic and optical properties of metal complexes and organic molecules has seen the realisation of a number of remarkable physical properties within porous hosts for the first time. These notably include the generation of porous magnets, porous hosts that are able to switch between multiple spin states, and porous luminescent materials, for each of which the host–guest chemistry and magnetic/electronic/optical functions are intertwined in interesting and potentially useful ways. Investigation of the pronounced structural flexibilities of MOFs has led to the achievement of a range of unprecedented materials properties, both relating to host–guest chemistry and mechanical properties, with the latter notably including the discovery of the highest negative thermal expansivities and nonlinear compressibilities known.

In reflecting on the immensely rich host–guest chemistry of MOFs it is encouraging to note that the numerous achievements highlighted above have emerged almost entirely this century. This extraordinarily rapid development has been made possible by the establishment of many important synthetic design principles in the 1990s and, more generally, has built on more than 100 years of coordination chemistry research. Given the rapid current expansion of the field, with particular focus both on porosity and on the targeted incorporation of other functional properties, it is reasonable to believe that the broad range of impressive materials properties outlined above are only the tip of the iceberg when considering the future scope for functionality in porous MOFs. In addition to further projected improvements in host–guest properties, considerable scope exists for the combination of multiple properties within individual systems to achieve a diverse array of further unique materials properties, in particular through the control of electron mobility and excitation. Armed with the considerable versatility of coordination chemistry, an ever improving eye for ligand and framework design, and an increasingly sophisticated arsenal of structural and physical characterisation techniques, we can look forward to further rapid developments in the future.

ACKNOWLEDGEMENTS

The author wishes to express his sincere thanks to his students and research fellows, past and present, who have worked tirelessly in MOF chemistry. He is grateful also to the Australian Research Council for providing ARC Discovery and Fellowship Grant funding to study these materials.

REFERENCES

[1] B.F. Hoskins and R. Robson, *J. Am. Chem. Soc.*, **112**, 1546 (1990).
[2] J.R. Long and O.M. Yaghi, *Chem. Soc. Rev.*, **38**, 1213 (2009).
[3] D.J. Tranchemontagne, J.L. Mendoza-Cortes, M. O'Keeffe and O.M. Yaghi, *Chem. Soc. Rev.*, **38**, 1257 (2009).
[4] M. O'Keeffe, *Chem. Soc. Rev.*, **38**, 1215 (2009).
[5] S. Kitagawa, R. Kitaura and S. Noro, *Angew. Chem. Int. Ed.*, **43**, 2334 (2004).
[6] G. Férey, *Chem. Soc. Rev.*, **37**, 191 (2008).
[7] M.J. Rosseinsky, *Microporous Mesoporous Mater.*, **73**, 15 (2004).
[8] C.J. Kepert, *Chem. Commun.*, **7**, 695 (2006).
[9] C.N.R. Rao, A.K. Cheetham and A. Thirumurugan, *J. Phys.: Condens. Matter*, **20**, 083202 (2008).
[10] C. Janiak, *Dalton Trans.*, **14**, 2781 (2003).
[11] B. Moulton and M.J. Zaworotko, *Chem. Rev.*, **101**, 1629 (2001).

[12] R.J. Hill, D.L. Long, N.R. Champness, P. Hubberstey and M. Schröder, *Acc. Chem. Res.*, **38**, 335 (2005).

[13] D. Maspoch, D. Ruiz-Molina and J. Veciana, *Chem. Soc. Rev.*, **36**, 770 (2007).

[14] S.L. James, *Chem. Soc. Rev.*, **32**, 276 (2003).

[15] J.L. Atwood, J.E.D. Davies, D.D. MacNicol and F. Vögtle (Eds), *Comprehensive Supramolecular Chemistry*, Vols 6–10, Pergamon Press, New York, 1996.

[16] J.L. Atwood, J.E.D. Davies and D.D. MacNicol (Eds), *Inclusion Compounds*, Vols 1–4, Academic Press, London, 1984.

[17] O.M. Yaghi, G. Li and H. Li, *Nature*, **378**, 703 (1995).

[18] M. Fujita, J. Yazaki and K. Ogura, *J. Am. Chem. Soc.*, **112**, 5645 (1990).

[19] R. Robson, *J. Chem. Soc., Dalton Trans.*, 3735 (2000).

[20] R. Robson, *Dalton Trans.*, 5113 (2008).

[21] M. Eddaoudi, D.B. Moler, H.L. Li, B.L. Chen, T.M. Reineke, M. O'Keeffe and O.M. Yaghi, *Acc. Chem. Res.*, **34**, 319 (2001).

[22] O.M. Yaghi, M. O'Keeffe, N.W. Ockwig, H.K. Chae, M. Eddaoudi and J. Kim, *Nature*, **423**, 705 (2003).

[23] A.F. Wells, *Three-Dimensional Nets and Polyhedra*, Wiley-Interscience, New York, 1977.

[24] M. O'Keeffe, M.A. Peskov, S.J. Ramsden and O.M. Yaghi, *Acc. Chem. Res.*, **41**, 1782 (2008).

[25] S. Hyde, O.D. Friedrichs, S.J. Ramsden and V. Robins, *Solid State Sci.*, **8**, 740 (2006).

[26] L. Ohrström and K. Larsson, *Molecule-Based Materials: The Structural Network Approach*, Elsevier, Amsterdam, 2005.

[27] S.R. Batten and R. Robson, *Angew. Chem. Int. Ed.*, **37**, 1460 (1998).

[28] J.J. Perry, J.A. Perman and M.J. Zaworotko, *Chem. Soc. Rev.*, **38**, 1400 (2009).

[29] J.S. Seo, D. Whang, H. Lee, S.I. Jun, J. Oh, Y.J. Jeon and K. Kim, *Nature*, **404**, 982 (2000).

[30] R. Vaidhyanathan, D. Bradshaw, J.N. Rebilly, J.P. Barrio, J.A. Gould, N.G. Berry and M.J. Rosseinsky, *Angew. Chem. Int. Ed.*, **45**, 6495 (2006).

[31] B.F. Abrahams, M. Moylan, S.D. Orchard and R. Robson, *Angew. Chem. Int. Ed.*, **42**, 1848 (2003).

[32] B. Kesanli and W.B. Lin, *Coord. Chem. Rev.*, **246**, 305 (2003).

[33] W.B. Lin, *J. Solid State Chem.*, **178**, 2486 (2005).

[34] D.N. Dybtsev, A.L. Nuzhdin, H. Chun, K.P. Bryliakov, E.P. Talsi, V.P. Fedin and K. Kim, *Angew. Chem. Int. Ed.*, **45**, 916 (2006).

[35] C.D. Wu, A. Hu, L. Zhang and W.B. Lin, *J. Am. Chem. Soc.*, **127**, 8940 (2005).

[36] C.D. Wu and W.B. Lin, *Angew. Chem. Int. Ed.*, **46**, 1075 (2007).

[37] S.S. Iremonger, P.D. Southon and C.J. Kepert, *Dalton Trans.*, 6103 (2008).

[38] T. Ezuhara, K. Endo and Y. Aoyama, *J. Am. Chem. Soc.*, **121**, 3279 (1999).

[39] M.J. Ingleson, J.P. Barrio, J. Bacsa, C. Dickinson, H. Park and M.J. Rosseinsky, *Chem. Commun.*, 1287 (2008).

[40] D. Bradshaw, J.B. Claridge, E.J. Cussen, T.J. Prior and M.J. Rosseinsky, *Acc. Chem. Res.*, **38**, 273 (2005).

[41] D. Bradshaw, T.J. Prior, E.J. Cussen, J.B. Claridge and M.J. Rosseinsky, *J. Am. Chem. Soc.*, **126**, 6106 (2004).

[42] C.J. Kepert, T.J. Prior and M.J. Rosseinsky, *J. Am. Chem. Soc.*, **122**, 5158 (2000).

[43] C.J. Kepert and M.J. Rosseinsky, *Chem. Commun.*, 31 (1998).

[44] G. Férey and A.K. Cheetham, *Science*, **283**, 1125 (1999).
[45] G. Férey, C. Mellot-Draznieks, C. Serre and F. Millange, *Acc. Chem. Res.*, **38**, 217 (2005).
[46] C.T. Kresge, M.E. Leonowicz, W.J. Roth, J.C. Vartuli and J.S. Beck, *Nature*, **359**, 710 (1992).
[47] Y.K. Park, S.B. Choi, H. Kim, K. Kim, B.H. Won, K. Choi, J.S. Choi, W.S. Ahn, N. Won, S. Kim, D.H. Jung, S.H. Choi, G.H. Kim, S.S. Cha, Y.H. Jhon, J.K. Yang and J. Kim, *Angew. Chem. Int. Ed.*, **46**, 8230 (2007).
[48] D. Britt, D. Tranchemontagne and O.M. Yaghi, *Proc. Natl. Acad. Sci. U.S.A*, **105**, 11623 (2008).
[49] H.K. Chae, D.Y. Siberio-Perez, J. Kim, Y. Go, M. Eddaoudi, A.J. Matzger, M. O'Keeffe and O.M. Yaghi, *Nature*, **427**, 523 (2004).
[50] H. Li, M. Eddaoudi, M. O'Keeffe and O.M. Yaghi, *Nature*, **402**, 276 (1999).
[51] M. Eddaoudi, J. Kim, N. Rosi, D. Vodak, J. Wachter, M. O'Keefe and O.M. Yaghi, *Science*, **295**, 469 (2002).
[52] G. Férey, C. Mellot-Draznieks, C. Serre, F. Millange, J. Dutour, S. Surblé and I. Margiolaki, *Science*, **309**, 2040 (2005).
[53] M. Latroche, S. Surble, C. Serre, C. Mellot-Draznieks, P.L. Llewellyn, J.H. Lee, J.S. Chang, S.H. Jhung and G. Férey, *Angew. Chem. Int. Ed.*, **45**, 8227 (2006).
[54] B. Wang, A.P. Cote, H. Furukawa, M. O'Keeffe and O.M. Yaghi, *Nature*, **453**, 207 (2008).
[55] K.S. Park, Z. Ni, A.P. Cote, J.Y. Choi, R.D. Huang, F.J. Uribe-Romo, H.K. Chae, M. O'Keeffe and O.M. Yaghi, *Proc. Natl. Acad. Sci. USA*, **103**, 10186 (2006).
[56] X. Lin, I. Telepeni, A.J. Blake, A. Dailly, C.M. Brown, J.M. Simmons, M. Zoppi, G.S. Walker, K.M. Thomas, T.J. Mays, P. Hubberstey, N.R. Champness and M. Schröder, *J. Am. Chem. Soc.*, **131**, 2159 (2009).
[57] X. Lin, J.H. Jia, X.B. Zhao, K.M. Thomas, A.J. Blake, G.S. Walker, N.R. Champness, P. Hubberstey and M. Schröder, *Angew. Chem. Int. Ed.*, **45**, 7358 (2006).
[58] H. Furukawa, M.A. Miller and O.M. Yaghi, *J. Mater. Chem.*, **17**, 3197 (2007).
[59] Z.Q. Wang and S.M. Cohen, *Chem. Soc. Rev.*, **38**, 1315 (2009).
[60] J.L.C. Rowsell and O.M. Yaghi, *J. Am. Chem. Soc.*, **128**, 1304 (2006).
[61] A.P. Nelson, O.K. Farha, K.L. Mulfort and J.T. Hupp, *J. Am. Chem. Soc.*, **131**, 458 (2009).
[62] S.S.Y. Chui, S.M.F. Lo, J.P.H. Charmant, A.G. Orpen and I.D. Williams, *Science*, **283**, 1148 (1999).
[63] Y. Liu, G. Li, X. Li and Y. Cui, *Angew. Chem. Int. Ed.*, **46**, 6301 (2007).
[64] M. Oh and C.A. Mirkin, *Angew. Chem. Int. Ed.*, **45**, 5492 (2006).
[65] F. Nouar, J. Eckert, J.F. Eubank, P. Forster and M. Eddaoudi, *J. Am. Chem. Soc.*, **131**, 2864 (2009).
[66] S.R. Halper, L. Do, J.R. Stork and S.M. Cohen, *J. Am. Chem. Soc.*, **128**, 15255 (2006).
[67] K.S. Min and M.P. Suh, *J. Am. Chem. Soc.*, **122**, 6834 (2000).
[68] S. Das, H. Kim and K. Kim, *J. Am. Chem. Soc.*, **131**, 3814 (2009).
[69] S. Bureekaew, S. Horike, M. Higuchi, M. Mizuno, T. Kawamura, D. Tanaka, N. Yanai and S. Kitagawa, *Nat. Mater.*, **8**, 831 (2009).
[70] H. Kitagawa, Y. Nagao, M. Fujishima, R. Ikeda and S. Kanda, *Inorg. Chem. Commun.*, **6**, 346 (2003).
[71] L.Q. Ma, C. Abney and W.B. Lin, *Chem. Soc. Rev.*, **38**, 1248 (2009).

[72] Y.K. Hwang, D.Y. Hong, J.S. Chang, S.H. Jhung, Y.K. Seo, J. Kim, A. Vimont, M. Daturi, C. Serre and G. Férey, *Angew. Chem. Int. Ed.*, **47**, 4144 (2008).

[73] S.S. Kaye and J.R. Long, *J. Am. Chem. Soc.*, **130**, 806 (2008).

[74] H.J. Choi and M.P. Suh, *J. Am. Chem. Soc.*, **126**, 15844 (2004).

[75] K.L. Mulfort and J.T. Hupp, *Inorg. Chem.*, **47**, 7936 (2008).

[76] K.L. Mulfort, T.M. Wilson, M.R. Wasielewski and J.T. Hupp, *Langmuir*, **25**, 503 (2009).

[77] K.K. Tanabe, Z.Q. Wang and S.M. Cohen, *J. Am. Chem. Soc.*, **130**, 8508 (2008).

[78] M.J. Ingleson, J.P. Barrio, J.B. Guilbaud, Y.Z. Khimyak and M.J. Rosseinsky, *Chem. Commun.*, 2680 (2008).

[79] G.J. Halder and C.J. Kepert, *J. Am. Chem. Soc.*, **127**, 7891 (2005).

[80] R. Kitaura, S. Kitagawa, Y. Kubota, T.C. Kobayashi, K. Kindo, Y. Mita, A. Matsuo, M. Kobayashi, H.C. Chang, T.C. Ozawa, M. Suzuki, M. Sakata and M. Takata, *Science*, **298**, 2358 (2002).

[81] P.V. Ganesan and C.J. Kepert, *Chem. Commun.*, 2168 (2004).

[82] C.J. Kepert and M.J. Rosseinsky, *Chem. Commun.*, 375 (1999).

[83] B.F. Abrahams, P.A. Jackson and R. Robson, *Angew. Chem. Int. Ed.*, **37**, 2656 (1998).

[84] B. Rather and M.J. Zaworotko, *Chem. Commun.*, 830 (2003).

[85] E.Y. Lee and M.P. Suh, *Angew. Chem. Int. Ed.*, **43**, 2798 (2004).

[86] K. Biradha, Y. Hongo and M. Fujita, *Angew. Chem. Int. Ed.*, **39**, 3843 (2000).

[87] A.J. Fletcher, E.J. Cussen, T.J. Prior, M.J. Rosseinsky, C.J. Kepert and K.M. Thomas, *J. Am. Chem. Soc.*, **123**, 10001 (2001).

[88] M. Kondo, M. Shimamura, S.-Noro, S. Minakoshi, A. Asami, K. Seki and S. Kitagawa, *Chem. Mater.*, **12**, 1288 (2000).

[89] M. Kondo, T. Yoshitomi, K. Seki, H. Matsuzaka and S. Kitagawa, *Angew. Chem. Int. Ed.*, **36**, 1725 (1997).

[90] H. Li, M. Eddaoudi, T.L. Groy and O.M. Yaghi, *J. Am. Chem. Soc.*, **120**, 8571 (1998).

[91] T. Duren, Y.S. Bae and R.Q. Snurr, *Chem. Soc. Rev.*, **38**, 1237 (2009).

[92] S.S. Han, J.L. Mendoza-Cortes and W.A. Goddard, *Chem. Soc. Rev.*, **38**, 1460 (2009).

[93] E.J. Cussen, J.B. Claridge, M.J. Rosseinsky and C.J. Kepert, *J. Am. Chem. Soc.*, **124**, 9574 (2002).

[94] M.P. Suh, J.W. Ko and H.J. Choi, *J. Am. Chem. Soc.*, **124**, 10976 (2002).

[95] K. Biradha and M. Fujita, *Angew. Chem. Int. Ed.*, **41**, 3392 (2002).

[96] H. Kumagai, K.W. Chapman, C.J. Kepert and M. Kurmoo, *Polyhedron*, **22**, 1921 (2003).

[97] G.J. Halder, C.J. Kepert, B. Moubaraki, K.S. Murray and J.D. Cashion, *Science*, **298**, 1762 (2002).

[98] R. Kitaura, K. Fujimoto, S. Noro, M. Kondo and S. Kitagawa, *Angew. Chem. Int. Ed.*, **41**, 133 (2002).

[99] G. Férey and C. Serre, *Chem. Soc. Rev.*, **38**, 1380 (2009).

[100] R. Kitaura, K. Seki, G. Akiyama and S. Kitagawa, *Angew. Chem. Int. Ed.*, **42**, 428 (2003).

[101] A.J. Fletcher, E.J. Cussen, T.J. Prior, M.J. Rosseinsky, C.J. Kepert and K.M. Thomas, *J. Am. Chem. Soc.*, **123**, 10001 (2001).

[102] F. Millange, C. Serre, N. Guillou, G. Férey and R.I. Walton, *Angew. Chem. Int. Ed.*, **47**, 4100 (2008).

[103] Y. Liu, J.H. Her, A. Dailly, A.J. Ramirez-Cuesta, D.A. Neumann and C.M. Brown, *J. Am. Chem. Soc.*, **130**, 11813 (2008).

[104] S. Bourrelly, P.L. Llewellyn, C. Serre, F. Millange, T. Loiseau and G. Férey, *J. Am. Chem. Soc.*, **127**, 13519 (2005).

[105] A. Pichon, A. Lazuen-Garay and S.L. James, *CrystEngComm*, **8**, 211 (2006).

[106] V. Niel, A.L. Thompson, M.C. Muñoz, A. Galet, A.S.E. Goeta and J.A. Real, *Angew. Chem. Int. Ed.*, **42**, 3760 (2003).

[107] A.V. Nossov, D.V. Soldatov and J.A. Ripmeester, *J. Am. Chem. Soc.*, **123**, 3563 (2001).

[108] M. Edgar, R. Mitchell, A.M.Z. Slawin, P. Lightfoot and P.A. Wright, *Chem. Eur. J.*, **7**, 5168 (2001).

[109] J.J. Vittal, *Coord. Chem. Rev.*, **251**, 1781 (2007).

[110] G.S. Papaefstathiou, Z. Zhong, L. Geng and L.R. MacGillivray, *J. Am. Chem. Soc.*, **126**, 9158 (2004).

[111] G.S. Papaefstathiou, I.G. Georgiev, T. Friscic and L.R. MacGillivray, *Chem. Commun.*, 3974 (2005).

[112] N.L. Toh, M. Nagarathinam and J.J. Vittal, *Angew. Chem. Int. Ed.*, **44**, 2237 (2005).

[113] M.J. Vela, B.B. Snider and B.M. Foxman, *Chem. Mater.*, **10**, 3167 (1998).

[114] T. Uemura, N. Yanai and S. Kitagawa, *Chem. Soc. Rev.*, **38**, 1228 (2009).

[115] R.E. Morris and P.S. Wheatley, *Angew. Chem. Int. Ed.*, **47**, 4966 (2008).

[116] L.J. Murray, M. Dincă and J.R. Long, *Chem. Soc. Rev.*, **38**, 1294 (2009).

[117] K.M. Thomas, *Dalton Trans.*, 1487 (2009).

[118] J.L.C. Rowsell and O.M. Yaghi, *Angew. Chem. Int. Ed.*, **44**, 4670 (2005).

[119] D.J. Collins and H.C. Zhou, *J. Mater. Chem.*, **17**, 3154 (2007).

[120] H. Wu, W. Zhou and T. Yildirim, *J. Am. Chem. Soc.*, **131**, 4995 (2009).

[121] R. Matsuda, R. Kitaura, S. Kitagawa, Y. Kubota, R.V. Belosludov, T.C. Kobayashi, H. Sakamoto, T. Chiba, M. Takata, Y. Kawazoe and Y. Mita, *Nature*, **436**, 238 (2005).

[122] S.C. Xiang, W. Zhou, J.M. Gallegos, Y. Liu and B.L. Chen, *J. Am. Chem. Soc.*, **131**, 12415 (2009).

[123] P. Horcajada, C. Serre, G. Maurin, N.A. Ramsahye, F. Balas, M. Vallet-Regi, M. Sebban, F. Taulelle and G. Férey, *J. Am. Chem. Soc.*, **130**, 6774 (2008).

[124] P. Horcajada, C. Serre, M. Vallet-Regi, M. Sebban, F. Taulelle and G. Férey, *Angew. Chem. Int. Ed.*, **45**, 5974 (2006).

[125] W.J. Rieter, K.M. Pott, K.M.L. Taylor and W.B. Lin, *J. Am. Chem. Soc.*, **130**, 11584 (2008).

[126] J.L.C. Rowsell, A.R. Millward, K.S. Park and O.M. Yaghi, *J. Am. Chem. Soc.*, **126**, 5666 (2004).

[127] S.S. Kaye, A. Dailly, O.M. Yaghi and J.R. Long, *J. Am. Chem. Soc.*, **129**, 14176 (2007).

[128] A.G. Wong-Foy, A.J. Matzger and O.M. Yaghi, *J. Am. Chem. Soc.*, **128**, 3494 (2006).

[129] M. Dincă, A. Dailly, Y. Liu, C.M. Brown, D.A. Neumann and J.R. Long, *J. Am. Chem. Soc.*, **128**, 16876 (2006).

[130] V.K. Peterson, Y. Liu, C.M. Brown and C.J. Kepert, *J. Am. Chem. Soc.*, **128**, 15578 (2006).

[131] P.M. Forster, J. Eckert, B.D. Heiken, J.B. Parise, J.W. Yoon, S.H. Jhung, J.S. Chang and A.K. Cheetham, *J. Am. Chem. Soc.*, **128**, 16846 (2006).

[132] S.S. Kaye and J.R. Long, *Chem. Commun.*, 4486 (2007).

[133] J.G. Vitillo, L. Regli, S. Chavan, G. Ricchiardi, G. Spoto, P.D.C. Dietzel, S. Bordiga and A. Zecchina, *J. Am. Chem. Soc.*, 130, 8386 (2008).

[134] B. Chen, X. Zhao, A. Putkham, K. Hong, E.B. Lobkovsky, E.J. Hurtado, A.J. Fletcher and K.M. Thomas, *J. Am. Chem. Soc.*, 130, 6411 (2008).

[135] X.B. Zhao, B. Xiao, A.J. Fletcher, K.M. Thomas, D. Bradshaw and M.J. Rosseinsky, *Science*, 306, 1012 (2004).

[136] J.R. Li, R.J. Kuppler and H.C. Zhou, *Chem. Soc. Rev.*, 38, 1477 (2009).

[137] C. Radu and A.M. Bruce, *Eur. J. Inorg. Chem.*, 2007, 1321 (2007).

[138] S. Ahuja, In *Chiral Separations: An Overview*, Vol. 471, S. Ahuja (Ed), American Chemical Society, Washington, DC, 1991, p. 1.

[139] S. Allenmark, *Chromatographic Enantioseparations*, Ellis Horwood, New York, 1991.

[140] D.N. Dybtsev, H. Chun, S.H. Yoon, D. Kim and K. Kim, *J. Am. Chem. Soc.*, 126, 32 (2004).

[141] M. Dincă and J.R. Long, *J. Am. Chem. Soc.*, 127, 9376 (2005).

[142] H.R. Li, Y. Tao, Q. Yu, X.H. Bu, H. Sakamoto and S. Kitagawa, *Chem. Eur. J.*, 14, 2771 (2008).

[143] S.Q. Ma, X.S. Wang, C.D. Collier, E.S. Manis and H.C. Zhou, *Inorg. Chem.*, 46, 8499 (2007).

[144] R. Banerjee, A. Phan, B. Wang, C. Knobler, H. Furukawa, M. O'Keeffe and O.M. Yaghi, *Science*, 319, 939 (2008).

[145] H.L. Guo, G.S. Zhu, I.J. Hewitt and S.L. Qiu, *J. Am. Chem. Soc.*, 131, 1646 (2009).

[146] T.K. Maji, K. Uemura, H.C. Chang, R. Matsuda and S. Kitagawa, *Angew. Chem. Int. Ed.*, 43, 3269 (2004).

[147] S.Q. Ma, D.F. Sun, X.S. Wang and H.C. Zhou, *Angew. Chem. Int. Ed.*, 46, 2458 (2007).

[148] H.M. Powell and J.H. Rayner, *Nature*, 163, 566 (1949).

[149] T. Iwamoto, in *Inclusion Compounds I*, J.L. Atwood, J.E.D. Davies and D.D. MacNicol (Eds), Academic Press, London, 1984, p. 29.

[150] J.S. Seo, D. Whang, H. Lee, S.I. Jun, J. Oh, Y.J. Jeon and K. Kim, *Nature*, 404, 982 (2000).

[151] O.R. Evans, H.L. Ngo and W.B. Lin, *J. Am. Chem. Soc.*, 123, 10395 (2001).

[152] Y. Cui, O.R. Evans, H.L. Ngo, P.S. White and W.B. Lin, *Angew. Chem. Int. Ed.*, 41, 1159 (2002).

[153] Y. Cui, S.J. Lee and W.B. Lin, *J. Am. Chem. Soc.*, 125, 6014 (2003).

[154] Y. Cui, H.L. Ngo, P.S. White and W.B. Lin, *Inorg. Chem.*, 42, 652 (2003).

[155] R.G. Xiong, X.Z. You, B.F. Abrahams, Z.L. Xue and C.M. Che, *Angew. Chem. Int. Ed.*, 40, 4422 (2001).

[156] M. Fujita, Y.J. Kwon, S. Washizu and K. Ogura, *J. Am. Chem. Soc.*, 116, 1151 (1994).

[157] J. Lee, O.K. Farha, J. Roberts, K.A. Scheidt, S.T. Nguyen and J.T. Hupp, *Chem. Soc. Rev.*, 38, 1450 (2009).

[158] A.U. Czaja, N. Trukhan and U. Müller, *Chem. Soc. Rev.*, 38, 1284 (2009).

[159] S. Kitagawa, S. Noro and T. Nakamura, *Chem. Commun.*, 701 (2006).

[160] L. Alaerts, E. Seguin, H. Poelman, F. Thibault-Starzyk, P.A. Jacobs and D.E. DeVos, *Chem. Eur. J.*, 12, 7353 (2006).

[161] K. Schlichte, T. Kratzke and S. Kaskel, *Microporous Mesoporous Mater.*, 73, 81 (2004).

[162] A. Henschel, K. Gedrich, R. Kraehnert and S. Kaskel, *Chem. Commun.*, 4192 (2008).

[163] S. Horike, M. Dincă, K. Tamaki and J.R. Long, *J. Am. Chem. Soc.*, 130, 5854 (2008).

[164] S.H. Cho, B.Q. Ma, S.T. Nguyen, J.T. Hupp and T.E. Albrecht-Schmitt, *Chem. Commun.*, 2563 (2006).

[165] M.H. Alkordi, Y.L. Liu, R.W. Larsen, J.F. Eubank and M. Eddaoudi, *J. Am. Chem. Soc.*, 130, 12639 (2008).

[166] S. Hermes, M.K. Schroter, R. Schmid, L. Khodeir, M. Muhler, A. Tissler, R.W. Fischer and R.A. Fischer, *Angew. Chem. Int. Ed.*, 44, 6237 (2005).

[167] O. Kahn, *Molecular Magnetism*, VCH, New York, 1993.

[168] R.L. Carlin and A.J. van Duyneveldt, *Magnetic Properties of Transition Metal Compounds*, Vol. 2, Springer-Verlag, New York, 1977.

[169] R.L. Carlin, *Magnetochemistry*, Springer-Verlag, Berlin, 1986.

[170] H.O. Stumpf, L. Ouahab, Y. Pei, D. Grandjean and O. Kahn, *Science*, 261, 447 (1993).

[171] X.Y. Wang, Z.M. Wang and S. Gao, *Chem. Commun.*, 281 (2008).

[172] M. Kurmoo, *Chem. Soc. Rev.*, 38, 1353 (2009).

[173] S. Decurtins, H.W. Schmalle, P. Schneuwly, J. Ensling and P. Gütlich, *J. Am. Chem. Soc.*, 116, 9521 (1994).

[174] S.G. Carling, C. Mathoniere, P. Day, K.M.A. Malik, S.J. Coles and M.B. Hursthouse, *J. Chem. Soc., Dalton Trans.*, 1839 (1996).

[175] H.Z. Kou and O. Sato, *Inorg. Chem.*, 46, 9513 (2007).

[176] X.Y. Wang, L. Gan, S.W. Zhang and S. Gao, *Inorg. Chem.*, 43, 4615 (2004).

[177] Z.M. Wang, B. Zhang, T. Otsuka, K. Inoue, H. Kobayashi and M. Kurmoo, *Dalton Trans.*, 2209 (2004).

[178] O. Kahn, J. Larionova and J.V. Yakhmi, *Chem. Eur. J.*, 5, 3443 (1999).

[179] A. Rujiwatra, C.J. Kepert and M.J. Rosseinsky, *Chem. Commun.*, 2307 (1999).

[180] A. Rujiwatra, C.J. Kepert, J.B. Claridge, M.J. Rosseinsky, H. Kumagai and M. Kurmoo, *J. Am. Chem. Soc.*, 123, 10584 (2001).

[181] M. Kurmoo, H. Kumagai, S.M. Hughes and C.J. Kepert, *Inorg. Chem.*, 42, 6709 (2003).

[182] D. Maspoch, D. Ruiz-Molina, K. Wurst, N. Domingo, M. Cavallini, F. Biscarini, J. Tejada, C. Rovira and J. Veciana, *Nat. Mater.*, 2, 190 (2003).

[183] M. Kurmoo, H. Kumagai, K.W. Chapman and C.J. Kepert, *Chem. Commun.*, 3012 (2005).

[184] S.S. Kaye, H.J. Choi and J.R. Long, *J. Am. Chem. Soc.*, 130, 16921 (2008).

[185] X.-M. Zhang, Z.-M. Hao, W.-X. Zhang and X.-M. Chen, *Angew. Chem. Int. Ed.*, 46, 3456 (2007).

[186] S. Xiang, X. Wu, J. Zhang, R. Fu, S. Hu and X. Zhang, *J. Am. Chem. Soc.*, 127, 16352 (2005).

[187] R.D. Poulsen, A. Bentien, M. Chevalier and B.B. Iversen, *J. Am. Chem. Soc.*, 127, 9156 (2005).

[188] Z. Wang, X. Zhang, S.R. Batten, M. Kurmoo and S. Gao, *Inorg. Chem.*, 46, 8439 (2007).

[189] R. Kuhlman, G.L. Schimek and J.W. Kolis, *Inorg. Chem.*, 38, 194 (1999).

[190] S.O.H. Gutschke, D.J. Price, A.K. Powell and P.T. Wood, *Angew. Chem. Int. Ed.*, **38**, 1088 (1999).

[191] M. Kurmoo, H. Kumagai, K.W. Chapman and C.J. Kepert, *Chem. Commun.*, 3012 (2005).

[192] M. Kurmoo, H. Kumagai, M. Akita-Tanaka, K. Inoue and S. Takagi, *Inorg. Chem.*, **45**, 1627 (2006).

[193] K. Barthelet, J. Marrot, D. Riou and G. Férey, *Angew. Chem. Int. Ed.*, **41**, 281 (2002).

[194] C. Serre, F. Millange, C. Thouvenot, M. Nogues, G. Marsolier, D. Louer and G. Férey, *J. Am. Chem. Soc.*, **124**, 13519 (2002).

[195] M. Kurmoo, H. Kumagai, S.M. Hughes and C.J. Kepert, *Inorg. Chem.*, **42**, 6709 (2003).

[196] M. Riou-Cavellec, C. Albinet, C. Livage, N. Guillou, M. Noguès, J.M. Grenèche and G. Férey, *Solid State Sci.*, **4**, 267 (2002).

[197] M. Viertelhaus, H. Henke, C.E. Anson and A.K. Powell, *Eur. J. Inorg. Chem.*, 2283 (2003).

[198] Z. Wang, B. Zhang, H. Fujiwara, H. Kobayashi and M. Kurmoo, *Chem. Commun.*, 416 (2004).

[199] Z. Wang, B. Zhang, M. Kurmoo, M.A. Green, H. Fujiwara, T. Otsuka and H. Kobayashi, *Inorg. Chem.*, **44**, 1230 (2005).

[200] Z. Wang, Y. Zhang, M. Kurmoo, T. Liu, S. Vilminot, B. Zhao and S. Gao, *Aust. J. Chem.*, **59**, 617 (2006).

[201] Z. Wang, B. Zhang, Y. Zhang, M. Kurmoo, T. Liu, S. Gao and H. Kobayashi, *Polyhedron*, **26**, 2207 (2007).

[202] Z. Wang, Y. Zhang, T. Liu, M. Kurmoo and S. Gao, *Adv. Funct. Mater.*, **17**, 1523 (2007).

[203] B. Zhang, Z. Wang, M. Kurmoo, S. Gao, K. Inoue and H. Kobayashi, *Adv. Funct. Mater.*, **17**, 577 (2007).

[204] K.R. Dunbar and R.A. Heintz, in *Progress in Inorganic Chemistry*, Vol. 45, John Wiley & Sons Inc., New York, 1997, p. 283.

[205] M. Verdaguer, A. Bleuzen, V. Marvaud, J. Vaissermann, M. Seuleiman, C. Desplanches, A. Sciuller, C. Train, R. Garde, G. Gelly, C. Lomenech, I. Rosenman, P. Veillet, C. Cartier and F. Villain, *Coord. Chem. Rev.*, **192**, 1023 (1999).

[206] M. Verdaguer, A. Bleuzen, C. Train, R. Garde, F.F. de Biani and C. Desplanches, *Philos. Trans. R. Soc. London, Ser. A*, **357**, 2959 (1999).

[207] D. Davidson and L.A. Welo, *J. Phys. Chem.*, **32**, 1191 (1928).

[208] V. Gadet, T. Mallah, I. Castro and M. Verdaguer, *J. Am. Chem. Soc.*, **114**, 9213 (1992).

[209] T. Mallah, S. Thiebaut, M. Verdaguer and P. Veillet, *Science*, **262**, 1554 (1993).

[210] S. Ferlay, T. Mallah, R. Ouahes, P. Veillet and M. Verdaguer, *Nature*, **378**, 701 (1995).

[211] Ø. Hatlevik, W.E. Buschmann, J. Zhang, J.L. Manson and J.S. Miller, *Adv. Mater.*, **11**, 914 (1999).

[212] S.M. Holmes and G.S. Girolami, *J. Am. Chem. Soc.*, **121**, 5593 (1999).

[213] M. Ohba and H. Ōkawa, *Coord. Chem. Rev.*, **198**, 313 (2000).

[214] M. Ohba, N. Maruono, H. Ōkawa, T. Enoki and J.M. Latour, *J. Am. Chem. Soc.*, **116**, 11566 (1994).

[215] M. Ohba, N. Usuki, N. Fukita and H. Ōkawa, *Angew. Chem. Int. Ed.*, **38**, 1795 (1999).

[216] G. Boxhoorn, J. Moolhuysen, J.G.F. Coolegem and R.A. van Santen, *J. Chem. Soc., Chem. Commun.*, 1305 (1985).

[217] S.S. Kaye and J.R. Long, *J. Am. Chem. Soc.*, **127**, 6506 (2005).

[218] K.W. Chapman, P.D. Southon, C.L. Weeks and C.J. Kepert, *Chem. Commun.*, 3322 (2005).

[219] K.W. Chapman, P.J. Chupas and C.J. Kepert, *J. Am. Chem. Soc.*, **127**, 11232 (2005).

[220] Y. Sato, S. Ohkoshi, K. Arai, M. Tozawa and K. Hashimoto, *J. Am. Chem. Soc.*, **125**, 14590 (2003).

[221] S. Ohkoshi, K.I. Arai, Y. Sato and K. Hashimoto, *Nat. Mat.*, **3**, 857 (2004).

[222] Z. Lu, X. Wang, Z. Liu, F. Liao, S. Gao, R. Xiong, H. Ma, D. Zhang and D. Zhu, *Inorg. Chem.*, **45**, 999 (2006).

[223] W. Kaneko, M. Ohba and S. Kitagawa, *J. Am. Chem. Soc.*, **129**, 13706 (2007).

[224] B. Nowicka, M. Rams, K. Stadnicka and B. Sieklucka, *Inorg. Chem.*, **46**, 8123 (2007).

[225] S. Ohkoshi, Y. Tsunobuchi, H. Takahashi, T. Hozumi, M. Shiro and K. Hashimoto, *J. Am. Chem. Soc.*, **129**, 3084 (2007).

[226] W. Kaim and M. Moscherosch, *Coord. Chem. Rev.*, **129**, 157 (1994).

[227] J.-H. Her, P.W. Stephens, K.I. Pokhodnya, M. Bonner and J.S. Miller, *Angew. Chem. Int. Ed.*, **46**, 1521 (2007).

[228] K.I. Pokhodnya, M. Bonner, J.-H. Her, P.W. Stephens and J.S. Miller, *J. Am. Chem. Soc.*, **128**, 15592 (2006).

[229] H. Miyasaka, T. Izawa, N. Takahashi, M. Yamashita and K.R. Dunbar, *J. Am. Chem. Soc.*, **128**, 11358 (2006).

[230] N. Motokawa, H. Miyasaka, M. Yamashita and K.R. Dunbar, *Angew. Chem. Int. Ed.*, **47**, 7760 (2008).

[231] S. Shimomura, R. Matsuda, T. Tsujino, T. Kawamura and S. Kitagawa, *J. Am. Chem. Soc.*, **128**, 16416 (2006).

[232] D. Maspoch, N. Domingo, D.R. Molina, K. Wurst, J.M. Hernandez, G. Vaughan, C. Rovira, F. Lloret, J. Tejada and J. Veciana, *Chem. Commun.*, 5035 (2005).

[233] D. Maspoch, D. Ruiz-Molina, K. Wurst, C. Rovira and J. Veciana, *Chem. Commun.*, 1164 (2004).

[234] P. Gütlich, A. Hauser and H. Spiering, *Angew. Chem. Int. Ed.*, **33**, 2024 (1994).

[235] O. Kahn and C.J. Martinez, *Science*, **279**, 44 (1998).

[236] Y. Garcia, P.J. van Koningsbruggen, R. Lapouyade, L. Fournes, L. Rabardel, O. Kahn, V. Ksenofontov, G. Levchenko and P. Gütlich, *Chem. Mater.*, **10**, 2426 (1998).

[237] J.G. Haasnoot, *Coord. Chem. Rev.*, **200**, 131 (2000).

[238] L.G. Lavrenova, N.G. Yudina, V.N. Ikorskii, V.A. Varnek, I.M. Oglezneva and S.V. Larionov, *Polyhedron*, **14**, 1333 (1995).

[239] J.A. Real, E. Andres, M.C. Muñoz, M. Julve, T. Granier, A. Bousseksou and F. Varret, *Science*, **268**, 265 (1995).

[240] V. Niel, J.M. Martinez-Agudo, M.C. Muñoz, A.B. Gaspar and J.A. Real, *Inorg. Chem.*, **40**, 3838 (2001).

[241] P. Gütlich, Y. Garcia and H.A. Goodwin, *Chem. Soc. Rev.*, **29**, 419 (2000).

[242] Y. Garcia, P.J. van Koningsbruggen, R. Lapouyade, L. Rabardel, O. Kahn, M. Wieczorek, R. Bronisz, Z. Ciunik and M.F. Rudolf, *C. R. Acad. Sci.*, **1**, 523 (1998).

[243] W. Vreugdenhil, J.H. van Diemen, R.A.G. de Graaff, J.G. Haasnoot, J. Reedijk, A.M. van der Kraan, O. Kahn and J. Zarembowitch, *Polyhedron*, **9**, 2971 (1990).

[244] Y. Garcia, O. Kahn, L. Rabardel, C. Benoit, L. Salmon and J.P. Tuchagues, *Inorg. Chem.*, **38**, 4663 (1999).

[245] P.J. van Koningsbruggen, Y. Garcia, G. Bravic, D. Chasseau and O. Kahn, *Inorg. Chim. Acta*, **326**, 101 (2001).

[246] J.A. Real, A.B. Gaspar, V. Niel and M.C. Muñoz, *Coord. Chem. Rev.*, **236**, 121 (2003).

[247] P.J. van Koningsbruggen, Y. Garcia, O. Kahn, L. Fournes, H. Kooijman, A.L. Spek, J.G. Haasnoot, J. Moscovici, K. Provost, A. Michalowicz, F. Renz and P. Gütlich, *Inorg. Chem.*, **39**, 1891 (2000).

[248] C.M. Grunert, J. Schweifer, P. Weinberger, W. Linert, K. Mereiter, G. Hilscher, M. Muller, G. Wiesinger and P.J. van Koningsbruggen, *Inorg. Chem.*, **43**, 155 (2004).

[249] T. Kitazawa, Y. Gomi, M. Takahashi, M. Takeda, M. Enomoto, A. Miyazaki and T. Enoki, *J. Mater. Chem.*, **6**, 119 (1996).

[250] W. Kosaka, K. Nomura, K. Hashimoto and S. Ohkoshi, *J. Am. Chem. Soc.*, **127**, 8590 (2005).

[251] D. Papanikolaou, S. Margadonna, W. Kosaka, S. Ohkoshi, M. Brunelli and K. Prassides, *J. Am. Chem. Soc.*, **128**, 8358 (2006).

[252] K.S. Murray and C.J. Kepert, in *Topics in Current Chemistry*, Vol. **233**, P. Gütlich and H.A. Goodwin (Eds), Springer-Verlag, Heidelberg, 2004, p. 195.

[253] G.J. Halder, K.W. Chapman, S.M. Neville, B. Moubaraki, K.S. Murray, J.F. Létard and C.J. Kepert, *J. Am. Chem. Soc.*, **130**, 17552 (2008).

[254] S.M. Neville, G.J. Halder, K.W. Chapman, M.B. Duriska, P.D. Southon, J.D. Cashion, J.-F. Létard, B. Moubaraki, K.S. Murray and C.J. Kepert, *J. Am. Chem. Soc.*, **130**, 2869 (2008).

[255] S.M. Neville, B. Moubaraki, K.S. Murray and C.J. Kepert, *Angew. Chem. Int. Ed.*, **46**, 2059 (2007).

[256] S.M. Neville, G.J. Halder, K.W. Chapman, M.B. Duriska, B. Moubaraki, K.S. Murray and C.J. Kepert, *J. Am. Chem. Soc.*, **131**, 12106 (2009).

[257] S. Bonhommeau, G. Molnar, A. Galet, A. Zwick, J.A. Real, J.J. McGarvey and A. Bousseksou, *Angew. Chem. Int. Ed.*, **44**, 4069 (2005).

[258] P.D. Southon, L. Liu, E.A. Fellows, D.J. Price, G.J. Halder, K.W. Chapman, B. Moubaraki, K.S. Murray, J.F. Létard and C.J. Kepert, *J. Am. Chem. Soc.*, **131**, 10998 (2009).

[259] M. Ohba, K. Yoneda, G. Agusti, M.C. Muñoz, A.B. Gaspar, J.A. Real, M. Yamasaki, H. Ando, Y. Nakao, S. Sakaki and S. Kitagawa, *Angew. Chem. Int. Ed.*, **48**, 4767 (2009).

[260] P. Day, N.S. Hush and R.J.H. Clark, *Philos. Trans. R. Soc. London, Ser. A*, **366**, 5 (2008).

[261] O. Sato, T. Iyoda, A. Fujishima and K. Hashimoto, *Science*, **272**, 704 (1996).

[262] O. Sato, Y. Einaga, A. Fujishima and K. Hashimoto, *Inorg. Chem.*, **38**, 4405 (1999).

[263] H. Tokoro, T. Matsuda, T. Nuida, Y. Moritomo, K. Ohoyama, E.D.L. Dangui, K. Boukheddaden and S. Ohkoshi, *Chem. Mater.*, **20**, 423 (2008).

[264] H. Tokoro, S. Ohkoshi and K. Hashimoto, *Appl. Phys. Lett.*, **82**, 1245 (2003).

[265] S. Ohkoshi and K. Hashimoto, *J. Am. Chem. Soc.*, **121**, 10591 (1999).

[266] S. Ohkoshi, H. Tokoro, T. Hozumi, Y. Zhang, K. Hashimoto, C. Mathoniere, I. Bord, G. Rombaut, M. Verelst, C.C.D. Moulin and F. Villain, *J. Am. Chem. Soc.*, **128**, 270 (2006).

[267] Y. Arimoto, S. Ohkoshi, Z.J. Zhong, H. Seino, Y. Mizobe and K. Hashimoto, *J. Am. Chem. Soc.*, **125**, 9240 (2003).

[268] S. Ohkoshi, S. Ikeda, T. Hozumi, T. Kashiwagi and K. Hashimoto, *J. Am. Chem. Soc.*, **128**, 5320 (2006).

[269] D.N. Hendrickson and C.G. Pierpont, in *Spin Crossover in Transition Metal Compounds II, Topics in Current Chemistry*, Vol. 234, P. Gütlich and H.A. Goodwin (Eds), Springer-Verlag, Berlin, 2004, p. 63.

[270] D.M. Adams, A. Dei, A.L. Rheingold and D.N. Hendrickson, *Angew. Chem. Int. Ed.*, **32**, 880 (1993).

[271] O.S. Jung and C.G. Pierpont, *J. Am. Chem. Soc.*, **116**, 2229 (1994).

[272] C.W. Lange, M. Foldeaki, V.I. Nevodchikov, V.K. Cherkasov, G.A. Abakumov and C.G. Pierpont, *J. Am. Chem. Soc.*, **114**, 4220 (1992).

[273] P.Y. Feng, X.H. Bu and N.F. Zheng, *Acc. Chem. Res.*, **38**, 293 (2005).

[274] A. Aumüller, P. Erk, G. Klebe, S. Hünig, J.U. von Schütz and H.P. Werner, *Angew. Chem. Int. Ed.*, **25**, 740 (1986).

[275] J.M. Williams, J.R. Ferraro and R.J. Thorn, Organic Superconductors: Synthesis, Structure, Properties and Theory (Includes Fullerenes), Prentice Hall, Englewood Cliffs, NJ, 1991.

[276] M. Kurmoo, A.W. Graham, P. Day, S.J. Coles, M.B. Hursthouse, J.L. Caulfield, J. Singleton, F.L. Pratt, W. Hayes, L. Ducasse and P. Guionneau, *J. Am. Chem. Soc.*, **117**, 12209 (1995).

[277] E. Coronado, J.R. Galan-Mascaros, C.J. Gomez-Garcia and V. Laukhin, *Nature*, **408**, 447 (2000).

[278] M.D. Allendorf, C.A. Bauer, R.K. Bhakta and R.J.T. Houk, *Chem. Soc. Rev.*, **38**, 1330 (2009).

[279] M.P. Suh, Y.E. Cheon and E.Y. Lee, *Coord. Chem. Rev.*, **252**, 1007 (2008).

[280] B.T.N. Pham, L.M. Lund and D.T. Song, *Inorg. Chem.*, **47**, 6329 (2008).

[281] J. Pang, E.J.P. Marcotte, C. Seward, R.S. Brown and S.N. Wang, *Angew. Chem. Int. Ed.*, **40**, 4042 (2001).

[282] Y.Q. Huang, B. Ding, H.B. Song, B. Zhao, P. Ren, P. Cheng, H.G. Wang, D.Z. Liao and S.P. Yan, *Chem. Commun.*, 4906 (2006).

[283] W.H. Zhu, Z.M. Wang and S. Gao, *Inorg. Chem.*, **46**, 1337 (2007).

[284] T.M. Reineke, M. Eddaoudi, M. Fehr, D. Kelley and O.M. Yaghi, *J. Am. Chem. Soc.*, **121**, 1651 (1999).

[285] B.L. Chen, Y. Yang, F. Zapata, G.N. Lin, G.D. Qian and E.B. Lobkovsky, *Adv. Mater.*, **19**, 1693 (2007).

[286] B.V. Harbuzaru, A. Corma, F. Rey, P. Atienzar, J.L. Jorda, H. Garcia, D. Ananias, L.D. Carlos and J. Rocha, *Angew. Chem. Int. Ed.*, **47**, 1080 (2008).

[287] B.L. Chen, L.B. Wang, F. Zapata, G.D. Qian and E.B. Lobkovsky, *J. Am. Chem. Soc.*, **130**, 6718 (2008).

[288] K.L. Wong, G.L. Law, Y.Y. Yang and W.T. Wong, *Adv. Mater.*, **18**, 1051 (2006).

[289] Y.C. Qiu, Z.H. Liu, Y.H. Li, H. Deng, R.H. Zeng and M. Zeller, *Inorg. Chem.*, **47**, 5122 (2008).

[290] B. Zhao, X.Y. Chen, P. Cheng, D.Z. Liao, S.P. Yan and Z.H. Jiang, *J. Am. Chem. Soc.*, **126**, 15394 (2004).

[291] W. Liu, T. Jiao, Y. Li, Q. Liu, M. Tan, H. Wang and L. Wang, *J. Am. Chem. Soc.*, **126**, 2280 (2004).

[292] E.Y. Lee, S.Y. Jang and M.P. Suh, *J. Am. Chem. Soc.*, **127**, 6374 (2005).

[293] C.A. Bauer, T.V. Timofeeva, T.B. Settersten, B.D. Patterson, V.H. Liu, B.A. Simmons and M.D. Allendorf, *J. Am. Chem. Soc.*, **129**, 7136 (2007).

[294] J.R. Salvador, F. Gu, T. Hogan and M.G. Kanatzidis, *Nature*, **425**, 702 (2003).

[295] J. Arvanitidis, K. Papagelis, S. Margadonna, K. Prassides and A.N. Fitch, *Nature*, **425**, 599 (2003).

[296] S. Margadonna, J. Arvanitidis, K. Papagelis and K. Prassides, *Chem. Mater.*, **17**, 4474 (2005).

[297] A.C. McLaughlin, F. Sher and J.P. Attfield, *Nature*, **436**, 829 (2005).

[298] A.W. Sleight, *Curr. Opin. Solid State Mater. Chem.*, **3**, 128 (1998).

[299] J.S.O. Evans, *J. Chem. Soc., Dalton Trans.*, 3317 (1999).

[300] M.G. Tucker, A.L. Goodwin, M.T. Dove, D.A. Keen, S.A. Wells and J.S.O. Evans, *Phys. Rev. Lett.*, **95**, 255501 (2005).

[301] T.A. Mary, J.S.O. Evans, T. Vogt and A.W. Sleight, *Science*, **272**, 90 (1996).

[302] S. Margadonna, K. Prassides and A.N. Fitch, *J. Am. Chem. Soc.*, **126**, 15390 (2004).

[303] A.L. Goodwin and C.J. Kepert, *Phys. Rev. B: Condens. Matter*, **71**, 140301/1 (2005).

[304] A.E. Phillips, A.L. Goodwin, G.J. Halder, P.D. Southon and C.J. Kepert, *Angew. Chem. Int. Ed.*, **47**, 1396 (2008).

[305] A.L. Goodwin, K.W. Chapman and C.J. Kepert, *J. Am. Chem. Soc.*, **127**, 17980 (2005).

[306] K.W. Chapman, P.J. Chupas and C.J. Kepert, *J. Am. Chem. Soc.*, **127**, 15630 (2005).

[307] K.W. Chapman, P.J. Chupas and C.J. Kepert, *J. Am. Chem. Soc.*, **128**, 7009 (2006).

[308] T. Pretsch, K.W. Chapman, G.J. Halder and C.J. Kepert, *Chem. Commun.*, 1857 (2006).

[309] A.L. Goodwin, M. Calleja, M.J. Conterio, M.T. Dove, J.S.O. Evans, D.A. Keen, L. Peters and M.G. Tucker, *Science*, **319**, 794 (2008).

[310] A.L. Goodwin, B.J. Kennedy and C.J. Kepert, *J. Am. Chem. Soc.*, **131**, 6334 (2009).

[311] S.J. Hibble, A.M. Chippindale, A.H. Pohl and A.C. Hannon, *Angew. Chem. Int. Ed.*, **46**, 7116 (2007).

[312] Y. Wu, A. Kobayashi, G.J. Halder, V.K. Peterson, K.W. Chapman, N. Lock, P.D. Southon and C.J. Kepert, *Angew. Chem. Int. Ed.*, **47**, 8929 (2008).

[313] D. Dubbeldam, K.S. Walton, D.E. Ellis and R.Q. Snurr, *Angew. Chem. Int. Ed.*, **46**, 4496 (2007).

[314] S.S. Han and W.A. Goddard, *J. Phys. Chem. C*, **111**, 15185 (2007).

[315] J.L. Korcok, M.J. Katz and D.B. Leznoff, *J. Am. Chem. Soc.*, **131**, 4866 (2009).

[316] A.L. Goodwin, D.A. Keen and M.G. Tucker, *Proc. Natl. Acad. Sci. USA*, **105**, 18708 (2008).

[317] K.W. Chapman, G.J. Halder and P.J. Chupas, *J. Am. Chem. Soc.*, **130**, 10524 (2008).

[318] K.W. Chapman and P.J. Chupas, *J. Am. Chem. Soc.*, **129**, 10090 (2007).

[319] G.B. Gardner, D. Venkataraman, J.S. Moore and S. Lee, *Nature*, **374**, 792 (1995).

[320] R.H. Baughman and D.S. Galvao, *Nature*, **365**, 735 (1993).

[321] K.E. Evans, M.A. Nkansah, I.J. Hutchinson and S.C. Rogers, *Nature*, **353**, 124 (1991).

2

Mesoporous Silicates

Karen J. Edler

Department of Chemistry, University of Bath, Bath, UK

2.1 INTRODUCTION

Mesoporous materials are generally defined by IUPAC as those having pore diameters between 20 Å and 500 Å.[1] There are several silicate materials which fulfil this description, including aerogels and xerogels, controlled pore or Vycor glasses, and surfactant templated mesoporous materials. All of these are amorphous forms of silica, while crystalline forms of silica, such as zeolites, tend to have smaller pores in the micropore size range. Of these mesoporous materials, aerogels and xerogels, produced *via* gelation of silica solutions and controlled pore glasses, produced *via* etching of microphase separated solid amorphous materials, have relatively large pore size distributions which can limit their use in applications such as size-selective catalysis, or optical waveguides which require uniform pore systems. This chapter will concentrate instead on the preparation of surfactant templated porous silicates which have highly uniform pore diameters, and a range of controllable pore structures through the use of surfactant micelles to determine the pore architecture.

Surfactant-templated mesoporous silicates have been the subject of intense research effort since the publication by Mobil scientists in 1992,[2,3] of the syntheses for MCM-41, MCM-48 and MCM-50. Some have suggested that similar materials were prepared earlier,[4] however

Porous Materials   Edited by Duncan W. Bruce, Dermot O'Hare and Richard I. Walton
© 2011 John Wiley & Sons, Ltd.

publication of the Mobil group's results in *Nature* sparked the current rapidly growing field of research which still continues to expand. To date, around 10 000 papers have been published on these materials, more than half of these since the beginning of 2005. Although initially the focus for applications was on catalysis, more recently these materials have been investigated for applications ranging from drug release to optical guides to low *k* dielectric layers in microelectronics. The general technique of surfactant templating has now been expanded far beyond silicates, with notable recent work producing mesostructured germanium,[5,6] and a range of metal oxides, including zinc[7] and tin[8] oxides, vanadia,[9] niobia[10] and ceria,[11] in both powder and thin film forms. The development of the technique of nanocasting, using a mesoporous silica as a sacrificial template upon which to build a negative but also mesoporous replica, has now made a huge range of mesoporous materials accessible to synthesis.[12] Electrodeposition has also been used to produce surfactant-templated mesoporous metals.[13] Given the breadth of this field, this chapter will focus solely on silicates, where the most work has been done on understanding the formation mechanism, as well as in producing a variety of mesoscale and macroscale morphologies. It will describe the synthesis of these materials, current evidence for their formation mechanisms, their properties, and briefly, some of the more recent work on applications.

2.2 NOMENCLATURE

The naming of materials in this field is far from systematic, and currently is under review by the Mesoporous Materials Commission of the International Zeolite Association. However, until a more rational system is devised, most surfactant templated materials have been named by their discoverers using a series of three letters, followed by a number. The letters refer to the research group or company that first published these materials, and the numbers refer to their internal numbering system for new materials. These have now developed into general terms for describing classes of mesoporous materials. Hence MCM-41, MCM-48 and MCM-50 are generally applied to materials synthesised using cationic surfactants under alkaline conditions with, respectively a two-dimensional (2D) hexagonal structure, an $Ia\bar{3}d$ cubic structure and a lamellar structure (details of structures are shown in Section 2.8). The term M41S is also used as a blanket term to cover any surfactant templated mesoporous silicates prepared from alkaline solutions.

Other commonly used names for mesoporous silicates include the SBA series, which are synthesised from acidic solutions. SBA-1, 2, 3 use cationic surfactants to form a $Pm\bar{3}n$ cubic phase, a 3D micellar hexagonal phase and a 2D hexagonal phase, respectively,[14,15] while SBA-15, 16 are 2D hexagonal and $Im\bar{3}m$ cubic phases, synthesised using nonionic triblock copolymer surfactants.[16,17] The MSU-X series are silicas templated in neutral solutions using nonionic polyoxyethylene surfactants (C_nEO_m) which have a wormlike structure,[18] while the FSM (Folded Sheet Material) series are formed using exfoliated sheets of the mineral kanemite with a cationic surfactant.[19] FSM-16 is the most commonly reported member of this family and has a 2D hexagonal structure. Other series include KIT,[20] FDU[21] and HOM.[22]

2.3 METHODS OF PREPARATION

The preparation of mesoporous surfactant-templated silicates is in principle straightforward, requiring a silica source, a catalyst for silica polymerisation, the template and a solvent. The templating reaction can be done under acidic, alkaline or neutral conditions. The solvent for these syntheses is usually water, although some work in solvent mixtures has been reported,[23–25] as well as a few syntheses in nonaqueous solvents.[26,27] The presence of cosolvents in aqueous synthesis solutions has a marked effect on the phase behaviour of the templating surfactant, resulting in changes to both nanoscale and macroscale structures.[28,29] In particular, the addition of ethanol has been found to favour particles with spherical or 'single crystal' morphologies and cubic phase structures from syntheses where otherwise 2D hexagonal powders would be formed.[30,31] In the case of evaporation-induced self-assembly (EISA) routes, a large amount of a volatile cosolvent, typically ethanol, is used, but in this route (discussed in more detail below) the evaporation of this solvent is crucial to formation of the mesophase structure, and the structures formed are controlled more by the residual water in the materials than by the cosolvent. The next sections discuss the nature of the surfactant template, and the silica sources generally used to prepare surfactant templated mesoporous silicates. Methods to remove the surfactant to open up the porosity are also mentioned.

Five specific routes to mesoporous templated materials exist, although only three of these are used to prepare mesoporous silicates. These are: true liquid crystal templating, cooperative self-assembly in solution and

EISA. The mechanism in each case has been the subject of much research and the current understanding of each mechanism is described below. The other two methods to produce mesoporous templated materials are electrodeposition,[13] which has been successfully used to produce surfactant templated porous metal films from high concentration surfactant solutions, and nanocasting,[12,32] where a surfactant templated silicate is used as a sacrificial template to generate further porous materials by coating the silica structure in another oxide or carbon precursors. The second material is sintered or solidified, and the silicate removed by HF or high pH solvent wash. This is discussed further in Section 2.10.

2.4 SURFACTANT AGGREGATION

Surfactant templating, as the name suggests, uses surfactant micelles in solution as a sacrificial template around which the inorganic species form. After the formation of the mesostructured inorganic-surfactant composite, which contains surfactant micelles encased in the inorganic phase, the surfactant is removed in order to open up the pores. Thus the size and shape of the pore network in the inorganic material is determined by the size and shape of the surfactant aggregates incorporated into the structure during synthesis.

Surfactant molecules aggregate in solution due to their amphiphilic character, as one part of the molecule is hydrophobic while the other part is hydrophilic. They therefore arrange themselves at interfaces and in solutions so that the hydrophobic part is removed from the water as far as possible, driving the formation of aggregates in the bulk solution above a certain concentration known as the critical micelle concentration (cmc). These aggregates, or micelles, are thermodynamically stable structures but are also dynamic, with surfactant molecules partitioning between the micelle and the solution. Thus the aggregates quickly achieve a fairly uniform size and shape for a given set of solution conditions (concentration, pH, ionic strength, temperature, etc.), making them ideal templates for producing uniform pores. The shape of the micelles just above the cmc, and thus some indication of the shape of eventual pores in the inorganic materials can be predicted by the packing parameter, g, devised by Israelachvili.[33,34] This parameter uses the headgroup area of the surfactant, a_0, molecule, the tail volume, v, and critical tail length, l_c, to calculate $g = v/a_0 l_c$ in order to predict micelle shape and explain trends in self-assembled surfactant structures with concentration, presence of screening counterions and other factors (see Table 2.1).

Table 2.1 Packing parameter and predicted surfactant aggregate shape[33,34]

Packing parameter, g	Predicted micelle shape
<1/3	Spherical
1/3–1/2	Cylindrical
1/2–1	Flexible bilayers or vesicles
~1	Planar bilayers
>1	Inverted micelles

The close-packed micellar structures formed by surfactants in concentrated surfactant–water solutions are closely mimicked by those found in surfactant templated inorganic materials, with the inorganic material filling the equivalent regions to the water in the structures. Thus changes in the size and shape of the headgroup of the surfactant relative to the tail volume can be used to tailor the pore shape and connectivity, while the overall size of the micelle hydrophobic region generally determines the pore size. (Note that the g parameter is not always an exact predictor of pore shapes in surfactant templated materials but it does explain observed trends as the surfactant molecular structure is altered.)[35] The pore diameter can also be increased by swelling the micelles with hydrophobic materials, typically oils, which are sequestered into the hydrophobic tail region of the micelle.[36–38] In this case however the shape of the micelles may also be altered, tending towards spherical, since this allows the most efficient solubilisation of the hydrophobic material.

The hydrophilic headgroup of the surfactant molecule for small molecule surfactants can be cationic (*e.g.* quaternary ammonium),[2,3] anionic (*e.g.* sulfate),[39–41] zwitterionic (*e.g.* phospholipids),[42,43] or neutral (*e.g.* amines).[23,44] Larger self-assembled structures are obtained using polymeric surfactants. In this case the water soluble headgroup can be either a charged polymer[45,46] or oligomer segment or a polar but neutral polymer or oligomer segment (*e.g.* oligoethylene oxides).[18] All of these, both small molecule and polymeric surfactants and their mixtures[47] have been used to template mesoporous materials (Table 2.2).

The hydrophobic (tail) part of the surfactant is typically a single alkyl chain in small-molecule surfactants, while polymeric surfactants have hydrophobic regions composed of less soluble polymers. Thus triblock copolymers of the type poly(ethylene oxide)-poly(propylene oxide)-poly(ethylene oxide) ($EO_nPO_mEO_n$) form micelles with poly(propylene oxide) in the hydrophobic core, and the poly(ethylene oxide) strands on the outside of the micelle. These triblock copolymers, particularly the

Table 2.2 Surfactants used for templating, conditions under which mesoporous silicas have been formed, and the interaction between surfactant (S) and inorganic species (I), from the charge-density matching model.[48] Note, X is the surfactant counterion, H is a hydrogen ion

Surfactant	Templating conditions	Interaction	Reference
Cationic e.g. hexadecyltrimethylammonium bromide 	Alkaline	S^+I^-	[2,3]
	Acidic	$S^+X^-I^+$	[49]
Anionic e.g. sodium dodecylsulfate 	Alkaline	$S^-X^+I^-$	[49]
	Acidic	S^-I^+	[49]
Neutral e.g. dodecylamine 	Neutral	S^0I^0	[44]
Nonionic e.g. alkyl polyoxyethylene ether 	Acidic	$S^0H^+X^-I^+$	[16]
triblock copolymers (e.g. Pluronic®) 	Neutral	S^0I^0	[18]

Pluronic® series, are widely used to template silica. The phase segregation between EO and PO is reasonably efficient, although it has been shown by small-angle neutron scattering experiments that significant amounts of water[50] and hydrophobic silica precursors[51] can also be found in the core of these micelles, which is not the case for micelles with more hydrophobic polymer segments or in the alkyl tail cores in micelles made from small surfactant molecules. However, in some preparations the degree of phase separation between EO and PO has been found to be insufficient to drive mesophase ordering, so a range of other block copolymers with more hydrophobic blocks have also been used to template inorganic oxides, including poly(styrene-b-ethylene oxide),[52] poly(isoprene-b-ethylene oxide)[26,53] and poly(ethylene-co-butylene)-b-(ethylene oxide).[54] More recently a range of triblock copolymer materials with three distinct polymeric blocks were made to prepare complex templated aluminosilicate structures, using poly(ethylene oxide) as the inorganic containing block, which phase separates from the other two polymer segments.[55,56]

The pore size produced in these materials depends on the size of the hydrophobic core of the micelles. Small molecule surfactants such as cetyltrimethylammonium bromide (CTAB) produce materials with pore sizes from 2 to 3 nm, and this can be controlled in steps of 2–3 Å by altering the hydrophobic tail length of the surfactant.[2] Polymeric surfactants produce much larger pores with pores up to 30 nm diameter reported for nonionic triblock copolymer surfactants.[16] Swelling the cores of the micelles with hydrophobic species such as trimethylbenzene[2,57] (mesitylene)[58] or alkanes[59] can also be used to increase pore size. Manipulation of the type of nanostructure of small molecule surfactant templated materials is largely done by variation of the headgroup area to tail volume ratio. Although surfactants with two or more alkyl tails on one charged headgroup exist, they are not commonly used in preparation of mesoporous materials since the large tail volume tends to result in formation of lamellar phases which collapse after surfactant removal. However, surfactants with two cationic headgroups, joined by a short spacer, and each with its own hydrophobic tail (known as Gemini surfactants) which self-assemble into a variety of 3D connected surfactant phases in water have been used to obtain specific silica pore geometries.[14,60] Generally however, in the drive towards applications, cheaper and less environmentally toxic surfactants are preferred, so that materials prepared using commercially available nonionic oligo-ethyleneoxide surfactants with an alkyl tail or the di- or triblock EO-PO copolymers are favoured.

2.5 SILICA SOURCE

The silica condensation reaction can be done under acidic, alkaline or neutral conditions and all three have been used to synthesise mesoporous materials. Under neutral conditions often the surfactant itself acts as a catalyst for silica condensation as formation of templated materials occurs much more quickly than precipitation or gel formation in silica-only solutions at the same pH.[61–63] Under neutral conditions addition of F^-, which also catalyses silica condensation, is also sometimes used to promote the reaction.[61,64] Under alkaline or acidic conditions the acid or base acts as a catalyst, particularly for the hydrolysis of alkoxysilane type silica precursors, but also to control the condensation rate and mechanism (Figure 2.1).

For alkaline syntheses the templating reaction occurs rapidly with precipitation observed immediately after addition of the silica source to the surfactant solution. Under neutral conditions precipitation of the

Hydrolysis:	$Si(OR)_4 + H_2O \rightarrow HO\text{-}Si(OR)_3 + ROH$
	$HO\text{-}Si(OR)_3 + H_2O \rightarrow (HO)_2\text{-}Si(OR)_2 + ROH\ldots\ldots\ etc.$
Condensation:	
Alkaline	$(OR)_3\text{-}Si\text{-}O^- + OH^- \rightarrow (OR)_3\text{-}Si\text{-}OH + H_2O$
	$(OR)_3\text{-}Si\text{-}OH + (OR)_3\text{-}Si\text{-}O^- \rightarrow (RO)_3\text{-}Si\text{-}O\text{-}Si\text{-}(OR)_3 + OH^-$
Acid	$(OR)_3\text{-}Si\text{-}O^- + H_3O^+ \rightarrow (RO)_3\text{-}Si\text{-}OH_2^+ + H_2O$
	$(RO)_3\text{-}Si\text{-}OH_2^+ + (OR)_3\text{-}Si\text{-}O^- \rightarrow (RO)_3\text{-}Si\text{-}O\text{-}Si\text{-}(OR)_3 + H_3O^+$

Figure 2.1 Acid and base catalysed reaction mechanisms,[65] R = H or alkyl group. The alkaline condensation mechanism operates above the isoelectric point of silica (pH 2)

silica-surfactant composite may occur immediately or be indefinitely stable until condensation is induced *via* addition of for instance, fluoride ions, depending on the surfactant used – oxyethylene based surfactants have longer induction times before precipitation than akyltrimethylammonium based surfactants.[61] In acidic conditions, below but close to the isoelectric point of silica at pH ∼2 the induction times can also be longer, up to several hours, again dependant upon the pH and surfactant used.[66] Very low pH solutions precipitate within minutes or tens of minutes at room temperature, although the final mesophase structure may take much longer to form.[63] In many syntheses a low or moderate temperature aging step (at room temperature up to about 40 °C) is frequently used for periods of up to 24 h. Although the silica condensation reaction occurs at room temperatures and templated materials can be recovered such preparations, it is slow, requiring about 7 days to obtain structures that are stable to calcination.[67] Therefore, most preparations are heated at 60–100 °C for a few hours to 3 days to complete the silica condensation process. Triblock copolymer surfactants containing EO hydrophilic units have an upper critical solution temperature and reach a cloud point, above which the surfactant forms larger aggregates in solution. When these are used to template silica, synthesis temperatures below the cloud point are recommended, as poorly ordered materials are produced above the cloud point.[68] However, higher temperatures around 95 °C have recently been found to improve ordering in low pH syntheses templated with triblock copolymer templates in the presence of ethanol, as the highly acidic solutions and ethanol both increase the cloud point of these surfactants.[69] Higher temperatures around 150 °C disrupt micelle formation in the synthesis solutions and microporous zeolitic materials templated on single surfactant molecules, such as ZSM-5 are produced, instead of the micelle templated mesoporous silicas.[70] Continued heating of the synthesis solutions can result in re-structuring of an initially formed mesophase composite, and has been used to increase pore diameters,[71,72] or induce phase transitions in the mesophase structure.[73]

Typically for materials prepared under alkaline conditions, the silica source has been either sodium silicate solution or colloidal sources such as Ludox or Aerosil fumed silicas which dissolve in alkaline synthesis solutions, since these are comparatively cheap sources which can give high quality materials.[74] There have also been various attempts to find cheaper silica sources with rice husk ash[75,76] geothermal silica[77] and coal fly ash[78,79] being used to produce mesoporous materials. However, these sources, which require dissolution of SiO_2, tend to form complex oligomeric silica species in solution which may impact upon their interactions with the organic template during formation of these materials. Batch variability may also result in some variation in the mesoporous materials produced. To overcome this in alkaline syntheses, and also in the majority of acidic syntheses, monomolecular sources of silica are typically used; either silicic acid but, or typically, alkoxysilanes of the form $Si(OR)_4$ which hydrolyse to produce reactive $Si(OR)_{4-n}(OH)_n$ species that can then condense to give SiO_2. The hydrolysis process produces small alcohols which can affect the surfactant phase behaviour. The production of the cubic phase MCM-48 for instance, is greatly assisted when small amounts of ethanol are present in the synthesis, thus it proved necessary to use tetraethoxysilane (TEOS) as a precursor or add ethanol when using other silica sources to produce this phase with the originally used surfactant template, cetyltrimethylammonium bromide.[73]

Alkoxysilanes [$R-Si(OR')_3$, where R and R' are different organic groups] also provide a straightforward method to introduce organic functionality into the pore walls of surfactant templated silicates.[81] Co-condensation of alkyltrialkoxysilanes such as aminopropyltriethoxysilane, methyltrimethoxysilane or phenyltriethoxysilane have been used in mixed syntheses with TEOS to produce materials containing these functionalities covalently linked to the silica walls.[82–86] The hydrophobicity of the Si-R group determines its position in the structure: hydrophobic groups insert into the palisade layer of the templating micelles resulting in the R groups lining the walls of the channels after surfactant removal by solvent wash.[87,88] In contrast, hydrophilic species can end up occluded in the wall if there is no attraction between the R group and the surface of the surfactant micelles.[87,88] These variable interactions have been used to locate dyes within the mesoporous structures, both within the channels,[89,90] and in the silica framework[91] (Figure 2.2). Several species can be simultaneously incorporated into different regions in mesostructured films *via* this method to study electron tunnelling between

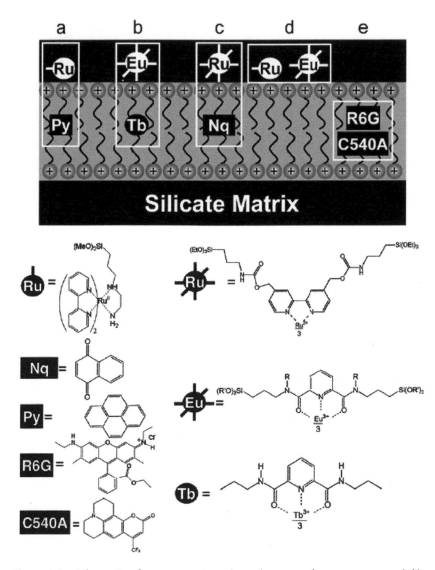

Figure 2.2 Schematic of a cross-section along the pore of a mesostructured film showing the functionalisation of surfactant templated silica in different locations using functional alkoxysilanes. White boxes show pairs of molecules incorporated simultaneously into silica films by Minoofar *et al.*[80] Reprinted with permission from Minoofar, P.N.; Hernandez, R.; Chia, S.; Dunn, B.; Zink, J.I.; Franville, A.-C., Placement and Characterization of Pairs of Luminescent Molecules in Spatially Separated Regions of Nanostructured Thin Films, *J. Am. Chem. Soc.*, **124**, 14388–14396. Copyright 2002 American Chemical Society

species,[92] and to develop active matrices for nanofunctional materials.[91,93,94] Specific interactions between the surfactant template and the R group have been used to help direct hydrophilic functionalities onto the channel surfaces rather than in the walls, for example use of an anionic surfactant, lauric acid sodium salt to direct condensation of aminopropyl functionalised silica precursors around the micelle surface so that the amine group lines the walls of the silica pores after surfactant removal.[95,96] Combinations of functional silanes have been used at different stages of the preparation in order to locate specific and different functionalities on the inside of the pores, or the external surface of the particles.[97]

More recently bifunctional species of the form $_3(RO)Si-R'-Si(OR)_3$ have been used to prepare a rapidly growing family of organosilica mesoporous materials.[98] Where the R' group has interactions with adjacent R' groups such as in $(C_2H_5O)_3Si-C_6H_4-Si(OC_2H_5)_3$ (1,4-bis (triethoxysilyl)-benzene), pore wall crystallinity, in this case due to π-stacking between adjacent phenyl groups, has been observed.[99]

2.6 TEMPLATE REMOVAL

Removal of the surfactant template from the initially formed composite inorganic-surfactant mesostructured material is required to open up the pores after synthesis of these materials. Typically this is done by burning out the organic template at high temperatures in an ashing furnace. This can be done in air,[100] although some groups prefer to calcine under flowing nitrogen or sequentially under nitrogen followed by oxygen.[2] Surfactant templated silicates have been shown to retain their structure upon heating in dry atmospheres up to 850 °C although in humid atmospheres pore collapse begins below 800 °C.[101] Calcination temperatures of 500–600 °C for periods of 3 h to overnight are generally used, with temperature ramps of 1–2 °C min^{-1}. Recent work suggests however that much faster heating rates could be used for some materials and that the heating rate can be used to fine-tune the porosity in the structures.[102] Several mechanistic studies of the thermal calcination process for both ionic surfactant templates and polymeric templates have been carried out showing that the template is removed in several steps, the majority below 200 °C, but with some organic species remaining up to 500 °C.[103–106] The surfactant degradation is consistent with a Hoffman degradation based mechanism,[103,106] although the surface bound species which remain at high temperature undergo side reactions which may include a

Cope elimination-like process.[103] For the triblock copolymer templated materials the higher degradation temperatures are associated with the oxidation of the significant amount of the ethylene oxide headgroup located in micropores within the silica walls.[104] The surfactant degradation is associated with some pore-blocking effects until the degradation is complete[103] and it has been suggested that this could be used to block unwanted micropores in the triblock copolymer templated silicates.[104]

The high temperature used to calcine these materials invariably leads to a contraction of the observed pore spacings due to the further condensation of the silicate species in the walls during this treatment. The amount of contraction varies with synthesis method, and can be up to 25 %[101] but even materials with good structures and well condensed walls undergo contractions of 2–3 Å.[107] Materials which were not initially well formed, where the degree of silica condensation in the walls was low, or the infiltration of inorganic species between surfactant micelles was not complete, will collapse during calcination. This is also especially problematic for oxides other than silica, where amorphous to crystalline transformations or transitions from one crystal phase to another can occur at relatively low temperatures. Crystal growth into the pore spaces tends to cause collapse of the pore structures in these cases unless the calcination conditions are tightly controlled. To overcome the shrinkage associated with calcination, and to remove templates from materials which have organic functionality or which have been prepared from inorganic oxides which re-crystallise at high temperatures, a number of other methods of template removal have been devised.

The second commonly used method of template removal is by solvent washing, which is of particular importance for template removal from organo-modified silicas prepared with either pendant organic groups on the pore walls or from bis-silanes to incorporate organic groups in the framework structure.[108–113] Often reflux or Soxhlet extraction for a few hours in acidic ethanol is used. In most cases the pore structures of solvent extracted materials are retained unchanged from those of the as-synthesised materials, and properties such as gas transport through membranes made from surfactant templated silica are improved over those of calcined materials.[114,115] The addition of acid to the solvent allows charge compensation during template removal in materials which have been templated with cationic surfactants under alkaline or neutral conditions where the silica species are negatively charged. The positively charged surfactant is electrostatically bound to the negatively charged silica walls and the addition of H^+ allows counterion exchange to occur during surfactant removal.[116] Ion exchange of the surfactant with an ethanolic solution of ammonium nitrate at 333 K followed by a low temperature thermal treatment, also produces highly ordered porous materials.[117] The ion exchange step can also be used to introduce new functionality into the materials simultaneously with surfactant removal.[118,119] For materials templated with cationic surfactants prepared from acidic syntheses, where the silica has a transient positive charge during condensation, the surfactant counterion is encapsulated into the pores along with the surfactant so the template is more easily removed without requiring ion exchange.[49] Amine templated materials from neutral syntheses can likewise be readily removed by solvent wash.[120]

Materials templated with polymeric surfactants can also be successfully solvent extracted using ethanol.[16] It has been shown that the templating surfactant can be recovered after solvent washing, and used to synthesise further mesoporous materials. This recycling of surfactant reduces the costs of making these materials, and would also be environmentally beneficial in industrial production processes.[16] The main drawback of solvent extraction is that excessive stirring of a suspension of mesoporous silica during solvent washing can reduce particle sizes by mechanical action, and some concerns over residual surfactant remaining in the materials have been raised, although it has been suggested this could be exploited to separately functionalise the meso- and micropores in these materials.[121]

Even milder conditions for template removal can be achieved using supercritical fluids, such as carbon dioxide.[122–125] Treatment with supercritical CO_2 alone is effective only in removing triblock copolymer templates from templated silicates, however inclusion of an alcohol improves the extraction efficiency for ionic surfactant templates.[125] Washing with methanol modified supercritical CO_2 was shown to remove 76–95 % of the surfactant template from MCM-41 and MCM-48 films, powders and SBA-1, SBA-3 powders at 85 °C.[126,127] Other groups prefer supercritical CO_2 modified with a methanol/dichloromethane mixture and showed that the template extracted in this manner could be re-used for further syntheses of mesoporous silica.[128] Treatment with supercritical CO_2 alone leads to pore swelling with retention of mesostructural ordering in 2D hexagonal phase silica triblock copolymer surfactant composites.[129] Pore swelling with supercritical CO_2 has also been observed for materials templated with fluorinated surfactants.[130]

The use of UV/ozone treatment to break down the surfactant at room temperature in air has also been shown to completely remove surfactant from the pores both for mesoporous powders,[131] and for thin films.[132] Silica condensation is promoted by the UV light, simultaneously with surfactant removal.[132] UV light in conjunction with dilute hydrogen peroxide solutions has also been shown to be effective in removing the template.[133] Use of ozone produced directly using an ozone generator, in humid air led to an uncontrolled reaction in mesoporous powders which degraded the materials, although ozone used on suspensions of mesoporous silica, either water or the original synthesis solution successfully removed template from MCM-41 materials after about 14 h.[134] Direct ozonolysis in suspensions however did not improve the silica condensation as in the case of UV/ozone treatment.

A more recent innovative approach to opening up the porosity in functionalised surfactant templated silicates is the use of 'lizard amphiphiles' which use either a chemically or photo-cleavable bond in the surfactant[135] or electrostatic interactions[136] to allow the surfactant to lose its 'tail' after templating is complete. The initial report used a dialkoxysilane linked quaternary ammonium ion 'headgroup', joined to a functional group, in that case alanine, which was joined *via* an ester bond to the alkyl tail.[135] Co-condensation of this species with tetraethoxysilane led to a mesostructured composite where the tail could be removed by acid hydrolysis. This opened the pores, and left densely grafted alanine groups on the pore walls. More recent work has used aminopropylsiloxane or quaternised aminopropylsiloxane as a co-structure-directing agent, along with a carboxylate anionic surfactant which links to them through electrostatic interactions in acidic solutions.[136] Surfactant removal is then done *via* a simple pH change in solution.

Other groups have used different strategies to degrade the surfactant templates. One group prepared surfactants containing regularly spaced unsaturation in their hydrophobic hydrocarbon tails which could be degraded *via* ring closing metathesis depolymerisation under mild conditions, although they have not yet demonstrated preparation of mesoporous materials using these surfactants.[137] Another group has used $KMnO_4/H_2SO_4$ solutions to chemically oxidise the triblock copolymer PEO-PPO-PEO template surfactant from SBA-15 materials.[138] The porosity of these materials was higher than equivalent calcined materials although the silica condensation was lower, as expected. Other chemical treatments have also been used, such as oxidation with hydrogen peroxide,[139] and decomposition of triblock copolymer templates *via* ether cleavage with sulfuric acid, followed by thermal decomposition of occluded EO chains in the pore walls, which are not accessible to the acid, to generate microporosity.[121,140]

Microwave digestion has also been shown to be effective for template removal from mesoporous materials.[141] Powder samples were suspended in aqueous solutions of mixed nitric acid and hydrogen peroxide and treated in a microwave system at operated at approximately 1200 W for 2 min. This resulted in porous materials with highly ordered inorganic frameworks with higher surface areas, larger pore volumes, lower structural shrinkage and a greater number of surface silanol groups compared with those from conventional template removal methods.

2.7 SYNTHETIC ROUTES AND FORMATION MECHANISMS

The synthetic route used determines not only the mesostructure but also the macroscopic structure of the materials produced, and this is due in large part to differences in the formation mechanism in each route. The synthetic routes for mesoporous silica divide into three main types, characterised by the surfactant concentration in the initial solution.

The three major routes are: (i) true liquid crystal templating at high surfactant concentrations, which is used for the formation of monoliths, thick layers or, *via* electrodeposition techniques, formation of thin films; (ii) cooperative self assembly at surfactant concentrations where micelles are present in solution, which can be used to make powders (with either well-defined particle shapes or random structures), fibres and thin films grown at interfaces from solution; and (iii) EISA at very low surfactant concentrations, where no micelles are initially present in solution, and solutions are in general prepared in nonaqueous solvents. This route is used to prepare thin films by dip or spin coating and powders *via* aerosol routes. The following sections will look at the current understanding of the mechanisms involved in each route to mesoporous materials.

2.7.1 True Liquid Crystal Templating

True liquid crystal templating refers to the replication of a high concentration water-surfactant phase in silica.[142–144] It is also sometimes referred to as 'nanocasting'.[145,146] At high concentrations surfactants in water self-assemble into liquid crystalline phases which are gels rather than low viscosity solutions. These phases can contain ordered nanostructures including the 2D hexagonal phase, the 3D hexagonal phase and a variety of cubic phases (see discussion of structures below). Ordered phases typically occur above 20 wt% surfactant in water so the solutions are very viscous. Replicas of these pre-assembled phases can be made by

(a) (b) (c)

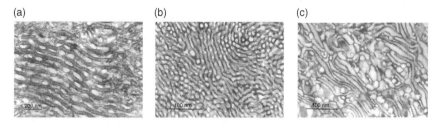

Figure 2.3 TEM micrographs showing pore structures in silicas templated with high concentrations of block poly(butadiene)-poly(ethylene oxide) copolymers: (a) 30 wt% with respect to water, transition between monodisperse vesicles and lamellae; (b) 50 wt% cylindrical pore structure, pore diameter 63 nm, wall thickness 31 nm; (c) 70 wt% lamellae coexisting with vesicles with average lamella thickness 39 nm. Reprinted from *Adv. Mat.*, **11**, 395–398, Nanoporous silicas by casting the aggregates of amphiphilic block co-polymers: the transition from cylinders to lamellae and vesicles by Göltner, C.G.; Berton, B.; Krämer, E.; Antonietti, M., Copyright (1999) Wiley-VCH Verlag GmbH & Co. KGaA. Reproduced with permission

the addition of a silica source, usually tetramethoxysilane (TMOS), to a viscous surfactant gel which has been prepared using dilute acid solution instead of water (Figure 2.3). A typical synthesis uses 50 wt% surfactant in 0.2 M HCl to which TMOS is added in amounts up to 0.25 mole equivalents of the water present in the mixture.[142] Thus the water content is at least equal to the amount stoichiometrically required for complete hydrolysis of the silica precursor. During the synthesis procedure the methanol generated by TMOS hydrolysis is generally removed by the application of a gentle vacuum, since if it is not removed only disordered phases result.

This method can be used to produce particles, thick films and monoliths as the gel will take on the shape of the container in which it is held, until silica condensation sets the final structure. Care must be taken in drying monolithic gels since, as for nontemplated silica sol-gel materials, the capillary forces due to receding water menisci pulling on the porous silica structure during evaporation will cause rapid shrinkage and cracking of the structure if they are exposed to a sudden drop in ambient humidity. With slow drying and/or solvent exchange (which also may remove some or all of the template) transparent mesoporous monoliths can be obtained. Particles can be formed by spreading the synthesis gel over a surface to solidify, during which cracking and contraction of the matrix will result in a coarse powder.

The advantage of this high surfactant concentration method is that the nanostructure obtained in the final materials matches that of the initial surfactant mesophase. The mechanism of this process is however not as straightforward as it may first seem. The methanol generated by silica hydrolysis has been shown to completely destroy the initial mesostructure during this synthesis step, and so the recovery of the structure during and after methanol removal relies upon the ability of the gel to reorder during this process. It has been shown that for very high weight fractions of surfactant template, particularly polymeric templates, the gel is too viscous to complete the re-ordering transition and only glass-like disordered phases result.[147]

One exception to the use of high viscosity block copolymer solutions for the true liquid crystal templating route is for materials templated on surfactant L_3 phases.[148–150] L_3 or sponge phases are formed in ternary systems containing oil and a small molecule surfactant often in salt solutions to regulate the charge density on the surfactant membrane. They consist of a multiply perforated bilayer of surfactant containing the minority oil phase, with large water-filled channels of uniform diameter (Figure 2.4). If HCl is used instead of salt, to regulate surfactant charge density, and TMOS is included in the water phase a replica silicate material can be generated. This material has significant advantages in that there is no need to calcine to open up the porosity. The silica polymerises at the surfactant headgroups and the large water pores

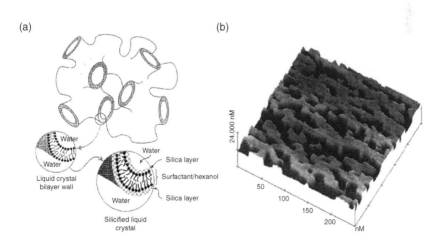

(a) (b)

Figure 2.4 (a) Schematic of the sponge phase and model of the silicification process and (b) a tapping mode AFM image of an L_3 silica film on glass substrate showing channels and walls at the surface of the film, prior to surfactant extraction. Reprinted with permission from McGrath, K.M.; Dabbs, D.M.; Yao, Y.; Edler, K.J.; Aksay, I.A.; Gruner, S.M., Silica Gels with Tunable Nanopores Through Templating of the L3 Phase, *Langmuir*, **16**, 398–406. Copyright (2000) American Chemical Society

remain open, while the surfactant and oil phase become occluded in the walls. Unfortunately the region of phase stability of the L_3 phase in surfactant phase diagrams is small, making templating using this phase difficult, despite the advantages of dispensing with the calcination step.

The true liquid crystal templating method has been successfully used for a number of materials besides silica,[151] primarily for the electrodeposition of metals[13,152] and metal oxides.[153] Pores up to 80 nm diameter have been obtained using high molecular weight block copolymer templates using this technique, substantially larger than the pores that are generally obtained via the other surfactant templating methods.[146,154] Larger molecular weight polymers have increasingly reduced mobility and this appears to be the limit at which strong segregation between hydrophobic and hydrophilic blocks, and thus templated porous materials, can be obtained. True liquid crystal templating is also useful to easily obtain large, millimetre sized (although irregular) particles of mesoporous materials, and for this reason was used to prepare core-shell particles having a core prepared with one template and a shell prepared with a second template giving a different mesopore structure.[155] Bimodal porosity has also been obtained via true liquid crystal templating by using two populations of noninteracting micelles. Fluorocarbon-ethylene oxide surfactants with a C_{6-16} fluorocarbon tail and EO_{4-5} or EO_{7-8} headgroup were mixed with either SE3030, a polystyrene-b-ethylene oxide block copolymer with a molecular weight of 3000 g mol^{-1} for each block, or KLE3935, a hydrogenated polybutadiene-b-ethylene oxide block copolymer with molecular weights of 3935 g mol^{-1} (the Kraton Liquid based hydrogenated polybutadiene) and 3460 g mol^{-1} [poly(ethylene oxide)]. This resulted in materials showing two distinct pore sizes with smaller pores intercalated between larger pores in both ordered and disordered phases depending on the relative concentration of small and large micelles.[156] Further pore hierarchy could be obtained via the inclusion of colloidal polymer spheres into the surfactant-silica gels to form macroporous materials having surfactant templated mesopores in the walls.[157]

Swelling of the micelles in these concentrated surfactant phases to form microemulsions has also been used to create monolithic silicas with large cage-like pores.[158,159] The use of aliphatic and aromatic hydrocarbon species to swell the hydrophobic interiors of the micelles and EO based surfactants with large EO headgroups resulted in rapid formation of monoliths with thick silica walls having high hydrothermal stability and a variety of ordered structures including $Ia\bar{3}d$ and $Fd\bar{3}m$ as well as worm-like phases.[160,161] The method of nanocasting has also been used to study the phase diagrams of block copolymer gels[145,156] and nanostructured ionic liquids[162,163] since the silica replica freezes in the nanostructures in these gels, and can be used to obtain solid casts of phases which are otherwise transient, or to assist in determining the structures of mixed phases.

A variation on true liquid crystal templating is the vapour phase synthesis route reported by Tanaka *et al.*[164–166] where the silica precursor TEOS, is vaporised and allowed to infiltrate into a surfactant film. The infiltration was carried out in a closed container which held a small amount of TEOS and a separate amount of HCl solution, along with the surfactant film which had been spin coated onto the substrate from an PEO-PPO-PEO triblock copolymer/water/ethanol mixture. This was heated at 90 °C for 1 h exposing the film to saturated TEOS vapour under autogenous pressure. Silica films with ordered mesostructures and good hydrothermal stability resulted from this method, which could be extended to other materials that form from volatile precursors.

2.7.2 Cooperative Self-Assembly

The cooperative self-assembly route is the most commonly used synthesis procedure for surfactant templated materials. It uses aqueous solutions at a much lower initial surfactant concentration than for the true liquid templating route, reducing the required amounts of expensive surfactant template. In these solutions, the surfactants are at a high enough concentration to form micelles, which may be spherical, elliptical, rod-like or vesicular, but do not form the ordered aggregates found in the final templated silica-surfactant composites. The solutions are in thermodynamic equilibrium so are stable at a given temperature until the silica precursors are added. Once added, a series of interactions between the inorganic species and the surfactant micelles occur, which involve simultaneous association of all components, hence the name, cooperative self-assembly. The result of the interactions is formation of the silica-surfactant composite, usually a precipitate, with an ordered nanoscale structure similar to those found in concentrated surfactant solutions.

Surfactant-inorganic systems of this type have been categorised based on the charge on each species (see Table 2.2), so S^+I^- and S^-I^+ denote the simple cases where surfactant (S) and inorganic species (I) have opposite charges. Syntheses using cationic surfactants with a range of negatively charged inorganic species (*e.g.* CTAB with silica in alkaline conditions) and anionic surfactants with cationic inorganic species (*e.g.* sodium dodecylsulfate with iron or lead oxides) were shown to produce meso-phase composite materials, although not all were stable to surfactant removal.[49] Mesophase materials can also be prepared where these species have the same charge, and this was originally explained as due to an anion mediated mechanism described as $S^+X^-I^+$ or $S^-X^+I^-$. Examples included silica or $[H_2ZnPO_4]^+$ with alkyltrimethylammonium surfactants under acidic conditions where the inorganic species both carry a positive charge. The use of neutral surfactants as templates was originally termed S^0I^0, referring to the use of either amine and poly(ethylene oxide) surfactants with silica at pH 7.[18,44] Later work on poly(ethylene oxide) surfactants has suggested that under acidic conditions this is more correctly described as $(S^0H^+)(X^-I^+)$ due to the protonation of the ether oxygen in the EO headgroup allowing hydrogen bonding to occur under these conditions.[16,17]

The mesophase formed for a given surfactant and silica source under particular conditions is hard to predict, and many studies have addressed the wide phase space available for different combinations of surfactant, silica source, pH, ionic strength, addition of hydrophobic species, temperature and post-synthesis treatments. Despite this lack of predictability, a given synthesis is usually readily reproducible, so the fact that much lower amounts of the surfactant template are required in this method makes it an attractive route to mesoporous materials both on grounds of cost and environmental considerations. Using this dilute surfactant route it is also possible to produce a number of different macroscale structures, from irregular mesostructured powders to well controlled shapes such as spheres, fibres, shell-like shapes and thin films on substrates[167] and at the air–water interface[168] (see discussion of macroscale structures below). Given the variety of interactions possible between the micelle surface and inorganic species, and the diversity of observed macroscale shapes, the evolution of the nanoscale solution structures into the final composite material has been the subject of numerous studies, which are now described in more detail.

Studies of the mechanism of formation for cooperative self-assembly divide into investigations on three types of system, depending on pH and the type of surfactant template. Most work has been done on formation in acidic systems since the reaction here is slower and more accessible to standard time-resolved measurement techniques. Within the acidic syntheses work has divided into that concerned with small molecule surfactants and with polymeric (usually Pluronic®) surfactants. For alkaline and neutral preparations the rapid precipitation has made measurement more difficult, requiring stop flow or other fast mixing devices to study the interaction between silica and surfactant on short timescales. These studies have focused on small molecule surfactants usually of the alkyl-trimethylammonium bromide family, and some work has been done using amine templates.[169] Although different mechanisms have been postulated at various times for these various categories of preparation, it seems that in the most recent work, the suggested formation mechanisms for all three types are again converging to involve similar steps in the pathway from free micelles to mesostructured composite.

The initial paper by the Mobil scientists postulated two possible mechanisms for the precipitation of mesophase silica surfactant composites from micellar solutions.[2] The first, which was equivalent to the true liquid templating pathway discussed above, suggested the initial formation of aggregates of micelles into the high concentration structures followed by infiltration of the inorganic precursor. This was quickly discarded since there was no evidence for such concentrated aggregates in the synthesis solutions prior to silica addition under the dilute surfactant conditions used in this route.[170,171] The second suggestion was that silica coated individual surfactant micelles in solution, leading to a reduction in the repulsion between micelles, allowing them to aggregate into the observed precipitates with lyotropic liquid crystalline ordering. This garnered some support from cryogenic transmission electron microscopy (cryo-TEM) imaging and small-angle X-ray scattering (SAXS) patterns of samples taken from alkaline preparations using the cationic surfactant CTAB with TEOS at low concentrations, to attempt to gain time-resolved information.[171] These were interpreted to show clusters of elongated silica coated micelles formed 3 min after addition of TEOS to the surfactant/alkali solution. Elongation of CTAB micelles in the presence of silica counterions was also observed under dilute conditions by Lee et al. using light scattering, although under similar conditions the $[AlO_2]^-$ counterion did not have the same effect.[172] This was ascribed to enhanced binding of silica counterions to the micelle surface due to their polydentate nature. Similar ultralong wormlike micelle formation is observed due to interactions between CTAB and counterions such as salicylate but since the silica oligomers can react with each other, concentrating them at the micelle surface causes further polymerisation catalysing the formation of larger oligomers and leading to the silica coating of the micelles.

This coated micelle mechanism remained a central concept in the mechanism of mesoporous material formation for several years. It was used by Yang et al.[173] to explain the formation of spectacular rope-like, disc, toroid and gyroid shaped particles and films at the solution surface from cationic surfactants in acidic solutions as being due to the sequential addition of silica coated micelles to a surfactant-silica seed central particle. The elongated silica coated micelles wrapped around the central core forming spiralling pores within the particle, and shapes that represented a nonequilibrium structure frozen at that point in time where the silica condensation prevented further rearrangement of the structure. Similarly Cai et al.[174] described the formation of mesoporous particles with pores radiating out from the particle centre as due to aggregation of stiff, silica coated micelles. Single silica coated micelles were also identified under specific conditions in a synthesis route involving nonionic surfactant micelles in pH 2 solutions, where the silica condensation was suppressed.[61] The silica coating developed around the micelles within minutes after addition of TEOS to the acidic micellar solution and formed a dilute weakly reticulated gel surrounding the micelle starting from the EO shell and growing into the medium. Addition of fluoride anions which catalyse silica condensation then caused precipitation of particles with a disordered wormlike micelle structure.

This coated micelle model does not however explain the formation of faceted 'single crystals' which were found in alkaline syntheses around the same time.[31] Very early experiments had shown that different mesostructures could be formed at the same surfactant concentration simply by varying the silica amounts, suggesting the inorganic ions played an important role.[175] Furthermore experiments using fluorescence to measure the binding of silica anions at the surface of micelles showed only fairly weak displacement of the surfactant counterion. This still concentrated silica at the surface of the micelle catalysing condensation, but was not considered to be sufficient to coat the micelle. Thus it was suggested that the micelles merely acted as a surfactant reservoir with assembly of the composite occurring elsewhere in the solution via surfactant monomer adsorption onto an oligomeric silica species.[176,177] More recent work on extremely dilute solutions reached similar conclusions about the degree of silica counterion binding but remained equivocal on the existence of silica coated micelles.[178] A competing mechanism also involving a cooperative process had been suggested earlier by Monnier et al. (Figure 2.5), where individual surfactant molecules interacted with small multidentate silica oligomers forming ion pairs in solution, and these species then cooperatively formed the concentrated mesophase, with headgroup volumes (and thus mesophase structure) determined by charge density matching between the inorganic and organic species.[48] Charge density matching is now commonly used to explain observed phase transitions in the solid state as silica condensation changes the charge on the

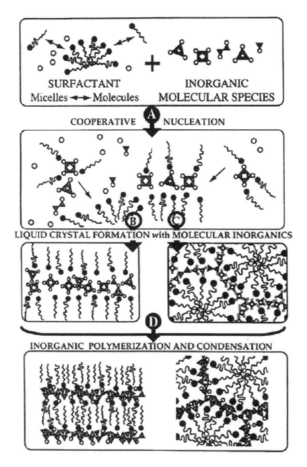

Figure 2.5 Cooperative templating model for silica-surfactant mesophase formation. (A) Single surfactant molecules react preferentially with silicate polyanions which displace the original surfactant mononanion. Micelles serve as a surfactant molecule source or are rearranged according to the anion charge density and shape requirements. (B and C) Organised arrays nucleate and precipitate in configurations determined by the cooperative interactions of ion pair charges, molecular geometries and van der Waals forces between surfactant tails since silica condensation at this stage is minimal. (D) Condensation of silica with increasing time and temperature, during which changes in framework charge density may lead to liquid crystal like phase transitions as the surfactant phase tries to reorganise the changing interface charge density. Reprinted with permission from Huo, Q.; Margolese, D.I.; Ciesla, U.; Demuth, D.G.; Feng, P.; Gier, T.E. *et. al.*, Organisation of organic molecules with inorganic molecular species into nanocomposite biphase arrays, *Chem. Mater.*, 6, 1176–1191. Copyright (1994) American Chemical Society

inorganic component and shrinks the structure altering the relative headgroup area of surfactants in the incorporated micelles.[179] Its role in the initial stages of synthesis *via* this route however has been questioned, particularly for acidic syntheses where the counterion mediated interaction between positively charged silica and surfactant species was originally mooted.

At this time, *in situ* experiments were also starting to be carried out which showed that the mechanism was more complex than the simple aggregation of silica-surfactant micelle building blocks and that phase separation played an important role in the process. Early experiments, by Firouzi *et al.*,[181] done under conditions where the silica does not condense (in highly alkaline solutions), showed that the association between silica counterions and cationic surfactant micelles changed the repulsive interaction between micelles into an attractive interaction. This caused formation of two-phase solutions, unlike the one-phase solution seen at similar concentrations in a surfactant/water only mixture. The two phases consisted of a concentrated phase with a lyotropic liquid crystalline structure (2D hexagonal or lamellar) and a dilute solution of isotropic micelles.

In situ EPR investigations by Galarneau *et al.*[182] using spin probes in an alkaline system with a CTAB template reported that MCM-41 assembled through a rapid initial step where mixed surfactants and silicates formed an amorphous gel. The hexagonal structure was then built within this gel by local reorganisation to form the ordered mesophase. Similarly in acidic solutions, also using CTAB as the template, studies of particle development showed initial growth of gel-like particles within which the mesostructure developed.[183] Chan *et al.*[183] proposed a model in which the growth of inorganic oligomers gives rise to a thermodynamically unstable mixture, which caused a phase separation to occur, producing droplets of a second liquid (or liquid crystal) phase rich in oligomer and surfactant. Within these colloidally stabilised droplets microphase separation occurs, giving rise to an ordered mesostructure which is then fixed by further polymerisation of the inorganic framework.

Around the same time *in situ* time-resolved reflectivity[184,185] and small-angle neutron scattering[66] experiments on acidic syntheses using a CTAB template which produced films at the air–solution interface (rather than spherical particles in solution) also found evidence for formation of particles in the subphase solution, which aggregated at the interface and underwent structural rearrangements to form the film. This particle-driven formation mechanism occurred when the solutions were at optimal conditions for the most rapid formation of films with the highest mesophase ordering. When more silica was added to the solutions, film formation became much slower, and the subphase solutions contained highly elongated CTAB micelles with a thin coating of silica. Entropic considerations prevented aggregation of these wormlike micelles to the growing film at the interface until the silica polymerisation had proceeded to a much greater extent.[185]

Studies by another group on formation of spherical particles found that as the concentration of silica and surfactant in the preparation increased, the number of spheres formed remained roughly constant, while the sphere size increased.[186] This also suggested that nucleation did not occur on micelles since increasing surfactant concentration would result in larger numbers of micelles and thus more nucleation sites. Instead nucleation was silica-driven with micelles or surfactant monomers adsorbing onto silica polymers or oligomers to form primary particles which could contain either spherical or rod-like micelles. The primary particles continued to grow *via* addition of either monomers or other primary particles, and reorganised into the final structure (Figure 2.6).

A model of colloidal phase separation was also proposed by Yu *et al.*[187] based on their work on formation of shapes ranging from

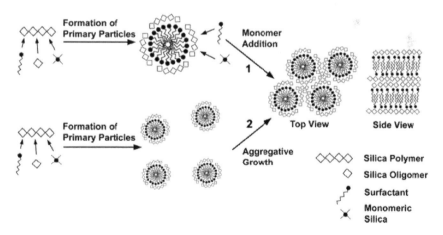

Figure 2.6 Scheme of the mechanism for the formation of mesoporous silica. Silica polymers formed initially from silica monomers, and associated with surfactant monomers, which form composite self-organised primary particles which can either continue to grow *via* monomer addition (path 1) or themselves aggregate in a directional fashion (path 2) to form the final mesophase composite. Nondirectional aggregation would cause formation of disordered pore structures. Reprinted with permission from Nooney, R.I.; Thirunavukkarasu, D.; Chen, Y.; Josephs, R.; Ostafin, A.E., Synthesis of Nanoscale Mesoporous Silica Spheres with Controlled Particle Size, *Chem. Mater.*, **14**, 4721–4728 . Copyright (2002) American Chemical Society

rod-like particles to 'single crystals' of nonionic triblock copolymer templated materials synthesised in acidic conditions both with and without salt. Synthesis conditions such as low temperature, low acidity, and low ionic strength which increased the induction time, also gave rise to mesoporous silica morphologies with increased curvatures suggesting surface energies played more of a role in materials which phase separated slowly from solution. Factors such as high ionic strength promoted phase separation from solution and also aggregation of the initially phase-separated particles, driving the formation of nonspherical and elongated particles. Their model involved three stages: (1) cooperative self-assembly of inorganic/organic composites in solution due to charge density matching between surfactant and silica oligomers; (2) formation of a new liquid crystal-like phase in spherical droplets which is rich in aggregates of block copolymer/silica species; and (3) continuing phase separation of these aggregates from the solution, and coalescence of aggregates coupled with growth of solid mesostructures within the aggregates driven by continued condensation of silica species and charge density matching between surfactant and framework. Macroscale morphologies of these materials developed after the phase separation stage so were influenced by the competition between the free energy of mesostructure self-assembly and the colloidal surface free energy, but also by other factors such as shear. When phase separation occurred early in the synthesis, the mesostructure self-assembly was dominant and caused the formation of single crystal structures.[187]

Further evidence for phase separation driven co-assembly was found by Flodström et al.[188–190] in work on triblock copolymer templated materials formed from acidic solutions. Diffuse flocs of unordered spherical micelles surrounded by a silica phase were observed in solution, prior to an ordering transition in the floc which formed the final ordered mesophase structures. The formation of particular mesophase structures depends upon the relative volumes of the hydrophobic region of the micelle and the headgroup area resulting after silica intercalation around the hydrophilic part of the micelle. This was observed to occur for materials both with final structures of 2D hexagonal, and micellar cubic ($Im\bar{3}m$) suggesting it is a general formation mechanism for triblock copolymer templated materials. A bicontinuous cubic ($Ia\bar{3}d$) phase however was found to assemble via a 'micro true liquid templating' mechanism, as in that case, phase-separated droplets of a bicontinuous phase of the polymeric surfactant template had already phase separated from solution prior to addition of the silica precursor.

Egger et al.[63] challenged both the formation of silica coated micelles and the idea of charge density matching in the case of acidic syntheses of SBA-1 using cationic cetyltriethylammonium bromide surfactant (*i.e.* the $S^+X^-I^+$ route). They suggested that simple Derjaguin and Landau, Verwey and Overbeek (DLVO) considerations would suggest that, at the high acid concentrations used in these preparations (around 4 M HCl), Coulombic interactions between species would be almost entirely screened, so that charge matching pathways leading to coated micelles would not work. Instead they observed a salting out effect coupled with a drastic change in the water activity as silica polymerisation progresses causing the precipitation of an amorphous silica-surfactant composite which then undergoes a substantial re-organisation in the precipitated phase to achieve the final structure (Figure 2.7). The silica oligomers bind weakly to the micelles and act as cross-linkers between micelles to provide an attractive force between micelles in the concentrated phase, however a layer of water remains between the condensing silica and the micelles allowing mobility of surfactant molecules within the micelles.

Several groups have also suggested that during the initial stages of the synthesis, rather than silica coating the micelles, hydrophobic silica precursors, such as TEOS, become solubilised in the hydrocarbon tail region

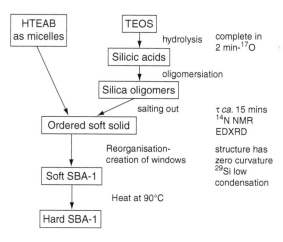

Figure 2.7 General kinetic picture representing the steps involved in forming SBA-1. TEOS is rapidly hydrolysed to silica acid species that start polymerising preferentially linearly. As they polymerise water molecules lose mobility and bind to their surfaces. This effect, coupled to the salting-out effect due to the presence of 4.4 M HCl induces the composite mesophase precipitation. An ordered soft solid is formed, which then rearranges to create windows between micelles in cages. Further condensation occurs to produce a hard solid upon heating to 90 °C. Reprinted from Egger, C.C.; Anderson, M.W.; Tiddy, G.J.T.; Casci, J.L., In situ NMR and XRD studies of the growth mechanism of SBA-1, *Phys. Chem. Chem. Phys.*, 7, 1845–1855. Copyright (2005). Reproduced by permission of the PCCP Owner Societies

of cationic surfactant micelles, forming a silica reservoir which, after phase separation occurs, can then polymerise at the water–micelle interface as the reaction proceeds.[191] TEOS has also been shown to penetrate the hydrophobic core of triblock copolymer micelles during early stages of synthesis[192] although that work suggests that this reservoir is responsible for generating coated micelles rather than phase separated particles.

Recent work by the Goldfarb group using cryo-TEM and EPR with a variety of spin probes to follow the dynamics and molecular scale interactions of micelles during the reaction with silica has however been interpreted again in terms of the coalescence of coated micelles but now also incorporates phase separation followed by rearrangement of the structure in the precipitate. In this work after the addition of silica, elongated micelles were observed to form, and coexist in solution with spherical micelles. These then formed disordered aggregates of threadlike micelles seen by TEM which rearranged to a 2D hexagonal structure, and then, if butanol was added, to a cubic structure, showing the ability of the precipitated phase to re-arrange after precipitation. Similar processes were observed in the synthesis of MCM-41[193–196] (cationic surfactants in alkaline solutions), SBA-15[51,197] (nonionic triblock copolymer surfactant under highly acidic conditions) and KIT-1[198] (nonionic triblock copolymer surfactant with butanol under dilute acidic conditions). A simple model of the formation process based on these data predicted coalescence of silica coated micelles.[199] In this model, silica monomer adsorption to surfactant micelles shields the charge on the surfactant headgroups, elongating the micelles and causing attraction between micelles due to bond formation between silica species on two different micelles. Micelles therefore form a given structure in solution (for a particular silica and surfactant concentration) and this coated micelle structure aggregates to produce particles (Figure 2.8). This model was shown to reproduce the results of the Goldfarb studies, producing elongated micelles due to silica monomer binding and aggregation of these micelles into ordered or disordered structures depending on silica and surfactant concentrations. It was also argued to be consistent with the phase separation seen by Flodström et al. since rearrangement of the structure within the particles was allowed in the model if the silica condensation had not proceeded sufficiently to render the particles solid.

Thus the experimental data appear to show that silica coated micelles can exist in solution and can either participate directly in forming mesostructures via coalescence or else play a minimal role as a surfactant/ silica oligomer reservoir while structure develops in a concentrated phase-separated gel phase which initially is formed from spherical or

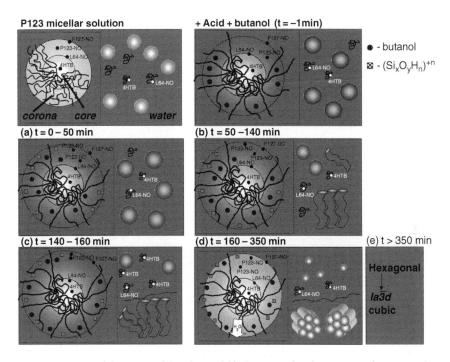

Figure 2.8 Model proposed by the Goldfarb group for formation of KIT-6. The right-hand side of each panel presents the nanostructures present in the reaction mixture at a given time. The left-hand side of each panel shows a cross-section of the micellar structure with the location of the different EPR spin probes within the micelles. Grey scale represents the amount of water in the corona region, dark grey denotes a large amount of water, light grey a region with less water. Reprinted with permission from Ruthstein, S.; Goldfarb, D., Evolution of solution structures during the formation of the cubic mesoporous material, KIT-6, determined by double electron-electron resonance, *J. Phys. Chem. C*, **112**, 7102–7109. Copyright (2008) American Chemical Society

close-to-spherical disordered micelles that rearranges internally to form the mesostructured materials. This dual behaviour dependant on synthesis conditions is similar to that observed in solutions of polyelectrolytes and surfactants where in this case the silica plays the role of a polyelectrolyte interacting in a multidentate manner with the surfactant micelles. This analogy was first suggested by Zana *et al.*[176] and also by Chan *et al.*[183] and in studies by Edler co-workers[185,200] The close analogy between silica and water soluble polar polymers was first confirmed experimentally by Edler co-workers,[200,201] where substitution of a polymer having similar characteristics to the polymerising silica under acidic conditions (*i.e.* branched, low but positive charge) with the same cationic surfactant template (CTAB) was shown to produce similar films at the air–solution interface with an identical nanoscale structure to that observed for the silica-surfactant case.[200] This analogy has also been explored by Linden co-workers[202] who carried out studies of oppositely charged polymers (linear polyacrylates of various molecular weights) with dodedcyltrimethylammonium chloride and showed formation of precipitates similar in nanoscale character to those formed in alkaline silica-surfactant solutions. Higher molecular weight polymers promoted both phase separation, into more concentrated 2D hexagonal phases at lower solution concentrations, and the formation of ordered mesophases after phase separation, and also allowed greater uptake of toluene showing stabilisation of the ordered hexagonal phase by polyelectrolyte binding to the micelles. These observations were in good agreement with similar indirect conclusions drawn from silica-surfactant studies.

Given the important role of silica oligomer binding to surfactants in this cooperative assembly formation route, significant effects on the structures formed are seen due to added salt. Early work by Ryoo et al.[107] on MCM-41 formed in alkaline syntheses with cationic surfactants showed improvements in mesoscale ordering if the solution pH was carefully controlled, and Edler and White[74] showed that the acid counterion used also affected the degree of ordering in MCM-41 in the order $[SO_4]^{2-} > [C_2H_3O_2]^- > Cl^- > Br^-$. Salt effects on the time required for silica precipitation in the synthesis of SBA-type materials with the non-ionic PEO surfactant Triton X-100 also depend both on $[H^+]$ and the anion used.[17] The trend in the precipitation time was found to be HBr $\sim$ HCl $<$ HI $<$ HNO$_3$ $<$ H$_2$SO$_4$ $<<$ H$_3$PO$_4$. In these nonionic templated syntheses the interaction between silica and surfactant was suggested to occur via the $(S^0H^+)(X^-I^+)$ mechanism and the assembly rate r was measured to increase with increasing concentration of $[H^+]$ and $[Cl^-]$, according to the kinetic expression $r = k[H^+]^{0.31}[Cl^-]^{0.31}$. However these observations could also be consistent with the salting out model of Egger et al.[63] since the change in water activity with increasing concentration of acid will also drive this effect and the different counterions bind water to a decreasing extent along the series measured.[203] The salting out effect is also important in the colloidal phase separation model proposed by Yu et al.[187] discussed above since high ionic strength favoured the rapid phase separation of silica-surfactant composites, and the coalescence of aggregated composite particles to form particles with low curvature, as opposed to spherical particles formed at low ionic strengths.

The effect of addition of NaCl to a nonionic triblock copolymer templated silica synthesis was investigated by Alfredsson et al.[204] who showed that NaCl made the aqueous solution more polar and thus a worse solvent for the polymer so that the EO chains of the polymer became solvophobic. As a result the micellar size and initial cell parameter of the 2D hexagonal phase were larger and the aggregation of the polymer-silica composite occurred faster. Baute and Goldfarb[205] noted a similar effect of NaNO₃ on nonionic triblock copolymer Pluronic P123 micelles in acidic SBA-15 synthesis. This effective dehydration of the EO groups decreased the curvature of the micelles, and led to formation of a final cubic phase rather than a hexagonal phase.

2.7.3 Evaporation-Induced Self-Assembly

The third route to surfactant templated mesoporous silicas is *via* EISA. This method, introduced initially by Ogawa in 1994,[206–209] relies upon the rapid evaporation of a volatile solvent to concentrate the silica-surfactant-water solution during dip or spin coating,[210,211] inkjet printing or dip pen writing methods[212,213] or aerosol drying.[214] The volatile solvent, normally ethanol, disperses the silica precursor and the surfactant, usually at surfactant concentrations below the cmc in that solvent, so no micelles are present in the initial solution. The pH of the solution is held around 2 so that silica condensation rates in the solution are minimised, reducing the formation of oligomeric species. During the application of this solution to a substrate, evaporation of the volatile component results in the rapid formation of micelles as the concentration in the solution increases above the cmc, and soluble inorganic species remain trapped in the interstices between micelles, where condensation takes place to fix the structure (Figure 2.9).[93,214] This method results in ordered thin films of controllable thicknesses of up to about 1 μm by changing the viscosity of the solution or by adjusting the spinning speed or withdrawal rate in dip coating.[215] Above 1 μm in thickness, films tend to crack or peel off the substrate, similar to other sol-gel films deposited *via* spin or dip coating methods.[206] After condensation has occurred, template removal by solvent wash, calcination or other techniques opens up the pores, as in the other routes to mesoporous materials.

Extensive work on the mechanism of formation of these films has been carried out by several groups.[39,211,214,216–224] It is generally agreed that it is the evaporation of the volatile solvent which triggers micelle assembly, once the concentration of the surfactant in the remaining solution increases above the cmc.[225] There was some debate over whether the mesostructure nucleated at the top and/or bottom surface of the film

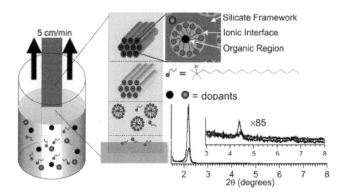

Figure 2.9 Diagram showing processes during surfactant templated film formation during dip coating (left). The film evolves continuously with the steps occurring in order from bottom to top as the substrate is withdrawn from a reservoir containing the surfactant, inorganic precursor and hydrophobic or hydrophilic dopants in ethanol. The magnification (right) shows the three distinct regions into which mesostructured surfactant-silica films may be subdivided and the location of hydrophilic (grey) and hydrophobic (black) dopant molecules. Representative X-ray diffraction patterns of the films are also shown (right). Reprinted with permission from Minoofar *et al.*, Copyright 2005 American Chemical Society

before propagating into the film interior,[219,222] however later work has shown that post-deposition conditions can have a much larger effect on structure than the initial deposition, so that the influence of the interfaces is small in comparison. However, at least one study has confirmed that the substrate type and roughness can affect mesostructural ordering in the films.[226,227] The EISA method can be used to deposit onto almost any type of surface including sacrificial polymer substrates to make free-standing films[228] however glass or silicon wafers have been used in most studies.

Using the EISA method, the mesostructure produced is influenced by a number of factors. The evaporation rate of the volatile solvent is crucial to the establishment of an ordered mesostructure and may induce changes between ordered mesostructures.[216] In general the surfactant structure (packing parameter)[229,230] and the ratio of inorganic species to surfactant in the initial solution determines the mesostructure of the films at a constant humidity;[211,221] however, by altering the relative humidity, rapid water uptake or loss by the film is observed and results in changes in structure. Once the volatile solvent has evaporated, the inorganic species is generally only partially condensed, leaving the film in a 'modulable steady state'.[231] Thus if the surrounding conditions, in particular the relative humidity, are altered, the film will adjust the mesostructure to the equilibrium structure favoured by the new conditions and this rearrangement in response to altered conditions can continue until the inorganic condensation process has progressed far enough to render the film too viscous to change. This high sensitivity to ambient conditions, although it allows the mesostructure to be tuned after deposition, is one of the main drawbacks of this method leading to problems with reproducibility of the nanostructures produced.

The ageing time of the synthesis solution prior to deposition is also an important variable for production of ordered structures.[233,234] Early work showed that at the optimum ageing time and temperature (4.5 h at 40 °C in that work) films had well-ordered 2D hexagonal structures throughout the entire film thickness while shorter or longer ageing times led to films with disordered interiors.[233] More recently the extent of oligomerisation in the synthesis solution has been used to adjust the curvature of the mesostructures formed using a triblock surfactant EO_{17}-PO_{12}-C_{14}.[232,234] The surfactant molecules in the ethanolic synthesis solution reduced the cluster–cluster growth rate of silica oligomers due to weak associative interactions, possibly hydrogen bonding, between the silica clusters and surfactant in solution (Figure 2.10); however, the primary structure of the silica oligomers of linked Si-O rings was not affected by the surfactant.[234] Preparation of films from these solutions showed that the ageing time (*i.e.* oligomer size) and ambient relative humidity during dip coating determined the phase formed in the films, with longer ageing times leading to formation of lower curvature structures (Figure 2.11).

These results however showed the opposite trend to similar studies using cationic surfactants such as CTAB.[236] For cationic surfactants in alkaline systems curvature is driven by charge density matching between the silica walls and surfactant headgroup. As the charge on the silica framework decreases, the headgroup area of the surfactant increases, leading to a change towards higher curvature structures, *i.e.* causing transition from lamellar to 2D hexagonal mesostructures. Similar arguments are proposed for acidic systems despite the difficulties in defining the charge on the silica walls at this pH.[236] In the case of nonionic surfactants, however, larger oligomers formed as ageing time increases do not mix as effectively with the EO headgroup, so reduction in curvature is observed, and the films alter structure from 2D hexagonal to bicontinuous phases to lamellar with increased ageing time of the solution before dip coating.[232]

Pore size in EISA materials can be adjusted by changing the size of the surfactant micelles with both small cationic surfactants[237] and larger nonionic triblock copolymer surfactants used to prepare films. While triblock copolymer surfactants of the EO-PO-EO family can be used to make

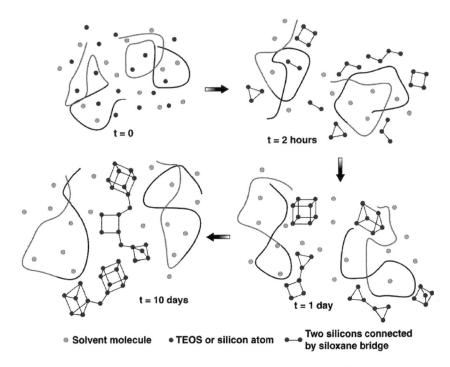

Figure 2.10 Schematic illustrating the evolution of silica species during ageing prior to coating and EISA. The times indicated are after mixing the silica and surfactant solutions. The EO_{17}-PO_{12}-C_{14} triblock copolymer surfactant is shown in light and dark grey because both the PO and C blocks are hydrophobic. Reprinted with permission from Urade, V.N.; Bollmann, L.; Kowalski, J.D.; Tate, M.P.; Hillhouse, H.W., Controlling interfacial curvature in nanoporous silica films formed by evaporation-induced selfassembly from nonionic surfactants. II. Effect of processing parameters on film structure, *Langmuir*, **23**, 4268–4278. Copyright (2007) American Chemical Society

these films,[230,238] the rapid assembly process and the use of a solvent, ethanol, which can solvate both blocks of this polymer mean that the phase separation responsible for forming the mesostructures can be hindered. Better reproducibility, the ability to use a wider range of solvents and more stable mesostructures for a wider range of different inorganic oxides have resulted from use of templates which have a stronger segregation between blocks.[217] The KLE copolymers poly(ethylene-*co*-butylene)-*b*-poly(ethylene oxide) in particular have been shown to produce stable well ordered films with large pores for a variety of inorganic oxide materials.[239]

Film deposition *via* EISA allows straightforward patterning methods to be used to determine the distribution of the mesostructured film on a

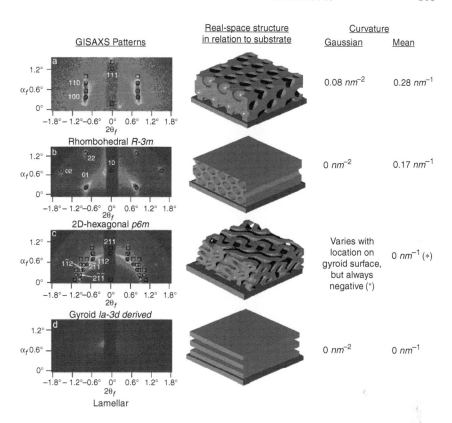

Figure 2.11 Well ordered silica films formed with EO_{17}-PO_{12}-C_{14} generated at various ageing times in the sol showing the different curvatures in the structures of the final films. Structures were determined by X-ray diffraction and modelling of the structure and orientation using NANOCELL.[235] The simulated Bragg peak positions are indicated on the experimental patterns with circles. The space group, orientation and lattice constants were determined to be (a) $R\bar{3}m$ structure, (111) oriented with $a = 87$Å and $\alpha = 80°$, (b) $c2m$ structure (10) oriented (distorted 2D hexagonal) with $a = 146$ Å and $b = 97$ Å, (c) double-gyroid topology (211) oriented (distorted $Ia\bar{3}d$) with $a = 172$Å, $b = c = 176.5$ Å, $\alpha = 91°$ and $= \gamma = 92°$, and (d) lamellar structure (the Bragg peak is partially covered by the beamstop). The 3D renderings on the right were based on the lattice constants and the interfacial curvatures calculated from observed lattice constants and rough estimates of the pore volume fractions. Reprinted with permission from Urade, V.N.; Bollmann, L.; Kowalski, J.D.; Tate, M.P.; Hillhouse, H.W., Controlling interfacial curvature in nanoporous silica films formed by evaporation-induced self-assembly from nonionic surfactants. II. Effect of processing parameters on film structure, *Langmuir*, **23**, 4268–4278. Copyright (2007) American Chemical Society

substrate.[240,241] Most simply, altering the hydrophobicity or hydrophilicity of the substrate by attachment of a self-assembled monolayer (SAM) leads to preferential formation of the mesophase on the hydrophilic areas.[213,214,240] The evaporation of ethanol from the deposition

Figure 2.12　Optical fluorescence micrograph of patterned rhodamine containing thin film mesophase formed by dip coating on a SAM-patterned substrate prepared by microcontact printing. (Inset) TEM micrograph of rhodamine containing 1D hexagonal mesostructure. Reprinted from 11/579–585 Evaporation-Induced Self-Assembly: Nanostructures made easy by Brinker, C.J.; Lu, Y.; Sellinger, A.; Fan, H. 9, 7 No 12, 13. Copyright (1999) Wiley-VCH Verlag GmbH & Co. KGaA. Reproduced with permission

solution leaves behind a water-rich phase causing selective dewetting from hydrophobic areas of the substrate (Figure 2.12). This has also been used as a preliminary patterning technique followed by pattern transfer to generate mesostructured silica patterns on flexible polymer substrates.[242] Films can also be patterned *via* lithographic techniques, since condensation of the silica can be driven by lowering or raising the pH. Thus by including a photoacid generator such as a pH sensitive dye, in the deposition solution, this species is co-deposited into the films and when exposed to light through a mask alters the local pH in the film causing silica condensation to occur in the patterned areas.[243,244] Depending on the degree of condensation and the deposition conditions of the film, this process may also result in a phase change in patterned areas allowing films with two different mesostructures in discrete regions to be formed.[243] UV light has also been used to photo-calcine deposited films leaving the calcined areas more highly condensed and thus less susceptible to attack by alkaline solutions, allowing selective removal of the areas which were not exposed to the light.[245,246] Other direct lithographic techniques which also rely on increased silica condensation and template degradation, such as deep X-ray lithography, have also been used to pattern films.[247]

The technique of micromoulding in capillaries has also been used to pattern mesostructures across a substrate.[248–250] In this case a polydimethylsiloxane (PDMS) stamp with channels is pressed against the substrate on top of a drop of synthesis solution[250] or the synthesis solution allowed to infiltrate *via* capillary action[250] or under the effect of an electric field.[248] The electric field also causes localised Joule heating causing silica condensation.[248] The walls of the channels in the PDMS stamp cause the surfactant micelles to align parallel to the substrate and the long axis of the mould, leading to highly aligned mesophases. Changing the shape of the mould leads to some control over channel orientation in the deposited silica-surfactant phase.[251] This technique has been used for patterning on several length scales simultaneously with a colloidal crystal deposited initially in the capillaries followed by infiltration of the silica-surfactant templating solution resulting in macropores from the colloidal spheres with mesoporous walls from the surfactant templating deposited in the pattern of the PDMS stamp.[250] A more rapid method to generate patterns of mesoporous materials involves deposition of very small amounts of solution. This can be achieved *via* inkjet printing[212,213] or dip-pen micro-[213,240] or nanolithography[252] where the synthesis solution acts as the ink in the printing or direct writing process, enabling fast deposition of complex patterns.

The EISA route to the formation of films has prompted a large number of publications by several groups since it has proved possible to use this method to prepare mesostructured films of many materials other than silica.[239,253–255] The method uses very small amounts of both inorganic precursor and surfactant template, making it economical, and the use of spin coating is already prevalent in industry for production of surface coatings. The incorporation of active molecules such as dyes has also been straightforward *via* this method.[256] Even very hydrophobic dyes can be easily solubilised in the organic solvents used in the synthesis solutions and can be directed to different locations within the structure depending on the hydrophobicity of the dye molecule, or the presence of - Si(OR)$_3$ functionality to allow binding to the silica walls.[80,93,94] Smooth porous silica coatings and patterned surfaces prepared by this method have been investigated as low *k* dielectric materials for use as insulating layers in microelectronics,[257] low refractive index layers for photonic crystal slab waveguides,[258] optical waveguides for mirrorless lasing,[249] pH sensors,[259,260] to guide formation of conjugated polymers into complex nanostructures *via* use of a diacetylene surfactant as template,[261] dye sensitised solar cells[262,263] and gas[264,265] or alcohol[266] sensors.

The formation of particles *via* the EISA route, using aerosol spraying followed by drying, in principle follows the same pathway, with solvent evaporation driving micellisation followed by condensation of the inorganic material in the spaces between micelles (Figure 2.13).[214,267] The

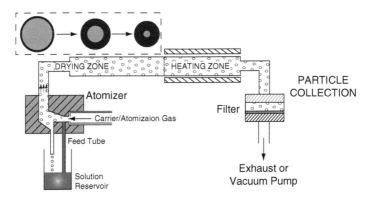

Figure 2.13 Apparatus for aerosol formation of particles *via* the EISA route. Particle evolution indicated in dashed box. Reprinted from 11/579-585 Evaporation-Induced Self-Assembly: Nanostructures made easy by Brinker, C.J.; Lu, Y.; Sellinger, A.; Fan, H. 9, 7 No 12, 13. Copyright (1999) Wiley-VCH Verlag GmbH & Co. KGaA. Reproduced with permission

surface evaporation from the droplet creates a radial concentration gradient and the presence of the liquid–vapour interface (which acts as a nucleating surface) causes ordered silica-surfactant liquid crystalline domains to grow radially inward, rather than outward from a seed as occurs in solution. The aerosol synthesis has been followed by *in situ* SAXS[268] showing that particles prepared from an ethanol-rich solution had more well ordered mesostructures than those from water-rich solutions due to the greater rate of silica condensation in water-rich solutions which prevents complete assembly into ordered structures. Similar to film formation, the rapid evaporation of the ethanol promotes formation of well-ordered mesophases and thus synthesis needed to be carried out above 80 °C in order to achieve almost total ethanol removal while retaining a small amount of water. More recent work has used TEM, SEM, SAXS and ^{29}Si NMR to show that after ethanol evaporation a disordered micellar phase initially forms within the silica-surfactant composite particle, which rearranges *via* micelle fusion to form the ordered mesostructures.[269] Domains at the outer edge of the particles were aligned with respect to the surface but domains in the interior of the particles were not oriented in the same directions as the surface domains (Figure 2.14). The mesophase formed depended both on the degree of condensation of the silica and the surfactant concentration since larger silica oligomers could cause disordering of the mesophase if they were too big to fit into the gaps between surfactant micelles in the ordered phases.

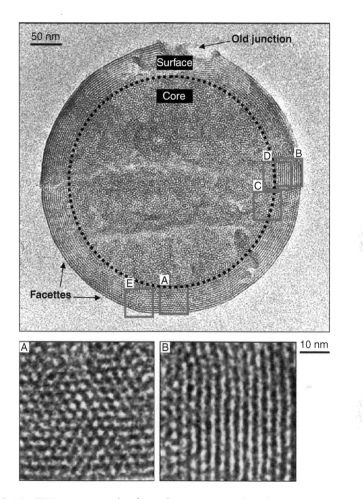

Figure 2.14 TEM micrograph of an ultramicrotomed surfactant templated silica sphere prepared *via* the EISA route using aerosol processing. The dashed circle represents the border between the ordered domain found at the surface and the nonordered domain found in the core of the sphere. The squares show the selected areas A and B which both have a hexagonal-like arrangement of pores, *P6mm* space group. Reprinted from Microporous Mesoporous Mater., Vol 106, B. Alonso, E. Véron, D. Durand, D. Massiot, C. Clinard, New insights into the formation of textures through spray-drying and self-assembly, p76–94, Copyright (2007), with permission from Elsevier

This method has been used to prepare core-shell particles by use of mixed surfactant solutions.[270] The core-shell structure was observed for certain pairs of surfactants which differed greatly in their surface activity in the initial ethanolic solution so that phase separation occurs during the drying process, as micelles for one surfactant form before those of the second amphiphile. Use of ethylene glycol as a swelling agent has also proved effective for pore size enlargement *via* this method.[271] As for film deposition, the incorporation of functionality in these materials *via* incorporation of functionalised siloxanes has been achieved,[272–274] and it has also been possible to prepare mesoporous spheres of various transition metal oxides,[275] or mesoporous silica spheres which incorporate nanoparticles of other transition metal oxides.[276] The incorporation of iron oxide colloids in these porous spheres is of particular interest for use in drug targeting applications since the porous silica has proved effective at encapsulation and release of therapeutic molecules while the magnetic particles incorporated in the silica matrix allow magnetic direction of these particles to the desired release location.[277]

2.8 PROPERTIES AND CHARACTERISATION

The important properties of mesoporous silica derive from the porosity. The attractive feature of surfactant templated materials is the highly uniform, controllable pore sizes, with well-determined pore shapes and connectivities, and high surface areas. For silicates the walls are normally amorphous but can be investigated to some extent by solid-state ^{29}Si MAS NMR to quantify the degree of silica condensation (Figure 2.15).[279] The signal from a central Si atom is referred to Q_n, depending on the degree of condensation in a purely inorganic SiO_x environment. Materials with more intensity in the Q_4 peak (-110 ppm) than in the Q_2 (at about -90 ppm) or Q_3 (at about -100 ppm) peak have more strongly cross-linked silica walls and thus tend to be more hydrothermally stable.[280,281] (Note, T_n is similarly defined for organosilanes where Si is linked to one organic group plus n other O-Si species.) This technique is also used to probe the surface chemistry of pores in these materials, usually in combination with a range of other techniques,[282,283] and also the incorporation of other metal oxides[284] or zeolitic units[285] into the silicate walls.

The acidity of sites in the walls for surfactant templated aluminosilicates, or silicates doped with other transition metals prepared *via* surfactant templating can be characterised using a range of techniques including temperature-programmed desorption (TPD) of ammonia,[101] solid-state NMR with probe molecules such as trimethylphosphine which bind to acid sites[286] or Fourier transform infrared and diffuse reflectance UV-Vis with ammonia or CO probes.[287,288] These techniques show that

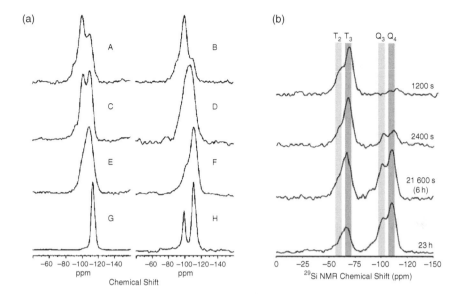

Figure 2.15 Typical ^{29}Si MAS NMR. (a) Silica-surfactant mesophase materials at various stages of preparation: (A) as-synthesied; (B) as-synthesised with cross-polarisation; (C) as-synthesised after removal of organic species using HCl-ethanol solution; (D) sample C after dehydration at 700 °C; (E) sample A calcined in air at 540 °C; (F) HiSil (an amorphous silica); (G) pure silica ZSM-5; and (H) magadiite. Reproduced with permission from Chen *et al.*, 1993 [101]. Copyright (1993) Elsevier. (b) Freeze-dried samples taken during formation of surfactant-organosilica-silica solids quenched at the times indicated, after TEOS addition to the surfactant template. The spectra reveal that the organo-siloxane (T_2, T_3) is initially highly condensed around the micelles, and only at later stages does inorganic silica form the outer shell of a three-layer hybrid material. Reprinted from Microporous Mater. Vol 2, Chen, C.-Y., H.-X. Li, and M.E. Davis Studies on Mesoporous Materials I. Synthesis and Characterisation of MCM-41., p17–26. Copyright (1993), with permission from Elsevier

surfactant templated materials have acidity comparable with other amorphous silicas and aluminosilicates,[101] and even when atoms such as aluminium or titanium are included in the framework, generally contain only weak acid sites rather than the strong Lewis acids generally required for catalysis. There has been some attempt to introduce zeolitic nanocrystals[289] or to crystallise parts of the pore walls in these materials to obtain stronger acid sites for catalytic applications but so far these have shown little improvement over standard silica gel catalysts. More promising is the use of mesoporous silica as an inert support for grafted organometallic complexes[290] or other species such as metal nanoparticles,[291] with the controlled pore sizes enabling size control, and restricted catalytic geometries which can help with improving the specificity of reactions. Post-synthesis grafting onto the walls of mesoporous silica materials is now widely used, including grafting of, for instance, phosphorus species to obtain Brønsted acids for dehydration of isopropanol,[292] or trimethylaluminium groups to create Lewis acid sites for catalyst mechanism studies.[293] Further discussion is beyond the scope of this chapter; however, these areas have been recently reviewed.[294,295]

The surface area and pore volumes for surfactant templated silicates are typically characterised by gas adsorption (Figure 2.16), although a range of other techniques including mercury porosimetry,[155,296] Xe NMR,[297,298] cryoporosimetry,[299] small-angle neutron scattering[300] and X-ray porosimetry[301] have also been used.

The interpretation of gas adsorption isotherms, typically nitrogen adsorption, for these materials is problematic due to the pore sizes, since for

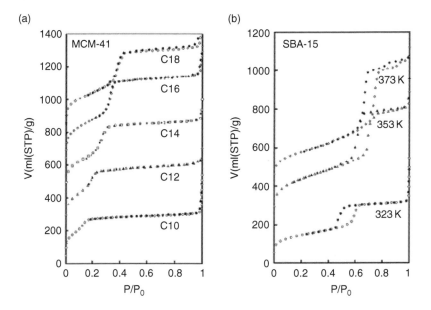

Figure 2.16 Typical nitrogen adsorption–desorption isotherms at 77 K for (a) MCM-41 materials templated with alkyltrimethylammonium bromide surfactants with hydrophobic tails of different lengths as indicated (volumes adsorbed for C12, C14, C16 and C18 were incremented by 200, 400, 600 and 800 ml(STP) g^{-1}, respectively) and (b) nonionic triblock copolymer templated SBA-15 silicas synthesised at different temperatures (volumes adsorbed for 353 K and 373 K were incremented by 200 and 400 ml(STP) g^{-1}, respectively). The hysteresis loops observed in this case are typical of the larger mesopores in these materials. Desorption points are represented by closed symbols. Reprinted with permission from Morishige, K.; Tateishi, M., Accurate relations between pore size and the pressure of capillary condensation and the evaporation of nitrogen in cylindrical pores, *Langmuir*, **22**, 4165–4169. Copyright (2006) American Chemical Society

pores smaller than ~10 nm the Kelvin equation is not a suitable model for the capillary condensation processes which are characteristic of the meso-porosity in these materials. Thus, the standard Brunauer–Emmett–Teller (BET) analysis of these isotherms will not give accurate pore sizes, parti-cularly for materials with noncylindrical pores. To attempt to improve accuracy of the calculated pore size distributions, independently calibrated, empirical variants of the Kelvin equation have been proposed[302–304] but apply only to a limited pore size range, and are strictly only valid for the case of cylindrical pores. Analysis using microscopic methods based on statistical mechanics such as nonlocal density functional theory (NLDFT) modelling is one alternate method of analysis which does not rely on the Kelvin equation, and these methods have been reviewed extensively.[305] However, research in this area to obtain accurate pore dimensions from adsorption data continues. There is an ongoing debate over whether the adsorption or the desorption branch of the isotherm hysteresis loop should be used in determining the pore size distribution. Simulations using NLDFT, which assume straight cylindrical pores, suggest that the desorption branch is the equilibrium branch while adsorption occurs at the spinodal point.[306] However, experimental data on the variation of the positions of the branches of the hysteresis loop with temperature and pore criticality for gases in cylindrical pores suggest that the adsorption branch corresponds to the equilibrium transition. Mean-field density functional theory (MFDFT) calculations for disordered materials suggest that neither branch of the hysteresis loop corresponds to the equilibrium transition.[307] Until this debate is settled and it becomes clear which isotherm branch and which model to use for a given material, there remains some uncertainty in pore sizes obtained from gas sorption. Nevertheless, BET analyses of pore size distributions for mesoporous silicates are common in the literature and continue to be used to allow comparison between newly synthesised materials and those reported in older publications.

For all of the pore geometries, materials prepared using surfactants that contain a PEO headgroup, may also contain significant microporosity due to the penetration of the PEO segments into the silica walls.[308,309] Around room temperatures the PEO block is soluble in water, but as the temperature is increased this polymer becomes increasingly insoluble due to entropic effects and collapses onto the micelle core.[310] Thus in materials synthesised at low temperatures, the interaction between silica and the PEO regions leads to PEO strands penetrating into the silica walls. Once the surfactant is removed, in addition to the primary meso-pores from the micelle cores, the walls also contain micropores less than

1 nm in size arising from the PEO chains.[308] The extent of micropore penetration and overall wall thickness can be controlled by the temperature of the synthesis, particularly in the high temperature ageing step, based on the temperature dependant solubility of the EO chain. As the synthesis temperature is increased the PEO chains become less soluble and collapse back against the hydrophobic micelle core, increasing the radius of the primary mesopores and decreasing the amount of micropores due to penetration of PEO into the silica walls.[16,310–312] At the highest synthesis temperatures, this decreasing solubility of PEO can also result in the development of significant mesoporosity in the walls between primary mesopores, possibly due to the EO self-segregating in regions of the wall, as well as collapsing onto the hydrophobic core of the micelle, or else due to the merging of adjacent micropores as the polymer retracts into the micelle core.[311]

The pore geometry in surfactant templated silicas is generally assigned by use of small-angle XRD, or imaging using electron microscopy and examples of these are shown for various pore geometries, discussed below. The pore systems in these materials are replicas of the micelle phases on which they are templated, so show the same variety of structures as seen in high concentration surfactant liquid crystalline phases. These are often imaged using TEM although care must be taken in using this technique to obtain a representative image of the sample. Many of these materials may be highly ordered in small regions but also contain significant areas of disordered pores or nonporous structures so TEM should be used in conjunction with other techniques to demonstrate well-ordered materials. TEM has however proved to be an extremely valuable technique to determine the pore connectivity in several of the mesoporous materials with complex structures, showing that these are frequently mixtures of structures or intergrowths of two phases.[313,314] A powerful technique of density reconstruction which gives a unique structural solution by using the Fourier sum of the 3D structure factors, both amplitudes and phases, obtained from Fourier analyses of a set of high-resolution electron microscope (HREM) images, has enabled the solution of several new cubic phase structures.[315,316]

The most commonly found surfactant templated silica mesophase is the 2D hexagonal phase (symmetry $P6m$), composed of long close packed cylinders aligned parallel to each other [Figure 2.17(a)]. This has been observed for many of the single tail small molecule surfactant templates, and the commonly used triblock copolymer surfactant P123 also usually produces 2D hexagonal phase materials. In cases where the hexagonal order is not fully established, the surfactant templated materials retain

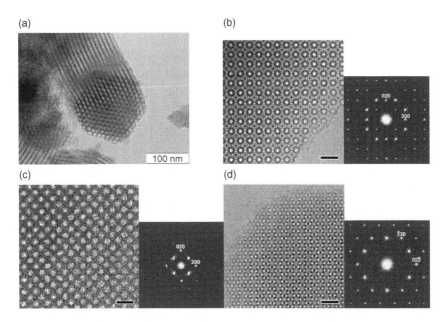

Figure 2.17 Transmission electron micrograph showing (a) SBA-15 silica with a 2D hexagonal phase. Both the side and the end of a particle can be seen showing the long axis and cross-section of the channels (with thanks to A. Lapkin and L. Torrente for kind permission to use this image). (b) $Pm\overline{3}n$ cubic phase silica [001] and the FFT transform of this image showing the electron diffraction pattern from this structure. (c) $Im\overline{3}m$ cubic phase silica [001] and FFT transform. (d) MCM-48 $Ia\overline{3}d$ cubic phase silica [111] and FFT transform. (with thanks to Y. Sakamoto for kind permission to use these images). The scale bar for (b,c,d) is 20 nm

the long tubular pores but these are twisted around each other rather than arranged in ordered arrays. These are known as wormlike phases, and contain a network of disordered pores, although with uniform diameter, which give a single broad peak in X-ray diffraction patterns.

More complex pore structures are produced in cubic phase materials, and these can have several types of interconnected pore structures. The earliest cubic pore structure reported was that of MCM-48, which has an $Ia\overline{3}d$ symmetry [Figure 2.17(d)], and high-resolution TEM studies have shown that this is templated on a bicontinuous cubic gyroid (G) minimal surface where the surfactant and silica phases form intertwined tubes through the structure.[317,318] Other cubic structures are due to the packing of mixtures of spherical and elliptical micelles where the silica phase between micelles lies on a minimal surface[315,319] [SBA-1 $Pm\overline{3}n$; Figure 2.17(b)] or are templated on different minimal surfaces such as the plumber's nightmare (P) surface.[320] The $Im\overline{3}m$ body centred cubic phase seen in SBA-16 and other materials consists of close packed spherical micelles surrounded by silica, with small pore openings in the silica walls at the points of closest approach of the spheres.[17,315] This results in a structure with large cages and small pore openings [Figure 2.17(c)]. Similar materials with pores formed by close packed spherical micelles include FDU-1 with a face centred cubic arrangement of micelles having $Fm\overline{3}m$ symmetry.[321,322] Aside from the cubic materials, materials have also been produced with an ordered array of 3D hexagonally close packed spherical micelles (symmetry $P6_3/mmc$).[17]

The final common characterisation method, used to determine the structure of surfactant templated materials, is SAXS [(Figure 2.18)] or, for films, grazing incidence XRD. Since the walls of these materials are normally amorphous, the diffraction peaks observed in X-ray scattering are due solely to the organisation of the pores in these materials. Given that the pores are typically greater than 2 nm in diameter with a wall of at least 1 nm between them, the first-order peaks occur at angles of $1.5°$ (2θ) or lower, making small-angle scattering the technique of choice to characterise these structures. Higher order reflections in well-ordered materials occur at higher angles, but, typical of all lyotropic liquid crystalline systems, it is rare to see more than four to five peaks for a 2D hexagonal phase, and often only the first three peaks are visible. The spacing of 2D hexagonal peaks follows the usual rule $1/d^2 = (4/3)(h^2+hk+k^2)/a^2$ so for the 100, 110 and 200 peaks that are normally seen, the peak spacings are in a ratio of $1:\sqrt{3}:2$, and the unit cell parameter, a (which corresponds to the centre-to-centre distance between adjacent channels) is $(2/\sqrt{3})d_{100}$. Lamellar phases have equally spaced peaks with spacings in the ratio $1:2:3...$ and the unit cell parameter is equal to the d-spacing of the first-order peak. For cubic phases a larger number of peaks occur in the small-angle region, and the peak spacings are complex, depending on which cubic structure is present.[323] Normally scattering can be seen out to about $5°$ (2θ), although after this the relatively low degree of very long-range order in these materials means reflections are damped out and cannot be measured.

By combining measurements of the repeat spacing from XRD measurements and pore diameters from one of the above techniques, or from modelling of the intensities of XRD peaks, the wall thickness of these materials can be calculated. The wall thicknesses of materials templated using neutral surfactants with EO headgroups, in particular the triblock copolymers such as the Pluronic® series are much greater than those templated using small molecule cationic or anionic surfactants. Early measurements of the wall thicknesses of MCM-41 and MCM-48 found a width of about 8 Å, equivalent to about two layers of tetrahedral SiO_2 units.[325]

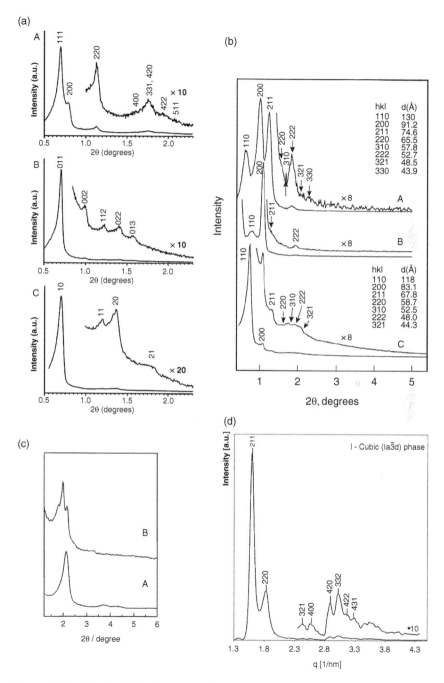

Figure 2.18 Typical SAXS patterns for mesoporous materials. (a) Mesophase silicates templated with F127 and different ratios of tetramethoxysilane and butanol: (A) $Fm\overline{3}m$ cubic phase; (B) $Im\overline{3}m$ cubic phase; (c) $P6mm$ 2D hexagonal

Materials templated using EO-containing surfactants can have walls that are several nanometres thick,[326] giving these materials much greater hydrothermal stability. The wall thickness in these materials depends on the extension of the EO group into solution, plus the distance maintained between micelles by the hydration shells surrounding the solubilised EO chains, so can range from ~2 to 10 nm.[17,309,326]

A wide range of other characterisation techniques have also been applied to study these materials by a variety of groups. These include Raman spectroscopy to study the condensation processes of the silica species in solution,[327] and silanol species on the walls of condensed materials,[328] EPR using spin probes,[169,198] small-angle neutron scattering[66,329,330] and conductivity[278] to investigate formation mechanisms and interactions between the inorganic wall materials and the surfactant templates, hyperpolarised ^{129}Xe-NMR to investigate porosity in the pore walls[297,331,332] and vibrational spectroscopies[333] to identify heterogeneity of hydroxyl groups on the pore surfaces. Positron annihilation spectroscopy has been used to directly measure pore sizes[334,335] particularly during template removal[105,336] and in thin films,[337,338] and quasielastic[339-341] and inelastic neutron scattering[333,342] have been used to characterise the dynamics of molecules in the pores. XANES and EXAFS are often used to characterise metal catalyst centres confined in the mesopores or within the walls of these materials.[343,344] Many other techniques are use to probe catalytic sites or activity but space does not permit a fuller examination of these investigations.

Figure 2.18 (*Continued*)   phase. Reprinted with permission from Kleitz, F.; Kim, T.W.; Ryoo, R., Phase domain of the cubic im3m mesoporous silica in the EO106PO70EO106-butanol-H2O system, *Langmuir*, **22**, 440–445. Copyright 2006 American Chemical Society. (b) (A) as-deposited and (B) calcined *Im3̄m* cubic mesoporous silica films prepared using amphiphilic $EO_{106}PO_{70}EO_{106}$ triblock copolymer surfactants. (C) A bulk sample of *Im3̄m* SBA-16 prepared using the same surfactant. Reprinted from Zhao, D.; Yang, P.; Melosh, N.; Feng, J.; Chmelka, B.F.; Stucky, G.D., Continuous Mesoporous Silica Films with Highly Ordered Large Pore Structures, *Adv. Mater.*, **10**, 1380–1385. Copyright (1998) Wiley-VCH Verlag GmbH & Co. KGaA. Reproduced with permission. (c) Silica-surfactant mesophases templated using cetyltrimethylammonium chloride at different HCl concentrations: (A) *P6mm* 2D hexagonal; (b) *Pm3̄n* cubic phase. Reprinted with permission from Kim, M.J.; Ryoo, R., Synthesis and Pore Size Control of Cubic Mesoporous Silica SBA-1, *Chem. Mater.*, **11**, 487–491, Copyright (1999) American Chemical Society. (d) *Ia3̄d* cubic phase. Reprinted from Microporous Mesoporous Mater., 38(2-3), S. Pevzner, O. Regev, The in situ phase transitions occurring during bicontinuous cubic phase formation, p413–421, Copyright (2000) with permission from Elsevier

2.9 MACROSCOPIC STRUCTURES

As well as control over the nanoscale structure of surfactant-templated silicas, it has been possible to prepare these materials in a variety of controlled macroscale structures. Initially the preparations of MCM-41 and MCM-48 produced fine powders with nonuniform particle shapes. However, it was found that by varying the synthesis conditions, in particular careful control of pH, allowed formation of 'single crystal' forms where faceted particles were observed [Figure 2.19(a)].[31] In the alkaline syntheses, combination of the surfactant-templating idea with the well established Stöber synthesis for production of uniform colloidal silica spheres, resulted in similar spheres of mesoporous silica.[345,346] Further experimentation with synthesis conditions for both alkaline and acidic preparations has led to the production of monodisperse mesoporous spheres in several size ranges from nanometres,[186] to micrometres[347] to millimetres[348] and even centimetres[349] [Figure 2.19(b)]. Interest in this area continues since uniform porous spheres are of interest for packing materials in separation columns, and improvements in particle strength, particularly for functionalised materials, are being targeted in order to withstand the pressures generated in HPLC and other similar separation techniques.[350–352]

In addition to these solution-based synthesis methods, uniform spherical particles of surfactant templated silicas and many other inorganic oxides have been prepared *via* drying of aerosol droplets of inorganic precursor with surfactant in a volatile solvent.[274] This formation method is discussed in more detail in Section 2.7.3.

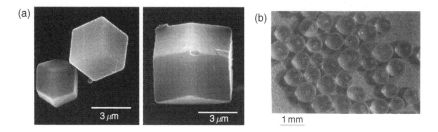

Figure 2.19 (a) SEM image of faceted 'single crystals' of SBA-16. Reprinted from Microporous Mesoporous Mater., Vol 81(1-3), B.-C. Chen, M.-C. Chao, H.-P. Lin, C.-Y. Mou, Faceted single crystals of mesoporous silica SBA-16 from a ternary surfactant system: surface roughening model. p. 241–249, Copyright (2005) with permission from Elsevier. (b) As-prepared millimetre sized hard mesoporous silica spheres templated on CTAB. Reprinted in part with permission from Huo, Q.; Feng, J.; Schüth, F.; Stucky, G.D., Preparation of Hard Mesoporous Silica Spheres, *Chem. Mater.*, 9, 14–17. Copyright (1997) American Chemical Society

Hollow particles of mesoporous silica, usually also spherical, are of interest for encapsulation and delivery applications, since silica is non-toxic and generally biocompatible (dependant on particle shape – silica fibres can induce silicosis in the lungs, but spherical particles are considered safer).[355] There is a large body of work on the preparation of vesicle-like hollow silica-surfactant structures either with multilamellar alternating silica/surfactant walls or a wormlike pore structure which allows loading and unloading of the silica capsules. These can be generated via complex solution phase self-assembly processes which require careful tuning of the synthesis parameters to avoid formation of solid mesostructured particles[356,357] or via use of a sacrificial core particle which is coated with mesoporous silica but removed by calcination or depolymerisation or dissolution during or after the surfactant template removal process. Templating of mesoporous silica shells can also be done onto emulsion droplets of either an oil droplet emulsion added to the synthesis solution[358] or on droplets of the silica precursor itself.[359,360] Using the silica source itself as a template relies on using a hydrophobic tetraalkoxysilane as the silica source which does not hydrolyse immediately in the synthesis solution. This droplet can then provide a surface where freshly hydrolysed silica species are available to react with surfactant that has diffused to the droplet interface, allowing a mesoporous shell to form. These shells are quite thin, depending on the ability of the silica to continue to escape from the central droplet as the reaction proceeds.

Other particle shapes have also been prepared via solution based syntheses, including flat hexagonal platelets,[362] rods[363] and elongated chiral particles with rope-like structures,[364,365] spheres which contain an inner pillar,[363,365] hollow tubes,[366,367] shell-like gyroid particles[368] and discs (Figure 2.20). Other more complex structures have been shown to result from the ordered aggregates of packed units such as hexagonal plates[362] or octahedral prisms[361] (Figure 2.21). Hierarchical structures with more than one size or shape of pores are of interest to control diffusion through these materials for instance to control the rate of reactions or to allow facile diffusion through macropores to mesopores where an active site is located. Core-shell particles having a core with one pore size and geometry, and a shell with a different pore structure have been prepared via true liquid crystal templating.[155] Bimodal structures can be created by using two templates simultaneously which can result in either ordered or randomly oriented pores. Silica-surfactant precursor solutions infiltrated into the gaps between spheres in colloidal crystals made from close packed polystyrene spheres result in an ordered

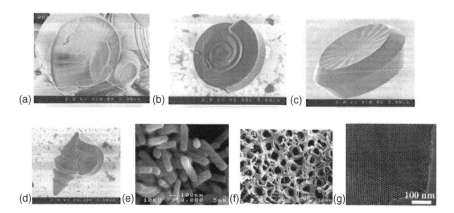

Figure 2.20 Surfactant templated silicas. (a,b) Spiral morphology. (c,d) Gyroid morphology. Reprinted from Ozin, G.A.; Yang, H.; Sokolov, I.; Coombs, N., Shell Mimetics, *Adv. Mater.*, **9**, 662–667 (1997). Copyright Wiley-VCH Verlag GmbH & Co. KGaA. Reproduced with permission. (e) Helical rod-like particles. Reprinted in part with permission from S. Yang, L.Z. Zhao, C.Z. Yu, X.F. Zhou, J.W. Tang, P. Yuan, D.Y. Chen and D.Y. Zhao, *J. Am. Chem. Soc.*, **128**, 10460. Copyright (2006) American Chemical Society. Hierarchical structures having (f) salt crystal templated macropores and (g) polymer micelle templated mesopores. Reprinted in part with permission from S. Yang, L.Z. Zhao, C.Z. Yu, X.F. Zhou, J.W. Tang, P. Yuan, D.Y. Chen and D.Y. Zhao, *J. Am. Chem. Soc.*, **128**, 10460. Copyright (2006) American Chemical Society

macroporous structure with mesoporous walls.[157,369] Disordered bimodal structures have been prepared *via* phase separation techniques using oil–water emulsions,[370,371] or solutions with high salt concentration[372] [Figure 2.20(f,g)] which undergo simultaneous liquid–liquid phase separation or crystallisation as the silica and surfactant self-assembly occurs.

Fibres of porous silica have also been synthesised both *via* EISA methods using fibre spinning techniques[376] and also from spontaneous growth in solution. Fibres were first found growing from an oil–water interface[358] where the fibres form from acidic solutions under static conditions. In the synthesis the silica source is located in the oil phase, while the surfactant is dissolved in the aqueous solution, and fibres form in the aqueous phase, along with a thin film which grows at the interface between the two phases. The fibres contain hexagonally packed mesopores which are oriented circularly around the long axis of the fibres[373,377] [Figure 2.22(a)]. Under certain conditions hollow fibres with a similar channel structure were observed.[367] Hollow fibres have also been prepared using strands of spider silk as a secondary template

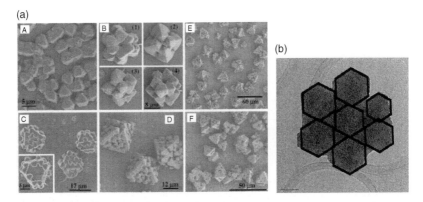

Figure 2.21 Hierarchical growth of surfactant templated silicas from well defined building blocks. (a) Octahedral open crystals: (A) primary octahedral crystals; (B) secondary octahedral crystals (1) face on, (2) top view from one corner, (3) edge on and (4) tilted edge on; (C) tertiary octahedral crystals (the insert shows small octahedral units nucleated on a secondary rosette crystal); (D) large open octahedral objects containing five primary octahedral units on each side; (E) large area view of high order open structures; (F) large open structures assembled from tertiary units. Reprinted from Tian, Z.R.; Liu, J.; Voigt, J.A.; Mckenzie, B.; Xu, H., Hierarchical and Self-Similar Growth of Self-Assembled Crystals, *Angew. Chem. Int. Ed.*, **42**, 413–417 (2003). Copyright Wiley-VCH Verlag GmbH & Co. KGaA. Reproduced with permission. (b) Large 2D hexagonal silica-surfactant platelet assembled from seven smaller hexagonal platelets. Scale bar 200 nm. Reprinted with permission from Linton et al. Copyright 2008 American Chemical Society

upon which the mesoporous silica was grown[378] or *via* gelation of silica around strands of sugar based polymers in an organogel.[379] Chiral sugar based surfactants produce organogel fibres with a helical twist, with frequently two fibres wrapping around each other to form double helical structures which can be coated in silica to form hollow tubular helices[374] [Figure 2.22(b)]. Nematic cholesteric liquid crystalline phases have also been frozen into silica *via* the use of cellulose derivatives as templates.[380]

Fibres have also been formed *via* the use of porous anodic alumina templates [Figure 2.22(c,d)]. Anodic alumina membranes contain uniform pores with micrometre scale diameters into which the precursor solution can be infiltrated.[381,382] The infiltration is done by a variety of methods including EISA,[375,383,384] soaking of the alumina membrane in the aqueous synthesis solution,[381,385] or *via* gentle aspiration of the synthesis solution through the membrane in a Buchner funnel-like set-up.[382,386] The effect of confinement in these pores, similar to effects observed in micromoulding in capillaries, causes micelles to orient along the capillary walls resulting in templated silica-surfactant phases with a

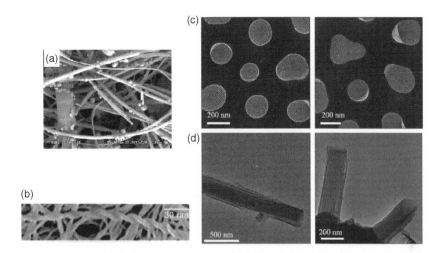

Figure 2.22 (a) SEM image of fibres synthesised from tetrapropoxysilane and CTAB in acidic solutions. Reprinted in part with permission from F. Kleitz, F. Marlow, G.D. Stucky and F. Schuth, *Chem. Mater.*, **13**, 3587. Copyright (2001) American Chemical Society. (b) SEM image of helical fibres templated on a chiral organogel. Reprinted in part with permission from J.H. Jung, K. Yoshida and T. Shimizu, *Langmuir*, **18**, 8724. Copyright (2002) American Chemical Society. TEM images showing (c, d) different pore morphologies of surfactant-templated silica in an anodic alumina template, and fibres protruding from the template where it was broken. Reprinted from Platschek, B.; Petkov, N.; Bein, T., Tuning the structure and orientation of hexagonally ordered mesoporous channels in anodic alumina membrane hosts: A 2D small-angle X-ray scattering study, *Angew. Chem. Intl. Ed.*, **45**, 1134–1138 (2006). Copyright Wiley-VCH Verlag GmbH & Co. KGaA. Reproduced with permission

variety of structures ranging from vertically oriented mesopores to doughnut like micelles to helical mesopores which spiral down the length of the fibre.[384,386–391] Hybrid membranes of porous anodic alumina filled with mesoporous silica are beginning to be investigated for applications in molecular separations.[392]

Thin films of mesoporous silica have many advantages for application of these materials as surface coatings, membranes for sieving and sensors, and patterning of porous materials on surfaces. As well as production of films *via* dip and spin coating discussed above it is possible to grow films at interfaces from solution *via* the cooperative self-assembly route.[167,168,393] Formation of free-standing films is possible *via* this route, however films formed from surfactant and silica alone tend to be brittle and fracture easily after removal from the solution surface and calcination. Thicker and more robust films with a range of mesostructures can be prepared using surfactant-polyelectrolyte complexes which themselves form films at the air–solution interface, in the absence of silica.[200] Addition of silica to the subphase results in ordered mesoporous films where the polyelectrolyte is threaded through the silica walls and the surfactant micelles act as porogens, as for standard silica-surfactant composite formation.[394]

Control of pore orientation in mesoporous films has been a long-standing goal to produce materials for sieving applications as well as to grow templated nanowires with known orientations. The co-assembly route allows orientation of the mesostructure *via* application of an external field in the presence of a substrate to achieve single-crystalline films where all of the mesopores follow a given direction. The group of Miyata *et al.* used substrates with surface treatments such as oriented Langmuir–Blodgett (LB)[395] films, mechanical rubbing[396] and restricted growth between two substrates to achieve uniform orientation of the mesopores. They achieved success with combined LB deposition with mechanical rubbing of polyimide layers which oriented the micelles allowing single crystal domains to be formed over large areas of the films.[397–399] Recently EISA films have also been prepared with uniformly oriented channels on freshly cleaved mica substrates where epitaxial interactions between the mica surface and the micelles guide micelle arrangement.[400,401] Similarly in-plane alignment of the mesopores has been achieved during EISA deposition by flowing hot air over the drying films.[402] Orientation of the mesophase within films has also been achieved *via* the use of epitaxial growth within channels on a substrate.[403] The pores of the film were filled with cobalt *via* a supercritical fluid method, to image the pore orientation and demonstrate production of metal nanowires in the films. The production of films with uniformly vertically oriented pores over the entire film has, however, proved more challenging.[254]

Early work attempted use of magnetic fields to orient silica-surfactant materials in materials grown from the concentrated part of phase-separated silica-surfactant solutions, achieving partial orientation.[404] Electric fields were used in micromoulding in capillaries to orient pores parallel to both the field and the capillary walls.[248] Changing the shape of the capillaries resulted in some control over the orientation of the mesophase formed within (Figure 2.23).[251] Recent work in capillaries also used electric fields to form films along the capillary walls which had a fibrous morphology along the capillary with the pores oriented along the long axis of the fibre.[405] Walcarius *et al.* also showed that perpendicular channels could be obtained by growing the silica-surfactant film from solution onto an electrode while applying a suitable cathodic potential.[406] The applied potential both generated hydroxyl ions that were

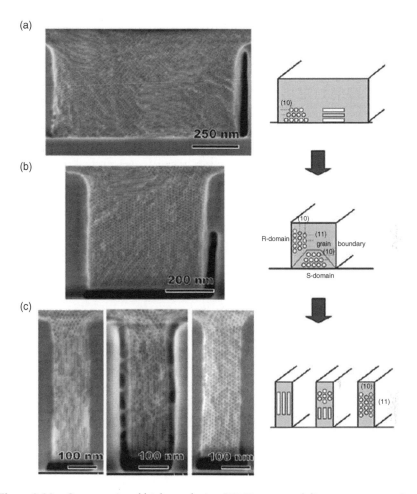

Figure 2.23 Cross-sectional high-resolution SEM images and diagrams representing the orientation of mesopores in confined linear nanospaces with different feature sizes: (a) >0.5 μm; (b) *ca* 0.5 μm; and (c) <0.5 μm. Reprinted from Wu, C.W.; Ohsuna, T.; Edura, T.; Kuroda, K., Orientational control of hexagonally packed silica mesochannels in lithographically designed confined nanospaces, *Angew. Chem. Intl. Ed.*, **46**, 5364–5368 (2007). Copyright Wiley-VCH Verlag GmbH & Co. KGaA. Reproduced with permission

necessary to catalyse condensation of the silica precursors and aligned the surfactant micelles perpendicularly to the electrode surface.

By comparing the self-assembly of cylindrical surfactant mesophases with that of diblock copolymer mesophases, Koganti *et al.*[407] generated mesoporous silicas with pores orthogonal to the substrate by dip coating then sandwiching the film between two substrates coated in a chemically neutral sacrificial PEO-PPO copolymer layer, favouring neither the EO-PO-EO surfactant nor the silica phase.[407] Tilted phases with the 2D hexagonally packed pores at a uniform angle to the substrate were generated in films which were thin enough (70–100 nm) for the orientation generated by the neutral substrate to propagate through the entire thickness of the film. Thicker films showed mixed orientations due to alignment of the micelles at the solution–air interface competing with the alignment imposed by the substrate.[408]

Vertically oriented mesopore films have also recently been obtained again using EISA methods *via* nanometre scale epitaxy.[409] With this route, a smooth cubic surfactant templated titania film which showed a hexagonal arrangement of spherical pores on its top surface was initially deposited. After calcination, a second deposition step coated a silica-surfactant mesophase. Silica has slow condensation kinetics allowing rearrangement and annealing of the structure to take place during deposition, resulting in the 2D hexagonal mesophase aligning on the hexagonal pore system exposed on the top surface of the titania film. Careful matching between the size of the pore system used as template and the 2D hexagonal mesophase generated during the second deposition was required to achieve epitaxial growth.[409] Deposition in polymer mesophases has also recently been reported as a means of preparing vertical channels in silica films.[410] In this method a diblock copolymer with hydrophobic and hydrophilic blocks was spin coated onto a neutral substrate and annealed to establish the desired structure of vertically oriented cylinders of hydrophobic polymer within the hydrophilic polymer matrix. This was then exposed to tetramethoxysilane in supercritical CO_2, allowing the hydrophilic regions to be selectively mineralised.[410] There are therefore now several routes to the generation of well aligned vertical mesopores, and it will be interesting to see which routes are sufficiently cheap and reproducible to be taken up by others for exploitation in a variety of applications.

2.10 APPLICATIONS

Although mesoporous silicas were originally synthesised in the hopes of obtaining a large pore analogue to zeolitic materials for catalysis, many other applications of these materials are now being pursued. Some, already mentioned above, include the use of mesoporous silicas in a variety of optical applications[260] such as waveguides, and, once doped with dyes or lanthanide metal ions, as optically active materials such as sensors or lasers.[89,411–413] The composite inorganic–organic materials have improved properties over those of either material alone including low thresholds of amplified spontaneous emission,[249] optical limiting properties,[414] fast/adjustable response times,[415] improved resistance to mechanical and photophysical damage,[414] and enhanced processability. Uniform pores lead to reduced aggregation of dyes[414] and the possibility to align dyes within anisotropic mesostructures.[416] The insulating nature of the silica walls also allow creation of molecular wires, either metallic[417] or polymeric[418,419] within the channels, and charge transfer along encapsulated polymer strands has been measured in mesoporous silica.[420,421] The manipulation of hydrophilic and hydrophobic regions within the structures allows precise placement of dyes and other functional species within the mesostructures, depending on their chemical natures.[93,94]

A second burgeoning area of application of mesoporous materials is in medically related areas for drug delivery, DNA transfection, and cell culture. The relative biocompatibility of silica, and the gradual dissolution of silica under physiological conditions lends itself to encapsulation for therapeutic delivery purposes.[422–424] Mesoporous silica nanoparticles have been developed for the controlled release of drugs,[425–428] as well as intracellular delivery of genes[429–431] and proteins.[432–435] Bioactive glass templated with surfactants with encapsulated osteogenic agents has been used for bone tissue regeneration.[436–440] Mesoporous silica as nanospheres has been coated with gadolinium compounds and used as a contrast agent for MRI scans,[441] multifunctional dye-containing mesoporous silicas have been used for cell imaging[442] and mesoporous silicas containing magnetic particles for stem cell labelling[443] as well as drug targeting.[277] Cell-directed assembly of silica in the presence of phospholipid templates resulted in cell encapsulation within multilamellar lipid vesicles inside a silica matrix, which prolonged cell viability by preventing excessive drying even in the absence of external buffer.[444,445] Investigations of the cytotoxicity of mesoporous silicas are also now beginning to appear in the literature, and the toxicity appears to depend on particle size, shape and surface functionalisation.[446]

Control over release rates from mesoporous surfactant templated particles has been obtained by several methods. The size and connectivity of the pore system as well as particle size and shape are of prime importance.[447] Simple organic functionalisation on the external and internal surfaces of the particles[448–450] or preparing hybrid particles in thermoresponsive hydrogels,[451] biodegradable polymers[452] or collagen[453] can affect release rates. Polymer layers have also been used to determine entrance of particular species into pores for sensor purposes.[454]

Magnetic nanoparticles[455] or CdS nanoparticles[456] have also been used to cap mesoporous silica to create a stimuli responsive delivery system. Coating the mesoporous particles with polyelectrolytes[457] or polyelectrolyte multilayers has been used to create stimuli responsive drug release via pH.[458] Particles have been also been coated with much more complex molecules to create light [91,459,460] or pH[461] or redox[462] activated molecular trap doors such as cucurbit[6]uril pseudorotaxanes[461] or azobenzene derivatives[460] which can be synthesised at the pore mouths to control diffusion into and out of the pores.[463,464] Nanovalves on mesoporous silicas have been prepared by covering the silica surface with a pseudorotaxane composed of two components: a long thread containing a 1,5-dioxnaphthalene donor unit, which is attached to the solid support, and a moving part, the tetracationic cyclophane acceptor/receptor, cyclobis(paraquat-p-phenylene), which controlled access to the interior of the nanopore. An external reducing reagent caused dethreading of the pseudorotaxane and opened the valve allowing guest molecules in the pores to be released.[465] More recently enzyme cleavable 'snap-top' mesoporous silica containers were prepared using [2]rotaxanes, consisting of tri(ethylene glycol) chains threaded by M-cyclodextrin tori, located on the surfaces of the nanoparticles and terminated by a large stoppering group (Figure 2.24). Enzyme cleavage of the bulky stopper group causes de-threading of the cyclodextrin and release of molecules transported in the pore system of the particles.[466]

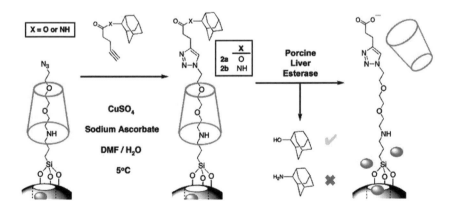

Figure 2.24 Synthesis and activation of enzyme responsive 'snap-top' cap to trigger release from mesoporous silica. Reprinted with permission from K. Patel, S. Angelos, W.R. Dichtel, A. Coskun, Y.W. Yang, J.I. Zink and J.F. Stoddart, *J. Am. Chem. Soc.*, **130**, 2382. Copyright (2008) American Chemical Society

Enzyme immobilisation on mesoporous silica has also proved attractive for biocatalysis applications, particularly in nonaqueous solvents,[467–472] since immobilisation of enzymes results in increased mechanical stability and potential for catalyst recycling.[473] Many factors, such as relative size of nanopore/enzyme, nanopore volume, surface characteristics of support/enzyme influence enzyme loading and activity.[467,474] Enzymes on mesoporous silicas have also been used for light harvesting[475] and biosensors.[476]

A final current application for mesoporous silicas is in the construction of other nanostructures. This is termed nanocasting,[12] and can involve direct, negative replication of the silica structure, or positive replicas *via* double nanocasting (Figure 2.25).[477] The initial step involves infilling

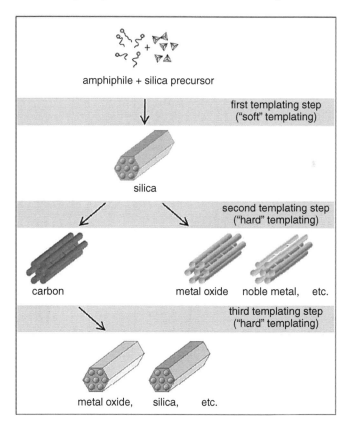

Figure 2.25 Scheme showing repeated templating (nanocasting) for the synthesis of ordered mesoporous materials. The first step is 'soft templating' using surfactants as structure directors. The following steps use a hard template. Reprinted with permission from M. Tiemann, *Chem. Mater.*, **20**, 961. Copyright (2008) American Chemical Society

the silica pores with a carbon precursor such as sucrose[32] or phenol/
formaldehyde resin,[478,479] an organic monomer or an inorganic precursor. Polymerisation or condensation results in a negative polymer[480]
metal,[481,482] metal sulfide[483] or metal oxide[484–487] replica from
which the silica can be removed. Otherwise, calcination produces a carbon
replica, completely filling the pore system, and after silica removal by
washing in NaOH, the negative carbon replica remains.[32,488] In the case
of 2D hexagonal phases such as MCM-41 this consists of separated rod-
like particles, however in the case of SBA-15 where micropores perforate
the channel walls, sufficient cross-shafts exist to retain the 2D hexagonally
close packed rod morphology.[488,489] This carbon negative replica can
then be used again for casting using metal salts, organic monomers or
metal precursor to deposit polymer, metal or metal oxide nanostructures.[7,490–493] Removal of the carbon then results in mesoporous metal
oxides and this route is now a major method to obtain porous metal oxides
which cannot be directly templated with surfactants, opening up further
major areas of application for mesoporous materials as battery materials,[494,495] catalysts[492] and gas sensors.[265] Efforts however continue to
directly template such materials using soft matter templates since this is a
much more efficient route to mesostructures.[10,496,497]

REFERENCES

[1] K.S.W. Sing, D.H. Everett, R.A.W. Haul, R.A. Pierotti, J. Rouquerol and T. Siemieniewska, *Pure Appl. Chem.*, **57**, 603 (1985).

[2] J.S. Beck, J.C. Vartuli, W.J. Roth, M.E. Leonowicz, C.T. Kresge, K.D. Schmitt,
C.T.-W. Chu, D.H. Olson, E.W. Sheppard, S.B. McCullen, J.B. Higgins and
J.L. Schlenker, *J. Am. Chem. Soc.*, **114**, 10834 (1992).

[3] C.T. Kresge, M.E. Leonowicz, W.J. Roth, J.C. Vartuli and J.S. Beck, *Nature*, **359**,
710 (1992).

[4] F. Di Renzo, H. Cambon and R. Dutartre, *Microporous Mater.*, **10**, 283 (1997).

[5] G.S. Armatas and M.G. Kanatzidis, *Nature*, **441**, 1122 (2006).

[6] D. Sun, A.E. Riley, A.J. Cadby, E.K. Richman, S.D. Korlann and S.H. Tolbert,
Nature, **441**, 1126 (2006).

[7] S. Polarz, A.V. Orlov, F. Schuth and A.H. Lu, *Chem. Eur. J.*, **13**, 592 (2007).

[8] J.H. Pan, S.Y. Chai, C. Lee, S.E. Park and W.I. Lee, *J. Phys. Chem. C*, **111**, 5582
(2007).

[9] H.T. Liu, P. He, Z.Y. Li, D.Z. Sun, H.P. Huang, J.H. Li and G.Y. Zhu, *Chem. Asian
J.*, **1**, 701 (2006).

[10] J. Lee, M.C. Orilall, S.C. Warren, M. Kamperman, F.J. Disalvo and U. Wiesner, *Nat.
Mater.*, **7**, 222 (2008).

[11] A. Corma, P. Atienzar, H. Garcìa and J.-Y. Chane-Ching, *Nat. Mater.*, **3**, 394 (2004).

[12] A.H. Lu and F. Schuth, *Adv. Mater.*, **18**, 1793 (2006).

[13] G.S. Attard, P.N. Bartlett, N.R.B. Coleman, J.M. Elliott, J.R. Owen and J.H. Wang, *Science*, **278**, 838 (1997).

[14] Q. Huo, R. Leon, P.M. Petroff and G.D. Stucky, *Science*, **268**, 1324 (1995).

[15] Q. Huo, D.I. Margolese and G.D. Stucky, *Chem. Mater.*, **8**, 1147 (1996).

[16] D. Zhao, J. Feng, Q. Huo, N. Melosh, G.H. Fredrickson, B.F. Chmelka and G.D. Stucky, *Science*, **279**, 548 (1998).

[17] D. Zhao, Q. Huo, J. Feng, B.F. Chmelka and G.D. Stucky, *J. Am. Chem. Soc.*, **120**, 6024 (1998).

[18] S.A. Bagshaw, E. Prouzet and T.J. Pinnavaia, *Science*, **269**, 1242 (1995).

[19] T. Yanagisawa, T. Shimizu, K. Kuroda and C. Kato, *Bull. Chem. Soc. Jpn.*, **63**, 988 (1990).

[20] R. Ryoo, J.M. Kim, C.H. Ko and C.H. Shin, *J. Phys. Chem.*, **100**, 17718 (1996).

[21] C. Yu, Y. Yu and D. Zhao, *Chem. Commun.*, 575 (2000).

[22] S.A. El-Safty and T. Hanaoka, *Chem. Mater.*, **15**, 2892 (2003).

[23] D.J. Macquarrie, B.C. Gilbert, L.J. Gilbey, A. Caragheorgheopol, F. Savonea, D.B. Jackson, B. Onida, E. Garrone and R. Luque, *J. Mater. Chem.*, **15**, 3946 (2005).

[24] Q. Cai, Y. Geng, X. Zhao, K. Cui, Q.Y. Sun, X.H. Chen, Q.L. Feng, H. Li and E.G. Vrieling, *Microporous Mesoporous Mater.*, **108**, 123 (2008).

[25] T.R. Pauly, Y. Liu, T.J. Pinnavaia, S.J.L. Billinge and T.P. Rieker, *J. Am. Chem. Soc.*, **121**, 8835 (1999).

[26] M. Templin, A. Franck, A. DuChesne, H. Leist, Y.M. Zhang, R. Ulrich, V. Schadler and U. Wiesner, *Science*, **278**, 1795 (1997).

[27] M. Tiemann, M. Schulz, C. Jäger and M. Fröba, *Chem. Mater.*, **13**, 2885 (2001).

[28] M.T. Anderson, J.E. Martin, J.G. Odinek and P.P. Newcomer, *Chem. Mater.*, **10**, 1490 (1998).

[29] M.T. Anderson, J.E. Martin, J.G. Odinek and P.P. Newcomer, *Chem. Mater.*, **10**, 311 (1998).

[30] O.I. Lebedev, G. Van Tendeloo, O. Collart, P. Cool and E.F. Vansant, *Solid State Sci.*, **6**, 489 (2004).

[31] J.M. Kim, S.K. Kim and R. Ryoo, *Chem. Commun.*, 259 (1998).

[32] R. Ryoo, S.H. Joo and S. Jun, *J. Phys. Chem. B*, **103**, 7743 (1999).

[33] J.N. Israelachvili, D.J. Mitchell and B.W. Ninham, *J. Chem. Soc., Faraday Trans. 2*, **72**, 1525 (1976).

[34] J.N. Israelachvili, *Intermolecular and Surface Forces*, 2nd Edn, Academic Press, San Diego, 1991.

[35] A.M. Lapena, A.F. Gross and S.H. Tolbert, *Langmuir*, **21**, 470 (2005).

[36] M.J. Kim and R. Ryoo, *Chem. Mater.*, **11**, 487 (1999).

[37] J.L. Blin, C. Otjacques, G. Herrier and B.-L. Su, *Langmuir*, **16**, 4229 (2000).

[38] J.L. Ruggles, E.P. Gilbert, S.A. Holt, P.A. Reynolds and J.W. White, *Langmuir*, **19**, 793 (2003).

[39] M.H. Huang, B.S. Dunn and J.I. Zink, *J. Am. Chem. Soc.*, **122**, 3739 (2000).

[40] L. Sicard, B. Lebeau, J. Patarin and R. Zana, *Langmuir*, **18**, 74 (2001).

[41] A. Mitra, A. Bhaumik and B.K. Paul, *Microporous Mesoporous Mater.*, **109**, 66 (2008).

[42] A. Galarneau, G. Renard, M. Mureseanu, A. Tourrette, C. Biolley, M. Choi, R. Ryoo, F. Di Renzo and F. Fajula, *Microporous Mesoporous Mater.*, **104**, 103 (2007).

[43] J. Garcia-Martinez, P. Brugarolas and S. Dominguez-Dominguez, *Microporous Mesoporous Mater.*, **100**, 63 (2007).

[44] P.T. Tanev and T.J. Pinnavaia A. Neutral Templating, *Science*, **267**, 865 (1995).

[45] S. Valkama, T. Ruotsalainen, H. Kosonen, J. Ruokolainen, M. Torkkeli, R. Serimaa, G. ten Brinke and O. Ikkala, *Macromolecules*, **36**, 3986 (2003).

[46] E. Krämer, S. Forster, C. Göltner and M. Antonietti, *Langmuir*, **14**, 2027 (1998).

[47] W. Guo, Y.W. Sun, G.S. Luo and Y.J. Wang, *Colloids Surf., A*, **252**, 71 (2005).

[48] A. Monnier, F. Schüth, Q. Huo, D. Kumar, D. Margolese, R.S. Maxwell, G.D. Stucky, M. Krishnamurty, P. Petroff, A. Firouzi, M. Janicke and B.F. Chmelka, *Science*, **261**, 1299 (1993).

[49] Q. Huo, D.I. Margolese, U. Ciesa, P. Feng, T.E. Gier, P. Sieger, R. Leon, P.M. Petroff, F. Schüth and G.D. Stucky, *Nature*, **368**, 317 (1994).

[50] J.S. Pedersen and M.C. Gerstenberg, *Colloids Surf., A*, **213**, 175 (2003).

[51] S. Ruthstein, V. Frydman and D. Goldfarb, *J. Phys. Chem. B*, **108**, 9016 (2004).

[52] C.G. Göltner, S. Henke, M.C. Weissenberger and M. Antonietti, *Angew. Chem. Int. Ed.*, **37**, 613 (1998).

[53] P.F.W. Simon, R. Ulrich, H.W. Spiess and U. Wiesner, *Chem. Mater.*, **13**, 3463 (2001).

[54] C.G. Göltner, B. Berton, E. Krämer and M. Antonietti, *Chem. Commun.*, 2287 (1998).

[55] G.E.S. Toombes, S. Mahajan, M. Thomas, P. Du, M.W. Tate, S.M. Gruner and U. Wiesner, *Chem. Mater.*, **20**, 3278 (2008).

[56] G.E.S. Toombes, S. Mahajan, M. Weyland, A. Jain, P. Du, M. Kamperman, S.M. Gruner, D.A. Muller and U. Wiesner, *Macromolecules*, **41**, 852 (2008).

[57] C. Boissière, M.A.U. Martines, M. Tokumoto, A. Larbot and E. Prouzet, *Chem. Mater.*, **15**, 509 (2003).

[58] B. Lindlar, A. Kogelbauer, P.J. Kooyman and R. Prins, *Microporous Mesoporous Mater.*, **44**, 89 (2001).

[59] J.L. Blin and B.L. Su, *Langmuir*, **18**, 5303 (2002).

[60] J. Hu, L.H. Zhou, J. Feng, H.L. Liu and Y. Hu, *J. Colloid Interface Sci.*, **315**, 761 (2007).

[61] C. Boissiére, A. Larbot, C. Bourgaux, E. Prouzet and C.A. Bunton, *Chem. Mater.*, **13**, 3580 (2001).

[62] M. Dubois, T. Gulik-Krzywicki and B. Cabane, *Langmuir*, **9**, 673 (1993).

[63] C.C. Egger, M.W. Anderson, G.J.T. Tiddy and J.L. Casci, *Phys. Chem. Chem. Phys.*, **7**, 1845 (2005).

[64] A.C. Voegtlin, F. Ruch, J.L. Guth, J. Patarin and L. Huve, *Microporous Mater.*, **9**, 95 (1997).

[65] C.J. Brinker and G.W. Scherer, *Sol-Gel Science. The Physics and Chemistry of Sol-Gel Processing*, Academic Press, San Diego, 1990.

[66] T. Brennan, S.J. Roser, S. Mann and K.J. Edler, *Chem. Mater.*, **14**, 4292 (2002).

[67] K.J. Edler and J.W. White, *Chem. Commun.*, 155 (1995).

[68] P. Kipkemboi, A. Fogden, V. Alfredsson and K. Flodstrom, *Langmuir*, **17**, 5398 (2001).

[69] M.J. Yuan, J.W. Tang, C.Z. Yu, Y.H. Chen, B. Tu and D.Y. Zhao, *Chem. Lett.*, **32**, 660 (2003).

[70] J.S. Beck, J.C. Vartuli, G.J. Kennedy, C.T. Kresge, W.J. Roth and S.E. Schramm, *Chem. Mater.*, **6**, 1816 (1994).

[71] A. Sayari, P. Liu, M. Kruk and M. Jaroniec, *Chem. Mater.*, **9**, 2499 (1997).

[72] D. Khushalani, A. Kuperman, G.A. Ozin, K. Tanaka, J. Garcés, M.M. Olken and N. Coombs, *Adv. Mater.*, **7**, 842 (1995).

[73] K.W. Gallis and C.C. Landry, *Chem. Mater.*, **9**, 2035 (1997).

[74] K.J. Edler and J.W. White, *Chem. Mater.*, **9**, 1226 (1997).

[75] M. Chareonpanich, A. Nanta-Ngern and J. Limtrakul, *Mater. Lett.*, **61**, 5153 (2007).

[76] K. Tungkananurak, S. Kerdsiri, D. Jadsadapattarakul and D.T. Burns, *Microchim. Acta*, **159**, 217 (2007).

[77] S.A. Bagshaw and F. Testa, *Microporous Mesoporous Mater.*, **39**, 67 (2000).

[78] H.L. Chang, C.M. Chun, I.A. Aksay and W.H. Shih, *Ind. Eng. Chem. Res.*, **38**, 973 (1999).

[79] P. Kumar, N. Mal, Y. Oumi, K. Yamana and T. Sano, *J. Mater. Chem.*, **11**, 3285 (2001).

[80] P.N. Minoofar, R. Hernandez, S. Chia, B. Dunn, J.I. Zink and A.-C. Franville, *J. Am. Chem. Soc.*, **124**, 14388 (2002).

[81] G.E. Fryxell, *Inorg. Chem. Commun.*, **9**, 1141 (2006).

[82] C.M. Bambrough, R.C.T. Slade and R.T. Williams, *J. Mater. Chem.*, **8**, 569 (1998).

[83] S.R. Hall, C.E. Fowler, B. Lebeau and S. Mann, *Chem. Commun.*, 201 (1999).

[84] D.J. Macquarrie, *Chem. Commun.*, 1961 (1996).

[85] C.E. Fowler, S.L. Burkett and S. Mann, *Chem. Commun.*, 1769 (1997).

[86] M. Kruk, T. Asefa, N. Coombs, M. Jaroniec and G.A. Ozin, *J. Mater. Chem.*, **12**, 3452 (2002).

[87] S. Huh, J.W. Wiench, J.-C. Yoo, M. Pruski and LinV.S.-Y. , *Chem. Mater.*, **15**, 4247 (2003).

[88] A.S.M. Chong, X.S. Zhao, A.T. Kustedjo and S.Z. Qiao, *Microporous Mesoporous Mater.*, **72**, 33 (2004).

[89] L. Borello, B. Onida, C. Barolo, K.J. Edler, C. Otero Areán and E. Garrone, *Sens. Actuators, B*, **100**, 107 (2004).

[90] B. Onida, L. Borello, S. Fiorilli, B. Bonelli, C. Barolo, G. Viscardi, D.J. Macquarrie and E. Garrone, *Comptes Rendus Chimie*, **8**, 655 (2005).

[91] E. Johansson, E. Choi, S. Angelos, M. Liong and J.I. Zink, *J. Sol-Gel Sci. Technol.*, **46**, 313 (2008).

[92] E. Johansson and J.I. Zink, *J. Am. Chem. Soc.*, **129**, 14437 (2007).

[93] P. Minoofar, R. Hernandez, A.-C. Franville, S. Chia, B. Dunn and J.I. Zink, *J. Sol-Gel Sci. Technol.*, **26**, 571 (2003).

[94] P.N. Minoofar, B.S. Dunn and J.I. Zink, *J. Am. Chem. Soc.*, **127**, 2656 (2005).

[95] T. Yokoi, H. Yoshitake, T. Yamada, Y. Kubota and T. Tatsumi, *J. Mater. Chem.*, **16**, 1125 (2006).

[96] C. Gao, Y. Sakamoto, O. Terasaki, K. Sakamoto and S. Che, *J. Mater. Chem.*, **17**, 3591 (2007).

[97] J. Kecht, A. Schlossbauer and T. Bein, *Chem. Mater.*, **20**, 7207 (2008).

[98] H. Zhu, D.J. Jones, J. Zajac, J. Rozière and R. Dutartre, *Chem. Commun.*, 2568 (2001)

[99] S. Inagaki, S.Y. Guan, T. Ohsuna and O. Terasaki, *Nature*, **416**, 304 (2002).

[100] P.T. Tanev and T.J. Pinnavaia, *Chem. Mater.*, **8**, 2068 (1996).

[101] C.-Y. Chen, H.-X. Li and M.E. Davis, *Microporous Mater.*, **2**, 17 (1993).

[102] S.A. Bagshaw and I.J. Bruce, *Microporous Mesoporous Mater.*, **109**, 199 (2008).

[103] M.T.J. Keene, R.D.M. Gougeon, R. Denoyel, R.K. Harris, J. Rouquerol and P.L. Llewellyn, *J. Mater. Chem.*, **9**, 2843 (1999).

[104] F. Berube and S. Kaliaguine, *Microporous Mesoporous Mater.*, **115**, 469 (2008).

[105] J. Goworek, A. Borowka, R. Zaleski and R. Kusak, *J. Therm. Anal. Calorim.*, **79**, 555 (2005).

[106] F. Kleitz, W. Schmidt and F. Schuth, *Microporous Mesoporous Mater.*, **65**, 1 (2003).

[107] R. Ryoo and J.M. Kim, *Chem. Commun.*, 711 (1995).

[108] G. Temtsin, T. Asefa, S. Bittner and G.A. Ozin, *J. Mater. Chem.*, **11**, 3202 (2001).

[109] L. Beaudet, K.-Z. Hossain and L. Mercier, *Chem. Mater.*, **15**, 327 (2002).

[110] A. Doyle and B.K. Hodnett, *Microporous Mesoporous Mater.*, **58**, 255 (2003).

[111] W.J. Hunks and G.A. Ozin, *Chem. Mater.*, **16**, 5465 (2004).

[112] M.A. Wahab, I. Kim and C.-S. Ha, *Microporous Mesoporous Mater.*, **69**, 19 (2004).

[113] P.C. Angelomé and G.J. Soler-Illia, *J. Mater. Chem.*, **15**, 3903 (2005).

[114] P. Kumar, J. Ida, S. Kim, V.V. Guliants and J.Y.S. Lin, *J. Membr. Sci.*, **279**, 539 (2006).

[115] H. Ji, Y.Q. Fan, W.Q. Jin, C.L. Chen and N.P. Xu, *J. Non-Cryst. Solids*, **354**, 2010 (2008).

[116] S. Hitz and R. Prins, *J. Catal.*, **168**, 194 (1997).

[117] N. Lang and A. Tuel, *Chem. Mater.*, **16**, 1961 (2004).

[118] V. Antochshuk and M. Jaroniec, *Chem. Commun.*, 2373 (1999)

[119] V. Antochshuk and M. Jaroniec, *Chem. Mater.*, **12**, 2496 (2000).

[120] A.G.S. Prado and C. Airoldi, *J. Mater. Chem.*, **12**, 3823 (2002).

[121] C.-M. Yang, B. Zibrowius, W. Schmidt and F. Schüth, *Chem. Mater.*, **16**, 2918 (2004).

[122] A.A. Sonin, J.B. Seon, M.H. Yang, H.J. Shin, H.D. Jeong, S.Y. Kim and K.P. Yoo, *Mol. Cryst. Liq. Cryst.*, **460**, 75 (2006).

[123] R. van Grieken, G. Calleja, G.D. Stucky, J.A. Melero, R.A. Garcia and J. Iglesias, *Langmuir*, **19**, 3966 (2003).

[124] S. Kawi and M.W. Lai, *Chem. Commun.*, 1407 (1998).

[125] S. Kawi and M.W. Lai, *AIChE J.*, **48**, 1572 (2002).

[126] L. Huang, S. Kawi, C. Poh, K. Hidajat and S.C. Ng, *Talanta*, **66**, 943 (2005).

[127] L. Huang, C. Poh, S.C. Ng, K. Hidajat and S. Kawi, *Langmuir*, **21**, 1171 (2005).

[128] X.B. Lu, W.H. Zhang, J.H. Xiu, R. He, L.G. Chen and X. Li, *Ind. Eng. Chem. Res.*, **42**, 653 (2003).

[129] J.P. Hanrahan, M.P. Copley, K.M. Ryan, T.R. Spalding, M.A. Morris and J.D. Holmes, *Chem. Mater.*, **16**, 424 (2004).

[130] K. Ghosh, S. Bashadi, H.J. Lehmler, S.E. Rankin and B.L. Knutson, *Microporous Mesoporous Mater.*, **113**, 106 (2008).

[131] M.T.J. Keene, R. Denoyel and P.L. Llewellyn, *Chem. Commun.*, 2203 (1998).

[132] A.M. Dattelbaum, M.L. Amweg, J.D. Ruiz, L.E. Ecke, A.P. Shreve and A.N. Parikh, *J. Phys. Chem. B*, **109**, 14551 (2005).

[133] L.P. Xiao, J.Y. Li, H.X. Jin and R.R. Xu, *Microporous Mesoporous Mater.*, **96**, 413 (2006).

[134] G. Buchel, R. Denoyel, P.L. Llewellyn and J. Rouquerol, *J. Mater. Chem.*, **11**, 589 (2001).

[135] Q. Zhang, K. Ariga, A. Okabe and T. Aida, *J. Am. Chem. Soc.*, **126**, 988 (2004).

[136] C. Gao, H. Qiu, W. Zeng, Y. Sakamoto, O. Terasaki, K. Sakamoto, Q. Chen and S. Che, *Chem. Mater.*, **18**, 3904 (2006).

[137] T.M. Long, B.A. Simmons, J.R. McElhanon, S.R. Kline, D.R. Wheeler, D.A. Loy, K. Rahimian, T. Zifer and G.M. Jamison, *Langmuir*, **21**, 9365 (2005).

[138] A.H. Lu, W.C. Li, W. Schmidt and F. Schuth, *J. Mater. Chem.*, **16**, 3396 (2006).

[139] L.M. Yang, Y.J. Wang, G.S. Luo and Y.Y. Dai, *Microporous Mesoporous Mater.*, **81**, 107 (2005).

[140] C.-M. Yang, B. Zibrowius, W. Schmidt and F. Schüth, *Chem. Mater.*, **15**, 3739 (2003).

[141] B. Tian, X. Liu, C. Yu, F. Gao, Q. Luo, S. Xie, B. Tu and D. Zhao, *Chem. Commun.*, 1186 (2002).

[142] G.S. Attard, J.C. Glyde and C.G. Göltner, *Nature*, **378**, 366 (1995).

[143] G.S. Attard, M. Edgar and C.G. Göltner, *Acta Mater.*, **46**, 751 (1998).

[144] M. Antonietti, *Philos. Trans. R. Soc. London, Ser. A*, **364**, 2817 (2006).

[145] B. Smarsly, S. Polarz and M. Antonietti, *J. Phys. Chem. B*, **105**, 10473 (2001).

[146] C.G. Göltner, B. Berton, E. Krämer and M. Antonietti, *Adv. Mater.*, **11**, 395 (1999).

[147] A. Thomas, H. Schlaad, B. Smarsly and M. Antonietti, *Langmuir*, **19**, 4455 (2003).

[148] D.M. Dabbs, N. Mulders and I.A. Aksay, *J. Nanoparticle Res.*, **8**, 603 (2006).

[149] K.M. McGrath, D.M. Dabbs, N. Yao, I.A. Aksay and S.M. Gruner, *Science*, **277**, 552 (1997).

[150] K.M. McGrath, D.M. Dabbs, Y. Yao, K.J. Edler, I.A. Aksay and S.M. Gruner, *Langmuir*, **16**, 398 (2000).

[151] D.G. Shchukin, A.A. Yaremchenko, M.G.S. Ferreira and V.V. Kharton, *Chem. Mater.*, **17**, 5124 (2005).

[152] A.H. Whitehead, J.M. Elliott, J.R. Owen and G.S. Attard, *Chem. Commun.*, 331 (1999).

[153] P.A. Nelson, J.M. Elliott, G.S. Attard and J.R. Owen, *Chem. Mater.*, **14**, 524 (2002).

[154] Y.J. Yuan, H.-P. Hentze, W.M. Arnold, B.K. Marlow and M. Antonietti, *Nano Lett.*, **2**, 1359 (2002).

[155] S.P. Rigby, K. Beanlands, M.J. Watt-Smith, K.J. Edler and R.S. Fletcher, *Chem. Eng. Sci.*, **59**, 5113 (2004).

[156] M. Groenewolt, M. Antonietti and S. Polarz, *Langmuir*, **20**, 7811 (2004).

[157] O. Sel, D.B. Kuang, M. Thommes and B. Smarsly, *Langmuir*, **22**, 2311 (2006).

[158] S.A. El-Safty, T. Hanaoka and F. Mizukami, *Chem. Mater.*, **17**, 3137 (2005).

[159] S.A. El-Safty, Y. Kiyozumi, T. Hanaoka and F. Mizukami, *J. Phys. Chem. C*, **112**, 5476 (2008).

[160] S.A. El-Safty, F. Mizukami and T. Hanaoka, *J. Mater. Chem.*, **15**, 2590 (2005).

[161] S.A. El-Safy, T. Hanaoka and F. Mizukami, *Adv. Mater.*, **17**, 47–+ (2005).

[162] Y. Zhou and M. Antonietti, *Chem. Mater.*, **16**, 544 (2004).

[163] Y. Zhou, J.H. Schattka and M. Antonietti, *Nano Lett.*, **4**, 477 (2004).

[164] S. Tanaka, N. Nishiyama, Y. Oku, Y. Egashira and K. Ueyama, *J. Am. Chem. Soc.*, **126**, 4854 (2004).

[165] S. Tanaka, H. Tada, T. Maruo, N. Nishiyama, Y. Egashira and K. Ueyama, *Thin Solid Films*, **495**, 186 (2006).

[166] S. Tanaka, M.P. Tate, N. Nishiyama, K. Ueyama and H.W. Hillhouse, *Chem. Mater.*, **18**, 5461 (2006).

[167] H. Yang, A. Kuperman, N. Coombs, S. Mamiche-Afara and G.A. Ozin, *Nature*, **379**, 703 (1996).

[168] H. Yang, N. Coombs, I. Sokolov and G.A. Ozin, *Nature*, **381**, 589 (1996).

[169] H. Caldararu, A. Caragheorgheopol, F. Savonea, D.J. Macquarrie and B.C. Gilbert, *J. Phys. Chem. B*, **107**, 6032 (2003).

[170] A. Firouzi, D. Kumar, L.M. Bull, T. Besier, P. Sieger, Q. Huo, S.A. Walker, J.A. Zasadzinski, C. Glinka, J. Nicol, D. Margolese, G.D. Stucky and B.F. Chmelka, *Science*, **267**, 1138 (1995).

[171] O. Regev, *Langmuir*, **12**, 4940 (1996).

[172] Y.S. Lee, D. Surjadi and J.F. Rathman, *Langmuir*, **12**, 6202 (1996).

[173] H. Yang, N. Coombs and G.A. Ozin, *Nature*, **386**, 692 (1997).

[174] Q. Cai, Z.-S. Luo, W.-Q. Pang, Y.-W. Fan, X.H. Chen and F.-Z. Cui, *Chem. Mater.*, **13**, 258 (2001).

[175] J.C. Vartuli, K.D. Schmitt, C.T. Kresge, W.J. Roth, M.E. Leonowicz, S.B. McCullen, S.D. Hellring, J.S. Beck, J.L. Schlenker, D.H. Olson and E.W. Sheppard, *Chem. Mater.*, **1994**, 2317 (1994).

[176] R. Zana, J. Frash, M. Soulard, B. Lebeau and J. Patarin, *Langmuir*, **15**, 2603 (1999).

[177] J. Frasch, B. Lebeau, M. Soulard, J. Patarin and R. Zana, *Langmuir*, **16**, 9049 (2000).

[178] C. Vautier-Giongo and H.O. Pastore, *J. Colloid Interface Sci.*, **299**, 874 (2006).

[179] M. Ogura, H. Miyoshi, S.P. Naik and T. Okubo, *J. Am. Chem. Soc.*, **126**, 10937 (2004).

[180] Q. Huo, D.I. Margolese, U. Ciesla, D.G. Demuth, P. Feng and T.E. Gier, *Chem. Mater.*, **6**, 1176 (1994).

[181] A. Firouzi, F. Atef, A.G. Oertli, G.D. Stucky and B.F. Chmelka, *J. Am. Chem. Soc.*, **119**, 3596 (1997).

[182] A. Galarneau, F. Di Renzo, F. Fajula, L. Mollo, B. Fubini and M.F. Ottaviani, *J. Colloid Interface Sci.*, **201**, 105 (1998).

[183] H.B.S. Chan, P.M. Budd and T. Naylor, *J. Mater. Chem.*, **11**, 951 (2001).

[184] K.J. Edler, A. Goldar, A.V. Hughes, S.J. Roser and S. Mann, *Microporous Mesoporous Mater.*, **44-45**, 661 (2001).

[185] T. Brennan, A.V. Hughes, S.J. Roser, S. Mann and K.J. Edler, *Langmuir*, **18**, 9838 (2002).

[186] R.I. Nooney, D. Thirunavukkarasu, Y. Chen, R. Josephs and A.E. Ostafin, *Chem. Mater.*, **14**, 4721 (2002).

[187] C. Yu, J. Fan, B. Tian and D. Zhao, *Chem. Mater.*, **16**, 889 (2004).

[188] K. Flodström, C.V. Teixeira, H. Amenitsch, V. Alfredsson and M. Linden, *Langmuir*, **20**, 4885 (2004).

[189] K. Flodström, H. Wennerström and V. Alfredsson, *Langmuir*, **20**, 680 (2004).

[190] K. Flodström, H. Wennerström, C.V. Teixeira, H. Amenitsch, M. Linden and V. Alfredsson, *Langmuir*, **20**, 10311 (2004).

[191] P. Ågren, M. Lindén, J.B. Rosenholm, R. Schwarzenbacher, M. Kriechbaum, H. Amenitsch, P. Laggner, J. Blanchard and F. Schüth, *J. Phys. Chem. B*, **103**, 5943 (1999).

[192] S. Ruthstein and D. Goldfarb, *J. Phys. Chem. C*, **112**, 7102 (2008).

[193] J. Zhang, P.J. Carl, H. Zimmermann and D. Goldfarb, *J. Phys. Chem. B*, **106**, 5382 (2002).

[194] J. Zhang and D. Goldfarb, *Microporous Mesoporous Mater.*, **48**, 143 (2001).

[195] J. Zhang, Z. Luz and D. Goldfarb, *J. Phys. Chem. B*, **101**, 7087 (1997).

[196] J. Zhang, Z. Luz, H. Zimmermann and D. Goldfarb, *J Phys. Chem. B*, **104**, 279 (2000).

[197] S. Ruthstein, J. Schmidt, E. Kesselman, Y. Talmon and D. Goldfarb, *J. Am. Chem. Soc.*, **128**, 3366 (2006).

[198] S. Ruthstein, J. Schmidt, E. Kesselman, R. Popovitz-Biro, L. Omer, V. Frydman, Y. Talmon and D. Goldfarb, *Chem. Mater.*, **20**, 2779 (2008).

[199] N. Gov, I. Borukhov and D. Goldfarb, *Langmuir*, **22**, 605 (2006).

[200] K.J. Edler, A. Goldar, T. Brennan and S.J. Roser, *Chem. Commun.*, 1724 (2003)

[201] B.M.D. O'Driscoll, E. Milsom, C. Fernandez-Martin, L. White, S.J. Roser and K.J. Edler, *Macromolecules*, **38**, 8785 (2005).

[202] R. Friman, S. Backlund, O. Hassan, V. Alfredsson and M. Linden, *Colloids Surf., A*, **291**, 148 (2006).

[203] A.L. Underwood and E.W. Anacker, *J. Colloid Interface Sci.*, **117**, 242 (1987).

[204] V. Alfredsson, H. Amenitsch, F. Kleitz, M. Linden, P. Linton and C.V. Teixeira, *Stud. Surface Sci. Catal.*, **158**, 97 (2005).

[205] D. Baute and D. Goldfarb, *J. Phys. Chem. C*, **111**, 10931 (2007).

[206] M. Ogawa, *J. Am. Chem. Soc.*, **116**, 7941 (1994).

[207] M. Ogawa, *Chem. Commun.*, 1149 (1996).

[208] M. Ogawa, T. Igarashi and K. Kuroda, *Bull. Chem. Soc. Jpn.*, **70**, 2833 (1997).

[209] M. Ogawa and N. Masukawa, *Microporous Mesoporous Mater.*, **38**, 35 (2000).

[210] J. Liu, J.R. Bontha, A.Y. Kim and S. Baskaran, in *Microporous and Macroporous Materials*, Vol. **431**, R.F. Lobo, J.S. Beck, S.L. Suib, D.R. Corbin, M.E. Davis, L.E. Iton and S.I. Zones (Eds), MRS, Pittsburgh, 1996, p. 245.

[211] Y. Lu, R. Ganguli, C.A. Drewien, M.T. Anderson, C.J. Brinker, W. Gong, Y. Guo, H. Soyez, B. Dunn, M.H. Huang and J.I. Zink, *Nature*, **389**, 364 (1997).

[212] M. Mougenot, M. Lejeune, J.F. Baumard, C. Boissiere, F. Ribot, D. Grosso, C. Sanchez and R. Noguera, *J. Am. Ceram. Soc.*, **89**, 1876 (2006).

[213] H. Fan, Y. Lu, A. Stump, S.T. Reed, T. Baer, R. Schunk, V. Perez-Luna, G.P. López and C.J. Brinker, *Nature*, **405**, 56 (2000).

[214] C.J. Brinker, Y. Lu, A. Sellinger and H. Fan, *Adv. Mater.*, **11**, 579 (1999).

[215] L.C. Huang, E.K. Richman, B.L. Kirsch and S.H. Tolbert, *Microporous Mesoporous Mater.*, **96**, 341 (2006).

[216] A. Gibaud, D. Grosso, B. Smarsly, A. Baptiste, J.F. Bardeau, F. Babonneau, D.A. Doshi, Z. Chen, C.J. Brinker and C. Sanchez, *J. Phys. Chem. B*, **107**, 6114 (2003).

[217] C. Sanchez, C. Boissiere, D. Grosso, C. Laberty and L. Nicole, *Chem. Mater.*, **20**, 682 (2008).

[218] M.H. Huang, B.S. Dunn, H. Soyez and J.I. Zink, *Langmuir*, **14**, 7331 (1998).

[219] A. Sellinger, P.M. Weiss, A. Nguyen, Y. Lu, R.A. Assink, W. Gong and J.C. Brinker, *Nature*, **394**, 256 (1998).

[220] D. Grosso, F. Babonneau, C. Sanchez, G. Soler-Illia, E.L. Crepaldi, P.A. Albouy, H. Amenitsch, A.R. Balkenende and A. Brunet-Bruneau, *J. Sol-Gel Sci. Technol.*, **26**, 561 (2003).

[221] P.C.A. Alberius, K.L. Frindell, R.C. Hayward, E.J. Kramer, G.D. Stucky and B.F. Chmelka, *Chem. Mater.*, **14**, 3284 (2002).

[222] D.A. Doshi, A. Gibaud, V. Goletto, M. Lu, H. Gerung, B. Ocko, S.M. Han and C.J. Brinker, *J. Am. Chem. Soc.*, **125**, 11646 (2003).

[223] P. Innocenzi, T. Kidchob, J.M. Bertolo, M. Piccinini, M.C. Guidi and C. Marcelli, *J. Phys. Chem. B*, **110**, 10837 (2006).

[224] P. Innocenzi, L. Malfatti, T. Kidchob, S. Costacurta, P. Falcaro, M. Piccinini, A. Marcelli, P. Morini, D. Sali and H. Amenitsch, *J. Phys. Chem. C*, **111**, 5345 (2007).

[225] D. Grosso, F. Cagnol, G.J. de A.A. Soler-Illia, E.L. Crepaldi, H. Amenitsch, A. Brunet-Bruneau, A. Bourgeois and C. Sanchez, *Adv. Funct. Mater.*, **14**, 309 (2004).

[226] A. Chougnet, C. Heitz, E. Sondergard, P.-A. Albouy and M. Klotz, *Thin Solid Films*, **495**, 40 (2006).

[227] A. Chougnet, C. Heitz, E. Søndergard, J.-M. Berquier, P.-A. Albouy and M. Klotz, *J. Mater. Chem.*, **15**, 3340 (2005).

[228] R. Bandyopadhyaya, E. Nativ-Roth, R. Yerushalmi-Rozen and O. Regev, *Chem. Mater.*, **15**, 3619 (2003).

[229] D. Zhao, P. Yang, D.I. Margolese, B.F. Chmelka and G.D. Stucky, *Chem. Commun.*, 2499 (1998).

[230] D. Zhao, P. Yang, N. Melosh, J. Feng, B.F. Chmelka and G.D. Stucky, *Adv. Mater.*, **10**, 1380 (1998).

[231] F. Cagnol, D. Grosso and G.J. Soler-Illia, *J. Mater. Chem.*, **13**, 61 (2003).

[232] V.N. Urade, L. Bollmann, J.D. Kowalski, M.P. Tate and H.W. Hillhouse, *Langmuir*, **23**, 4268 (2007).

[233] S. Besson, T. Gacoin, C. Jacquiod, C. Ricolleau, D. Babonneau and J.P. Boilot, *J. Mater. Chem.*, **10**, 1331 (2000).

[234] L. Bollmann, V.N. Urade and H.W. Hillhouse, *Langmuir*, **23**, 4257 (2007).

[235] M.P. Tate, V.N. Urade, J.D. Kowalski, T.-C. Wei, B.D. Hamilton, B.W. Eggiman and H.W. Hillhouse, *J. Phys. Chem. B*, **110**, 9882 (2006).

[236] G. Grosso, F. Babonneau, A.A. de, G.J. Soler-Illia, P.A. Albouy and H. Amenitsch, *Chem. Commun.*, 748 (2002).

[237] D. Grosso, A.R. Balkenende, P.A. Albouy, M. Lavergne, L. Mazerolles and F. Babonneau, *J. Mater. Chem.*, **10**, 2085 (2000).

[238] D. Grosso, A.R. Balkenende, P.A. Albouy, A. Ayral, H. Amenitsch and F. Babonneau, *Chem. Mater.*, **13**, 1848 (2001).

[239] D. Grosso, C. Boissiére, B. Smarsly, T. Brezesinski, N. Pinna, P.A. Albouy, H. Amenitsch, M. Antonietti and C. Sanchez, *Nat. Mater.*, **3**, 787 (2004).

[240] H.Y. Fan, S. Reed, T. Baer, R. Schunk, G.P. Lopez and C.J. Brinker, *Microporous Mesoporous Mater.*, **44**, 625 (2001).

[241] P. Innocenzi, T. Kidchob, P. Falcaro and M. Takahashi, *Chem. Mater.*, **20**, 607 (2008).

[242] A. Hozumi, T. Kizuki, M. Inagaki and N. Shirahata, *J. Vac. Sci. Technol., A*, **24**, 1494 (2006).

[243] D.A. Doshi, N.K. Huesing, M. Lu, H. Fan, Y. Lu, K. Simmons-Potter, B.G. Potter Jr., A.J. Hurd and C.J. Brinker, *Science*, **290**, 107 (2000).

[244] Y. Lu, Y. Yang, A. Sellinger, M. Lu, J. Huang, H. Fan, R. Haddad, G. Lopez, A.R. Burns, D.Y. Sasaki, J. Shelnutt and C.J. Brinker, *Nature*, **410**, 913 (2001).

[245] A.M. Dattelbaum, M.L. Amweg, L.E. Ecke, C.K. Yee, A.P. Shreve and A.N. Parikh, *Nano Lett.*, **3**, 719 (2003).

[246] T. Clark Jr, J.D. Ruiz, H. Fan and C.J. Brinker, *Chem. Mater.*, **12**, 3879 (2000).

[247] P. Falcaro, S. Costacurta, L. Malfatti, M. Takahashi, T. Kidchob, M.F. Casula, M. Piccinini, A. Marcelli, B. Marmiroli, H. Amenitsch, P. Schiavuta and P. Innocenzi, *Adv. Mater.*, **20**, 1864 (2008).

[248] M. Trau, N. Yao, E. Kim, Y. Xia, G.M. Whitesides and I.A. Aksay, *Nature*, **390**, 674 (1997).

[249] P.D. Yang, G. Wirnsberger, H.C. Huang, S.R. Cordero, M.D. McGehee, B. Scott, T. Deng, G.M. Whitesides, B.F. Chmelka, S.K. Buratto and G.D. Stucky, *Science*, **287**, 465 (2000).

[250] P. Yang, T. Deng, D. Zhao, P. Feng, D. Pine, B.F. Chmelka, G.M. Whitesides and G.D. Stucky, *Science*, **282**, 2244 (1998).

[251] C.W. Wu, T. Ohsuna, T. Edura and K. Kuroda, *Angew. Chem. Int. Ed.*, **46**, 5364 (2007).

[252] M. Su, X. Liu, S.Y. Li, V.P. Dravid and C.A. Mirkin, *J. Am. Chem. Soc.*, **124**, 1560 (2002).

[253] T. Brezesinski, A. Fischer, K.-i Iimura, C. Sanchez, D. Grosso, M. Antonietti and B.M. Smarsly, *Adv. Funct. Mater.*, **16**, 1433 (2006).

[254] C.J. Brinker and D.R. Dunphy, *Curr. Opin. Colloid Interface Sci.*, **11**, 126 (2006).

[255] P.D. Yang, D.Y. Zhao, D.I. Margolese, B.F. Chmelka and G.D. Stucky, *Chem. Mater.*, **11**, 2813 (1999).

[256] B. Lebeau, C.E. Fowler, S.R. Hall and S. Mann, *J. Mater. Chem.*, **9**, 2279 (1999).

[257] H. Fan, H.R. Bentley, K.R. Kathan, P. Clem, Y. Lu and C.J. Brinker, *J. Non-Cryst. Solids*, **285**, 79 (2001).

[258] M. Schmidt, G. Boettger, M. Eich, W. Morgenroth, U. Huebner, R. Boucher, H.G. Meyer, D. Konjhodzic, H. Bretinger and F. Marlow, *Appl. Phys. Lett.*, **85**, 16 (2004).

[259] G. Wirnsberger, B.J. Scott and G.D. Stucky, *Chem. Commun.*, 119 (2001)

[260] G. Wirnsberger, P.D. Yang, B.J. Scott, B.F. Chmelka and G.D. Stucky, *Spectrochim. Acta, Part A*, **57**, 2049 (2001).

[261] Y. Yang, Y. Lu, M. Lu, J. Huang, R. Haddad, G. Xomeritakis, N. Liu, A.P. Malanoski, D. Sturmayr, H. Fan, D.Y. Sasaki, R.A. Assink, J.A. Shelnutt, F. van Swol, G.P. Lopez, A.R. Burns and C.J. Brinker, *J. Am. Chem. Soc.*, **125**, 1269 (2003).

[262] K. Hou, B. Tian, F. Li, Z. Bian, D. Zhao and C. Huang, *J. Mater. Chem.*, **15**, 2414 (2005).

[263] M. Zukalová, A. Zukal, L. Kavan, M.K. Nazeeruddin, P. Liska and M. Grätzel, *Nano Lett.*, **5**, 1789 (2005).

[264] V.N. Urade and H.W. Hillhouse, *J. Phys. Chem. B*, **109**, 10538 (2005).

[265] T. Wagner, T. Waitz, J. Roggenbuck, M. Froba, C.D. Kohl and M. Tiemann, *Thin Solid Films*, **515**, 8360 (2007).

[266] P. Innocenzi, A. Martucci, M. Guglielmi, A. Bearzotti, E. Traversa and J.C. Pivin, *J. Eur. Ceram. Soc.*, **21**, 1985 (2001).

[267] Y. Lu, H. Fan, A. Stump, T.L. Ward, T. Rieker and C.J. Brinker, *Nature*, **398**, 223 (1999).

[268] C. Boissiere, D. Grosso, H. Amenitsch, A. Gibaud, A. Coupé, N. Baccile and C. Sanchez, *Chem. Commun.*, 2798 (2003).

[269] B. Alonso, E. Veron, D. Durand, D. Massiot and C. Clinard, *Microporous Mesoporous Mater.*, **106**, 76 (2007).

[270] S. Areva, C. Boissière, D. Grosso, T. Asakawa, C. Sanchez and M. Lindén, *Chem. Commun.*, 1630 (2004).

[271] M.H. Sorensen, R.W. Corkery, J.S. Pedersen, J. Rosenholm and P.C. Alberius, *Microporous Mesoporous Mater.*, **113**, 1 (2008).

[272] B. Alonso, A. Vrain, E. Beaubois and D. Massiot, *Prog. Solid State Chem.*, **33**, 153 (2005).

[273] X.L. Ji, Q.Y. Hu, J.E. Hampsey, X.P. Qiu, L.X. Gao, J.B. He and Y.F. Lu, *Chem. Mater.*, **18**, 2265 (2006).

[274] I.V. Melnyk, Y.L. Zub, E. Veron, D. Massiot, T. Cacciaguerra and B. Alonso, *J. Mater. Chem.*, **18**, 1368 (2008).

[275] D. Grosso, G. Illia, E.L. Crepaldi, B. Charleux and C. Sanchez, *Adv. Funct. Mater.*, **13**, 37 (2003).

[276] N. Baccile, A. Fischer, B. Julian-Lopez, D. Grosso and C. Sanchez, *J. Sol-Gel Sci. Technol.*, **47**, 119 (2008).

[277] E. Ruiz-Hernandez, A. Lopez-Noriega, D. Arcos and M. Vallet-Regi, *Solid State Sci.*, **10**, 421 (2008).

[278] A.E. Garcia-Bennett, U. Brohede, R.P. Hodgkins, N. Hedin, M. Stromme and A. Mechanistic, *Langmuir*, **23**, 9875 (2007).

[279] A. Steel, S.W. Carr and M.W. Anderson, *Chem. Mater.*, **7**, 1829 (1995).

[280] Y.C. Du, X.J. Lan, S. Liu, Y.Y. Ji, Y.L. Zhang, W.P. Zhang and F.S. Xiao, *Microporous Mesoporous Mater.*, **112**, 225 (2008).

[281] K.X. Wang, Y.J. Lin, M.A. Morris and J.D. Holmes, *J. Mater. Chem.*, **16**, 4051 (2006).

[282] P. Sutra and J.B. Nagy, *Colloids Surf., A*, **158**, 21 (1999).

[283] X.S. Zhao, G.Q. Lu, A.K. Whittaker, G.J. Millar and H.Y. Zhu, *J. Phys. Chem. B*, **101**, 6525 (1997).

[284] J.-Y. Piquemal, E. Briot, G. Chottard, P. Tougne, J.-M. Manoli and J.-M. Bregeault, *Microporous Mesoporous Mater.*, **58**, 279 (2003).

[285] H. Yang, R.I. Walton, S. Antonijevic, S. Wimperis and A.C. Hannon, *J. Phys. Chem. B*, **108**, 8208 (2004).

[286] Q. Luo, F. Deng, Z. Yuan, J. Yang, M. Zhang, Y. Yue and C. Ye, *J. Phys. Chem. B*, **107**, 2435 (2002).

[287] E. Gianotti, V. Dellarocca, L. Marchese, G. Martra, S. Coluccia and T. Maschmeyer, *Phys. Chem. Chem. Phys.*, **4**, 6109 (2002).

[288] B. Bonelli, B. Onida, J.D. Chen, A. Galarneau, F. Di Renzo, F. Fajula and E. Garrone, *Microporous Mesoporous Mater.*, **67**, 95 (2004).

[289] N. Petkov, S. Mintova, B. Jean, T.H. Metzger and T. Bein, *Chem. Mater.*, **15**, 2240 (2003).

[290] T. Maschmeyer, F. Rey, G. Sankar and J.M. Thomas, *Nature*, **378**, 159 (1995).

[291] D.S. Shephard, T. Maschmeyer, B.F.G. Johnson, J.M. Thomas, G. Sankar, D. Ozkaya, W. Zhou, R.D. Oldroyd and R.G. Bell, *Angew. Chem. Int. Ed.*, **36**, 2242 (1997).

[292] S. Kawi, S.C. Shen and P.L. Chew, *J. Mater. Chem.*, **12**, 1582 (2002).

[293] R. Anwander, C. Palm, O. Groeger and G. Engelhardt, *Organometallics*, **17**, 2027 (1998).

[294] J.M. Thomas and R. Raja, *Acc. Chem. Res.*, **41**, 708 (2008).

[295] J.M. Sun and X.H. Bao, *Chem. Eur. J.*, **14**, 7478 (2008).

[296] M. Thommes, R. Skudas, K.K. Unger and D. Lubda, *J. Chromatogr., A*, **1191**, 57 (2008).

[297] A. Nossov, E. Haddad, F. Guenneau and A. Gédéon, *Phys. Chem. Chem. Phys.*, **5**, 4473 (2003).

[298] M.A. Springuel-Huet, J.L. Bonardet, A. Gedeon, Y. Yue, V.N. Romannikov and J. Fraissard, *Microporous Mesoporous Mater.*, **44**, 775 (2001).

[299] J.C. Dore, J.B.W. Webber and J.H. Strange, *Colloids Surf., A*, **241**, 191 (2004).

[300] O. Sel, A. Brandt, D. Wallacher, M. Thommes and B. Smarsly, *Langmuir*, **23**, 4724 (2007).

[301] A. Gibaud, S. Dourdain and G. Vignaud, *Appl. Surf. Sci.*, **253**, 3 (2006).

[302] K. Morishige and M. Tateishi, *Langmuir*, **22**, 4165 (2006).

[303] M. Kruk, M. Jaroniec and A. Sayari, *Langmuir*, **13**, 6267 (1997).

[304] M. Kruk and M. Jaroniec, *Chem. Mater.*, **12**, 222 (2000).

[305] M. Thommes, in *Nanoporous Materials, Science and Engineering*, Vol. 4, G.Q. Lu, X.S. Zhao (Eds), Imperial College Press, London, 2004, p. 317.

[306] A.V. Neimark and P.I. Ravikovitch, *Microporous Mesoporous Mater.*, **44**, 697 (2001).

[307] E. Kierlik, P.A. Monson, M.L. Rosinberg and G. Tarjus, *J. Phys.: Condens. Matter*, **14**, 9295 (2002).

[308] M. Impéror-Clerc, P. Davidson and A. Davidson, *J. Am. Chem. Soc.*, **122**, 11925 (2000).

[309] B. Smarsly, S. Polarz and M. Antonietti, *J. Phys. Chem. B*, **105**, 10473 (2001).

[310] E. Prouzet and T.J. Pinnavaia, *Angew. Chem. Int. Ed.*, **36**, 516 (1997).

[311] A. Galarneau, H. Cambon, F. Di Renzo, R. Ryoo, M. Choi and F. Fajula, *New J. Chem.*, **27**, 73 (2003).

[312] M. Kruk, M. Jaroniec, C.H. Ko and R. Ryoo, *Chem. Mater.*, **12**, 1961 (2000).

[313] Y. Sakamoto, I. Dìaz, O. Terasaki, D. Zhao, J. Pérez-Pariente, J.M. Kim and G.D. Stucky, *J. Phys. Chem. B*, **106**, 3118 (2002).

[314] Y. Sakamoto, T.-W. Kim, R. Ryoo and O. Terasaki, *Angew. Chem. Int. Ed.*, **43**, 5231 (2004).

[315] Y. Sakamoto, M. Kaneda, O. Terasaki, D.Y. Zhao, J.M. Kim and G. Stucky, *Nature*, **408**, 449 (2000).

[316] M. Kaneda, T. Tsubakiyama, A. Carlsson, Y. Sakamoto, T. Ohsuna, O. Terasaki, S.H. Joo and R. Ryoo, *J. Phys. Chem. B*, **106**, 1256 (2002).

[317] V. Alfredsson and M.W. Anderson, *Chem. Mater.*, **8**, 1141 (1996).

[318] A. Carlsson, M. Kaneda, Y. Sakamoto, O. Terasaki, R. Ryoo and S.H. Joo, *J. Electron Microsc.*, **48**, 795 (1999).

[319] M.W. Anderson, C.C. Egger, G.J.T. Tiddy, J.L. Casci and K.A. Brakke, *Angew. Chem. Int. Ed.*, **44**, 3243 (2005).

[320] R.C. Hayward, P.C.A. Alberius, E.J. Kramer and B.F. Chmelka, *Langmuir*, **20**, 5998 (2004).

[321] J.R. Matos, M. Kruk, L.P. Mercuri, M. Jaroniec, L. Zhao, T. Kamiyama, O. Terasaki, T.J. Pinnavaia and Y. Liu, *J. Am. Chem. Soc.*, **125**, 821 (2003).

[322] T. Yu, H. Zhang, X.W. Yan, Z.X. Chen, X.D. Zou, P. Oleynikov and D.Y. Zhao, *J. Phys. Chem. B*, **110**, 21467 (2006).

[323] C.H. MacGillavry and G.R. Rieck (Eds), *International Tables for Crystallography*, Vol. 3, Kynock Press, Birmingham, 1968.

[324] F. Kleitz, T.W. Kim and R. Ryoo, *Langmuir*, **22**, 440 (2006).

[325] G.D. Stucky, A. Monnier, F. Schüth, Q. Huo, D. Margolese, D. Kumar, M. Krishnamurty, P. Petroff, A. Firouzi, M. Janicke and B.F. Chmelka, *Mol. Cryst. Liq. Cryst.*, **240**, 187 (1994).

[326] K. Flodström and V. Alfredsson, *Microporous Mesoporous Mater.*, **59**, 167 (2003).

[327] N. Baccile, C.V. Teixeira, H. Amenitsch, F. Villain, M. Linden and F. Babonneau, *Chem. Mater.*, **20**, 1161 (2008).

[328] J.L. Blin and C. Carteret, *J. Phys. Chem. C*, **111**, 14380 (2007).

[329] J.D.F. Ramsay and S. Kallus, *J. Non-Cryst. Solids*, **285**, 142 (2001).

[330] K.M. Ryan, N.R.B. Coleman, D.M. Lyons, J.P. Hanrahan, T.R. Spalding, M.A. Morris, D.C. Steytler and R.K. Heenan, *Langmuir*, **18**, 4996 (2002).

[331] A. Galarneau, M. Nader, F. Guenneau, F. Di Renzo and A. Gedeon, *J. Phys. Chem. C*, **111**, 8268 (2007).

[332] T. Pietrass, J.M. Kneller, R.A. Assink and M.T. Anderson, *J. Phys. Chem. B*, **103**, 8837 (1999).

[333] E. Geidel, H. Lechert, J. Dobler, H. Jobic, G. Calzaferri and F. Bauer, *Microporous Mesoporous Mater.*, **65**, 31 (2003).

[334] K. Ito, Y. Yagi, S. Hirano, M. Miyayama, T. Kudo, A. Kishimoto and Y. Ujihira, *J. Ceram. Soc. Jpn.*, **107**, 123 (1999).

[335] S. Boskovic, A.J. Hill, T.W. Turney, M.L. Gee, G.W. Stevens and A.J. O'Connor, *Prog. Solid State Chem.*, **34**, 67 (2006).

[336] R. Zaleski and J. Wawryszczuk, *Acta Phys. Pol., A*, **113**, 1543 (2008).

[337] M.P. Petkov, C.L. Wang, M.H. Weber, K.G. Lynn and K.P. Rodbell, *J. Phys. Chem. B*, **107**, 2725 (2003).

[338] C.Q. He, R. Suzuki, T. Ohdaira, N. Oshima, A. Kinomura, M. Muramatsu and Y. Kobayashi, *Chem. Phys.*, **331**, 213 (2007).

[339] K.J. Edler, P.A. Reynolds, F. Trouw and J.W. White, *Chem. Phys. Lett.*, **249**, 438 (1996).

[340] S. Takahara, N. Sumiyama, S. Kittaka, T. Yamaguchi and M.-C. Bellissent-Funel, *J. Phys. Chem. B*, **109**, 11231 (2005).

[341] S. Takahara, S. Kittaka, T. Mori, Y. Kuroda, T. Takamuku and T. Yamaguchi, *J. Phys. Chem. C*, **112**, 14385 (2008).

[342] K.J. Edler, P.A. Reynolds, P.J. Branton, F. Trouw and J.W. White, *J. Chem. Soc., Faraday Trans.*, **93**, 1667 (1997).

[343] F.G. Requejo, J.M. Ramallo-Lopez, R. Rosas-Salas, J.M. Dominguez, J.A. Rodriguez, J.-Y. Kim and R. Quijada, *Catal. Today*, **107-108**, 750 (2005).

[344] K. Ikeda, Y. Kawamura, T. Yamamoto and M. Iwamoto, *Catal. Commun.*, **9**, 106 (2008).

[345] S. Liu, L. Lu, Z. Yang, P. Cool and E.F. Vansant, *Mater. Chem. Phys.*, **97**, 203 (2006).

[346] M. Grun, I. Lauer and K.K. Unger, *Adv. Mater.*, **9**, 254 (1997).

[347] L. Qi, J. Ma, H. Cheng and Z. Zhao, *Chem. Mater.*, **10**, 1623 (1998).

[348] Q. Huo, J. Feng, F. Schüth and G.D. Stucky, *Chem. Mater.*, **9**, 14 (1997).

[349] C.P. Kao, H.P. Lin and C.Y. Mou, *J. Phys. Chem. Solids*, **62**, 1555 (2001).

[350] L.M. Yang, Y.J. Wang, G.S. Luo and Y.Y. Dai, *Particuology*, **6**, 143 (2008).

[351] T. Martin, A. Galarneau, F. Di Renzo, D. Brunel, F. Fajula, S. Heinisch, G. Crétier and J.-L. Rocca, *Chem. Mater.*, **16**, 1725 (2004).

[352] T. Prouzet and C. Boissiere, *C. R. Chim.*, **8**, 579 (2005).

[353] G.A. Ozin, H. Yang, I. Sokolov and N. Coombs, *Adv. Mater.*, **9**, 662 (1997).

[354] S. Yang, L.Z. Zhao, C.Z. Yu, X.F. Zhou, J.W. Tang, P. Yuan, D.Y. Chen and D.Y. Zhao, *J. Am. Chem. Soc.*, **128**, 10460 (2006).

[355] G.A. Hart, L.M. Kathman and T.W. Hesterberg, *Carcinogenesis*, **15**, 971 (1994).

[356] Z.G. Feng, Y.S. Li, D.C. Niu, L. Li, W.R. Zhao, H.R. Chen, J.H. Gao, M.L. Ruan and J.L. Shi, *Chem. Commun.*, 2629 (2008).

[357] H. Zhang, J. Wu, L. Zhou, D. Zhang and L. Qi, *Langmuir*, **23**, 1107 (2007).

[358] S. Schacht, Q. Huo, I.G. Voigt-Martin, G.D. Stucky and F. Schüth, *Science*, **273**, 768 (1996).

[359] C.E. Fowler, D. Khushalani and S. Mann, *J. Mater. Chem.*, **11**, 1968 (2001).

[360] C.E. Fowler, D. Khushalani and S. Mann, *Chem. Commun.*, 2028 (2001).

[361] Z.R. Tian, J. Liu, J.A. Voigt, B. Mckenzie and H. Xu, *Angew. Chem. Int. Ed.*, **42**, 413 (2003).

[362] P. Linton and V. Alfredsson, *Chem. Mater.*, **20**, 2878 (2008).
[363] C.Y. Mou and H.P. Lin, *Pure Appl. Chem.*, **72**, 137 (2000).
[364] H.-P. Lin, S.-B. Liu, C.-Y. Mou and C.-Y. Tan, *Chem. Commun.*, 583 (1999).
[365] H.-P. Lin and C.-Y. Mou, *Acc. Chem. Res.*, **35**, 927 (2002).
[366] H.-P. Lin, S. Cheng and C.-Y. Mou, *Chem. Mater.*, **10**, 581 (1998).
[367] F. Kleitz, U. Wilczok, F. Schuth and F. Marlow, *Phys. Chem., Chem. Phys.*, **3**, 3486 (2001).
[368] D. Zhao, J. Sun, Q. Li and G.D. Stucky, *Chem. Mater.*, **12**, 275 (2000).
[369] H. Maekawa, J. Esquena, S. Bishop, C. Solans and B.F. Chmelka, *Adv. Mater.*, **15**, 591 (2003).
[370] K. Nakanishi, Y. Kobayashi, T. Amatani, K. Hirao and T. Kodaira, *Chem. Mater.*, **16**, 3652 (2004).
[371] K. Nakanishi, Y. Sato, Y. Ruyat and K. Hirao, *J. Sol-Gel Sci. Technol.*, **26**, 567 (2003).
[372] D. Zhao, P. Yang, B.F. Chmelka and G.D. Stucky, *Chem. Mater.*, **11**, 1174 (1999).
[373] F. Kleitz, F. Marlow, G.D. Stucky and F. Schüth, *Chem. Mater.*, **13**, 3587 (2001).
[374] J.H. Jung, K. Yoshida and T. Shimizu, *Langmuir*, **18**, 8724 (2002).
[375] B. Platschek, N. Petkov and T. Bein, *Angew. Chem. Int. Ed.*, **45**, 1134 (2006).
[376] K.T. Jung, Y.H. Chu, S. Haam and Y.G. Shul, *J. Non-Cryst. Solids*, **298**, 193 (2002).
[377] F. Marlow and F. Kleitz, *Microporous Mesoporous Mater.*, **44**, 671 (2001).
[378] L. Huang, H. Wang, C.Y. Hayashi, B. Tian, D. Zhao and Y. Yan, *J. Mater. Chem.*, **13**, 666 (2003).
[379] J. Jung, M. Amaike and S. Shinkai, *Chem. Commun.*, 2343 (2000).
[380] A. Thomas and M. Antonietti, *Adv. Funct. Mater.*, **13**, 763 (2003).
[381] Q. Lu, F. Gao, S. Komarneni and T.E. Mallouk, *J. Am. Chem. Soc.*, **126**, 8650 (2004).
[382] A. Yamaguchi, F. Uejo, T. Yoda, T. Uchida, Y. Tanamura, T. Yamashita and N. Teramae, *Nat. Mater.*, **3**, 337 (2004).
[383] S.J. Yoo, D.A. Ford and D.F. Shantz, *Langmuir*, **22**, 1839 (2006).
[384] Y. Wu, G. Cheng, K. Katsov, S.W. Sides, J. Wang, J. Tang, G.H. Fredrickson, M. Moskovits and G.D. Stucky, *Nat. Mater.*, **3**, 816 (2004).
[385] Z. Yang, Z. Niu, X. Cao, Z. Yang, Y. Lu, Z. Hu and C.C. Han, *Angew. Chem. Int. Ed.*, **42**, 4201 (2003).
[386] A. Yamaguchi, H. Kaneda, W.S. Fu and N. Teramae, *Adv. Mater.*, **20**, 1034 (2008).
[387] B. Platschek, R. Kohn, M. Doblinger and T. Bein, *ChemPhysChem*, **9**, 2059 (2008).
[388] B. Platschek, N. Petkov and T. Bein, *Angew. Chem. Int. Ed.*, **45**, 1134 (2006).
[389] B. Platschek, R. Kohn, M. Doblinger and T. Bein, *Langmuir*, **24**, 5018 (2008).
[390] J.F. Wang, C.K. Tsung, W.B. Hong, Y.Y. Wu, J. Tang and G.D. Stucky, *Chem. Mater.*, **16**, 5169 (2004).
[391] K.W. Jin, B.D. Yao and N. Wang, *Chem. Phys. Lett.*, **409**, 172 (2005).
[392] A. Yamaguchi and N. Teramae, *Anal. Sci.*, **24**, 25 (2008).
[393] I.A. Aksay, M. Trau, S. Manne, I. Honma, N. Yao, L. Zhou, P. Fenter, P.M. Eisenberger and S.M. Gruner, *Science*, **273**, 892 (1996).
[394] B. Yang and K.J. Edler, *Chem. Mater.*, **21**, 1221 (2009).
[395] H. Miyata and K. Kuroda, *Adv. Mater.*, **11**, 1448 (1999).
[396] H. Miyata, T. Noma and K. Kuroda, *Chem. Mater.*, **14**, 766 (2002).
[397] H. Miyata, T. Suzuki, A. Fukuoka, T. Sawada, M. Watanabe, T. Noma, K. Takada, T. Mukaide and K. Kuroda, *Nat. Mater.*, **3**, 651 (2004).

[398] H. Miyata, *Microporous Mesoporous Mater.*, **101**, 296 (2007).
[399] H. Miyata, T. Suzuki, M. Watanabe, T. Noma, K. Takada, T. Mukaide and K. Kuroda, *Chem. Mater.*, **20**, 1082 (2008).
[400] T. Suzuki, Y. Kanno, Y. Morioka and K. Kuroda, *Chem. Commun.*, 3284 (2008)
[401] T. Suzuki, H. Miyata and K. Kuroda, *J. Mater. Chem.*, **18**, 1239 (2008).
[402] B. Su, X.M. Lu, Q.H. Lu and A. Facile, *J. Am. Chem. Soc.*, **130**, 14356 (2008).
[403] R.L. Rice, D.C. Arnold, M.T. Shaw, D. Iacopina, A.J. Quinn, H. Amenitsch, J.D. Holmes and M.A. Morris, *Adv. Funct. Mater.*, **17**, 133 (2007).
[404] S.H. Tolbert, A. Firouzi, G.D. Stucky and B.F. Chmelka, *Science*, **278**, 264 (1997).
[405] A.Y. Ku, D.A. Saville and I.A. Aksay, *Langmuir*, **23**, 8156 (2007).
[406] A. Walcarius, E. Sibottier, M. Etienne and J. Ghanbaja, *Nat. Mater.*, **6**, 602 (2007).
[407] V.R. Koganti and S.E. Rankin, *J. Phys. Chem. B*, **109**, 3279 (2005).
[408] V.R. Koganti, D. Dunphy, V. Gowrishankar, M.D. McGehee, X.F. Li, J. Wang and S.E. Rankin, *Nano Lett.*, **6**, 2567 (2006).
[409] E.K. Richman, T. Brezesinski and S.H. Tolbert, *Nat. Mater.*, **7**, 712 (2008).
[410] S. Nagarajan, M.Q. Li, R.A. Pai, J.K. Bosworth, P. Busch, D.M. Smilgies, C.K. Ober, T.P. Russell and J.J. Watkins, *Adv. Mater.*, **20**, 246 (2008).
[411] F. Marlow, M.D. McGehee, D. Zhao, B.F. Chmelka and G.D. Stucky, *Adv. Mater.*, **11**, 632 (1999).
[412] L.-N. Sun, H.-J. Zhang, C.-Y. Peng, J.-B. Yu, Q.-G. Meng, L.-S. Fu, F.-Y. Liu and X.-M. Guo, *J. Phys. Chem. B*, **110**, 7249 (2006).
[413] G. Wirnsberger and G.D. Stucky, *Chem. Mater.*, **12**, 2525 (2000).
[414] N.A. Melosh, C.A. Steinbeck, B.J. Scott, R.C. Hayward, P. Davidson, G.D. Stucky and B.F. Chmelka, *J. Phys. Chem. B*, **108**, 11909 (2004).
[415] G. Wirnsberger, B.J. Scott, B.F. Chmelka and G.D. Stucky, *Adv. Mater.*, **12**, 1450 (2000).
[416] C.A. Steinbeck, M. Ernst, B.H. Meier and B.F. Chmelka, *J. Phys. Chem. C*, **112**, 2565 (2008).
[417] A. Fukuoka, Y. Sakamoto, T. Higuchi, N. Shimomura and M. Ichikawa, *J. Porous Mater.*, **13**, 231 (2006).
[418] J.J. Wu, A.F. Gross and S.H. Tolbert, *J. Phys. Chem. B*, **103**, 2374 (1999).
[419] S.H. Tolbert, J.J. Wu, A.F. Gross, T.Q. Nguyen and B.J. Schwartz, *Microporous Mesoporous Mater.*, **44**, 445 (2001).
[420] T.-Q. Nguyen, J.J. Wu, V. Doan, B.J. Schwartz and S.H. Tolbert, *Science*, **288**, 652 (2000).
[421] T.-Q. Nguyen, J.J. Wu, B.J. Schwartz and S.H. Tolbert, *Adv. Mater.*, **13**, 609 (2001).
[422] S.P. Rigby, M. Fairhead and C.F. van der Walle, *Curr. Pharm. Design*, **14**, 1821 (2008).
[423] X.X. Li, S. Barua, K. Rege and B.D. Vogt, *Langmuir*, **24**, 11935 (2008).
[424] J.D. Bass, D. Grosso, C. Boissiere, E. Belamie, T. Coradin and C. Sanchez, *Chem. Mater.*, **19**, 4349 (2007).
[425] M. Vallet-Regi, F. Balas and D. Arcos, *Angew. Chem. Int. Ed.*, **46**, 7548 (2007).
[426] M. Manzano, V. Aina, C.O. Arean, F. Balas, V. Cauda, M. Colilla, M.R. Delgado and M. Vallet-Regi, *Chem. Eng. J.*, **137**, 30 (2008).
[427] M. Colilla, I. Izquierdo-Barba and M. Vallet-Regi, *Exp. Op. Therapeut. Patents*, **18**, 639 (2008).
[428] R. Mellaerts, C.A. Aerts, J. Van Humbeeck, P. Augustijns, G. Van den Mooter and J.A. Martens, *Chem. Commun.*, 1375 (2007).

[429] I.Y. Park, I.Y. Kim, M.K. Yoo, Y.J. Choi, M.H. Cho and C.S. Cho, *Int. J. Pharm.*, **359**, 280 (2008).

[430] F. Torney, B.G. Trewyn, LinV.S.-Y. and K. Wang, *Nat. Nanotechnol.*, **2**, 295 (2007).

[431] D.R. Radu, C.Y. Lai, K. Jeftinija, E.W. Rowe, S. Jeftinija and V.S.Y. Lin, *J. Am. Chem. Soc.*, **126**, 13216 (2004).

[432] Y.J. Han, G.D. Stucky and A. Butler, *J. Am. Chem. Soc.*, **121**, 9897 (1999).

[433] B.G. Trewyn, J.A. Nieweg, Y. Zhao and V.S.Y. Lin, *Chem. Eng. J.*, **137**, 23 (2008).

[434] B.G. Trewyn, I.I. Slowing, S. Giri, H.T. Chen and V.S.Y. Lin, *Acc. Chem. Res.*, **40**, 846 (2007).

[435] S. Giri, B.G. Trewyn and V.S.Y. Lin, *Nanomedicine*, **2**, 99 (2007).

[436] M. Vallet-Regi, M. Colilla and I. Izquierdo-Barba, *J. Biomed. Nanotechnol.*, **4**, 1 (2008).

[437] F. Balas, M. Manzano, M. Colilla and M. Vallet-Regi, *Acta Biomater.*, **4**, 514 (2008).

[438] A. Lopez-Noriega, D. Arcos, I. Izquierdo-Barba, Y. Sakamoto, O. Terasaki and M. Vallet-Regi, *Chem. Mater.*, **18**, 3137 (2006).

[439] Y.F. Zhu, C.T. Wu, Y. Ramaswamy, E. Kockrick, P. Simon, S. Kaskel and H. Zrelqat, *Microporous Mesoporous Mater.*, **112**, 494 (2008).

[440] A. El-Ghannam, K. Ahmed and M. Omran, *J. Biomed. Mater. Res. B*, **73B**, 277 (2005).

[441] K.M.L. Taylor, J.S. Kim, W.J. Rieter, H. An, W.L. Lin and W.B. Lin, *J. Am. Chem. Soc.*, **130**, 2154 (2008).

[442] C.P. Tsai, Y. Hung, Y.H. Chou, D.M. Huang, J.K. Hsiao, C. Chang, Y.C. Chen and C.Y. Mou, *Small*, **4**, 186 (2008).

[443] H.M. Liu, S.H. Wu, C.W. Lu, M. Yao, J.K. Hsiao, Y. Hung, Y.S. Lin, C.Y. Mou, C.S. Yang, D.M. Huang and Y.C. Chen, *Small*, **4**, 619 (2008).

[444] H.K. Baca, E. Carnes, S. Singh, C. Ashley, D. Lopez and C.J. Brinker, *Acc. Chem. Res.*, **40**, 836 (2007).

[445] H.K. Baca, C. Ashley, E. Carnes, D. Lopez, J. Flemming, D. Dunphy, S. Singh, Z. Chen, N.G. Liu, H.Y. Fan, G.P. Lopez, S.M. Brozik, M. Werner-Washburne and C.J. Brinker, *Science*, **313**, 337 (2006).

[446] A.J. Di Pasqua, K.K. Sharma, Y.L. Shi, B.B. Toms, W. Ouellette, J.C. Dabrowiak and T. Asefa, *J. Inorg. Biochem.*, **102**, 1416 (2008).

[447] F. Qu, G. Zhu, S. Huang, S. Li, J. Sun, D. Zhang and S. Qiu, *Microporous Mesoporous Mater.*, **92**, 1 (2006).

[448] Z. Wu, Y. Jiang, T. Kim and K. Lee, *J. Controlled Release*, **119**, 215 (2007).

[449] Q.L. Tang, Y. Xu, D. Wu and Y.H. Sun, *Chem. Lett.*, **35**, 474 (2006).

[450] J.C. Doadrio, E.M.B. Sousa, I. Izquierdo-Barba, A.L. Doadrio, J. Perez-Pariente and M. Vallet-Regi, *J. Mater. Chem.*, **16**, 462 (2006).

[451] Q. Fu, G.V.R. Rao, T.L. Ward, Y.F. Lu and G.P. Lopez, *Langmuir*, **23**, 170 (2007).

[452] J.M. Xue and M. Shi, *J. Controlled Release*, **98**, 209 (2004).

[453] L.B. Fagundes, T.G.F. Sousa, A. Sousa, V.V. Silva and E.M.B. Sousa, *J. Non-Cryst. Solids*, **352**, 3496 (2006).

[454] D.R. Radu, C.Y. Lai, J.W. Wiench, M. Pruski and V.S.Y. Lin, *J. Am. Chem. Soc.*, **126**, 1640 (2004).

[455] S. Giri, B.G. Trewyn, M.P. Stellmaker and V.S.Y. Lin, *Angew. Chem. Int. Ed.*, **44**, 5038 (2005).

[456] C.Y. Lai, B.G. Trewyn, D.M. Jeftinija, K. Jeftinija, S. Xu, S. Jeftinija and V.S.Y. Lin, *J. Am. Chem. Soc.*, **125**, 4451 (2003).
[457] Q. Yang, S. Wang, P. Fan, L. Wang, Y. Di, K. Lin and F.-S. Xiao, *Chem. Mater.*, **17**, 5999 (2005).
[458] Y.F. Zhu, J.L. Shi, W.H. Shen, X.P. Dong, J.W. Feng, M.L. Ruan and Y.S. Li, *Angew. Chem. Int. Ed.*, **44**, 5083 (2005).
[459] N.K. Mal, M. Fujiwara and Y. Tanaka, *Nature*, **421**, 350 (2003).
[460] S. Angelos, E. Choi, F. Vogtle, L. De Cola and J.I. Zink, *J. Phys. Chem. C*, **111**, 6589 (2007).
[461] S. Angelos, Y.W. Yang, K. Patel, J.F. Stoddart and J.I. Zink, *Angew. Chem. Int. Ed.*, **47**, 2222 (2008).
[462] T.D. Nguyen, H.R. Tseng, P.C. Celestre, A.H. Flood, Y. Liu, J.F. Stoddart and J.I. Zink, *Proc. Natl. Acad. Sci. USA*, **102**, 10029 (2005).
[463] S. Angelos, E. Johansson, J.F. Stoddart and J.I. Zink, *Adv. Funct. Mater.*, **17**, 2261 (2007).
[464] S. Angelos, M. Liong, E. Choi and J.I. Zink, *Chem. Eng. J.*, **137**, 4 (2008).
[465] R. Hernandez, H.-R. Tseng, J.W. Wong, J.F. Stoddart and J.I. Zink, *J. Am. Chem. Soc.*, **126**, 3370 (2004).
[466] K. Patel, S. Angelos, W.R. Dichtel, A. Coskun, Y.W. Yang, J.I. Zink and J.F. Stoddart, *J. Am. Chem. Soc.*, **130**, 2382 (2008).
[467] S. Hudson, E. Magner, J. Cooney and B.K. Hodnett, *J. Phys. Chem. B*, **109**, 19496 (2005).
[468] A. Salis, D. Meloni, S. Ligas, M.F. Casula, M. Monduzzi, V. Solinas and E. Dumitriu, *Langmuir*, **21**, 5511 (2005).
[469] M. Hartmann and C. Streb, *J. Porous Mater.*, **13**, 347 (2006).
[470] S. Jang, D. Kim, J. Choi, K. Row and W. Ahn, *J. Porous Mater.*, **13**, 385 (2006).
[471] D. Jung, C. Streb and M. Hartmann, *Microporous Mesoporous Mater.*, **113**, 523 (2008).
[472] P. Reis, T. Witula and K. Holmberg, *Microporous Mesoporous Mater.*, **110**, 355 (2008).
[473] S. Hudson, J. Cooney, B.K. Hodnett and E. Magner, *Chem. Mater.*, **19**, 2049 (2007).
[474] H. Essa, E. Magner, J. Cooney and B.K. Hodnett, *J. Mol. Catal., B*, **49**, 61 (2007).
[475] Y. Fukushima, T. Kajino and T. Itoh, *Curr. Nanosci.*, **2**, 211 (2006).
[476] A. Kumar, R.R. Pandey and B. Brantley, *Talanta*, **69**, 700 (2006).
[477] M. Tiemann, *Chem. Mater.*, **20**, 961 (2008).
[478] S. Han and T. Hyeon, *Chem. Commun.*, 1955 (1999).
[479] J. Lee, S. Yoon, T. Hyeon, S.M. Oh and K.B. Kim, *Chem. Commun.*, 2177 (1999)
[480] A.B. Fuertes, M. Sevilla, S. Alvarez and T. Valdes-Solis, *Microporous Mesoporous Mater.*, **112**, 319 (2008).
[481] Z. Liu, Y. Sakamoto, T. Ohsuna, K. Hiraga, O. Terasaki, C.H. Ko, H.J. Shin and R. Ryoo, *Angew. Chem. Int. Ed.*, **39**, 3107 (2000).
[482] K. Lee, Y.-H. Kim, S.B. Han, H. Kang, S. Park, W.S. Seo, J.T. Park, B. Kim and S. Chang, *J. Am. Chem. Soc.*, **125**, 6844 (2003).
[483] Y.F. Shi, Y. Wan, R.L. Liu, B. Tu and D.Y. Zhao, *J. Am. Chem. Soc.*, **129**, 9522 (2007).
[484] B.Z. Tian, X.Y. Liu, L.A. Solovyov, Z. Liu, H.F. Yang, Z.D. Zhang, S.H. Xie, F.Q. Zhang, B. Tu, C.Z. Yu, O. Terasaki and D.Y. Zhao, *J. Am. Chem. Soc.*, **126**, 865 (2004).

[485] B. Tian, X. Liu, H. Yang, S. Xie, C. Yu, B. Tu and D. Zhao, *Adv. Mater.*, **15**, 1370 (2003).

[486] G.S. Li, D.Q. Zhang and J.C. Yu, *Chem. Mater.*, **20**, 3983 (2008).

[487] H. Yang, Q. Shi, B. Tian, Q. Lu, F. Gao, S. Xie, J. Fan, C. Yu, B. Tu and D. Zhao, *J. Am. Chem. Soc.*, **125**, 4724 (2003).

[488] R. Ryoo, S.H. Joo, M. Kruk and M. Jaroniec, *Adv. Mater.*, **13**, 677 (2001).

[489] A.-H. Lu, J.-H. Smatt, S. Backlund and M. Linden, *Microporous Mesoporous Mater.*, **72**, 59 (2004).

[490] A.H. Lu, W. Schmidt, A. Taguchi, B. Spliethoff, B. Tesche and F. Schüth, *Angew. Chem. Int. Ed.*, **41**, 3489 (2002).

[491] A.-H. Lu, W. Schmidt, B. Spliethoff and F. Schüth, *Chem. Eur. J.*, **10**, 6085 (2004).

[492] J. Roggenbuck, H. Schafer, T. Tsoncheva, C. Minchev, J. Hanss and M. Tiemann, *Microporous Mesoporous Mater.*, **101**, 335 (2007).

[493] A. Rumplecker, F. Kleitz, E.L. Salabas and F. Schuth, *Chem. Mater.*, **19**, 485 (2007).

[494] F. Jiao and P.G. Bruce, *Adv. Mater.*, **19**, 657 (2007).

[495] K.M. Shaju, F. Jiao, A. Debart and P.G. Bruce, *Phys. Chem. Chem. Phys.*, **9**, 1837 (2007).

[496] Y. Wan, Y.F. Shi and D.Y. Zhao, *Chem. Mater.*, **20**, 932 (2008).

[497] Y. Deng, J. Liu, C. Liu, D. Gu, Z. Sun, J. Wei, J. Zhang, L. Zhang, B. Tu and D. Zhao, *Chem. Mater.*, **20**, 7281 (2008).

3

Ordered Porous Crystalline Transition Metal Oxides

Masahiro Sadakane[1] and Wataru Ueda[2]

[1]*Chemistry and Chemical Engineering, Graduate School of Engineering, Hiroshima University, Higashi-Hiroshima, Japan*
[2]*Catalysis Research Center, Hokkaido University, Sapporo, Japan*

3.1   INTRODUCTION

Humans are fascinated by the beauty of naturally and artificially ordered materials, and chemists are attracted by the beauty of ordered porous materials.[1] Ordered porous materials offer a wide variety of applications based on properties specific to pore size and arrangement. Material properties are also dominated by chemical and physical properties of the solid matrix.

The International Union of Pure and Applied Chemistry (IUPAC) has classified porous materials into three distinct categories (microporous, mesoporous and macroporous) according to their pore sizes.[2] This classification was based on gas adsorption isotherms. Pores with widths not exceeding 2 nm are called micropores, pores with widths between 2 nm and 50 nm are called mesopores, and those exceeding 50 nm are called macropores.

Natural zeolites with crystalline aluminosilicate frameworks to form ordered micropores have been known since the eighteenth century.

Porous Materials   Edited by Duncan W. Bruce, Dermot O'Hare and Richard I. Walton
© 2011 John Wiley & Sons, Ltd.

Methods for synthesising ordered porous materials have been developed since the 1950s following the creation of the first commercially important artificial zeolite.[3] The general procedure for preparation of porous materials involves the use of templates around which a solid wall structure forms. By removing the templates, porous materials are produced. The templates used to make zeolites are composed of alkaline cations and organic molecules, and pore size can be controlled by adjusting the molecular size of the templates.

In the early 1990s, a new synthetic strategy using self-assembling molecular arrays as templates for the fabrication of ordered mesoporous materials was developed.[4–7] Inorganic walls formed around micelles of an assembly of surfactants, and ordered mesoporous materials with cubic or hexagonal symmetry were obtained after removing the templates. Pore diameters up to 30 nm were attained by using long-chain surfactants.

In the late 1990s, a new synthetic strategy for producing larger ordered porous materials, macroporous materials, using submicrometre-ordered spheres was introduced. [8,9] In this strategy, a close-packed array (opal structure) of monodisperse spheres, *e.g.* polystyrene (PS), poly(methyl methacrylate) (PMMA) or silica (SiO_2), was used as template, and inorganic walls were formed in the voids of this opal structure. After removing the templates, three-dimensionally ordered macroporous materials were obtained.

A transition element containing an incomplete d subshell has many interesting properties[10] and its oxides form a series of compounds with various unique electronic properties. They have a variety of applications such as catalysis, photocatalysis, sensors and electrode materials because of their catalytic, optical and electronic properties. Recently, many attempts have been made to combine these chemical and physical properties and ordered porous properties in order to create novel functional materials. In this chapter, we summarise the synthetic procedures, structural characterisation and applications of ordered porous crystalline transition metal oxides.

3.2 SCOPE AND LIMITATIONS OF THIS REVIEW

We have selected ordered porous crystalline transition metal oxides that meet all three of the following criteria. The first criterion is that they are ordered porous materials, and that ordered pores were observed by transmission electron microscopy (TEM) or scanning electron microscopy (SEM). For mesoporous materials, order was also detected by small-angle X-ray diffraction (XRD). For microporous and mesoporous materials, ordered porosity was confirmed by the gas adsorption technique. Unordered porous materials were excluded. The second criterion is that the materials are transition metal oxides: metal oxides with porous walls containing transition metal oxide networks were selected. According to the IUPAC definition,[10] a transition metal is an element whose atom has an incomplete d subshell corresponding to Groups 3–11 elements in the periodic table. The third criterion is crystallinity. Only crystalline metal oxides that show a clear powder XRD pattern or electron diffraction (ED) spots were selected.

3.3 MICROPOROUS TRANSITION METAL OXIDE MATERIALS

Ordered micropores (<2 nm) in crystalline metal oxides are formed as part of their crystal structures (*i.e.* they are intracrystallite). The most popular ordered microporous metal oxides are the zeolites. Zeolites contain metals that prefer a tetrahedral coordination to four oxygen atoms, such as silicon, aluminium, phosphorus, sulfur, germanium, and arsenic (As). Other metals can be incorporated in the zeolite framework, and new zeolites have been produced by combining metals and organic structure-directing agent (SDA) molecules. Many transition metals having octahedral coordination are now also incorporated in zeolite-like frameworks, such as $(Me_4N)_{1.3}(H_2O)_{0.7}$ [Mo_4O_8 $(PO_4)_2$]·$2H_2O$ (pore diameter 0.28 nm).[11–13] These transition-metal-containing zeolites contain tetrahedrally coordinated non-transition metals in the wall framework. However, it is still a challenging issue to produce ordered microporous metal oxides with only transition metal elements.

There are two examples of transition metal oxides forming ordered microporous materials: (1) manganese oxide and (2) molybdenum vanadium mixed oxide, both of which contain MO_6 octahedra as the basic structural units to form one-dimensional channel pores.

The first ordered microporous transition metal oxides were microporous manganese oxides, known as 'octahedral molecular sieves' (OMS).[14–16] The manganese oxide OMS are classified into three families: the pyrolusite-ramsdellite family with a $(1 \times n)$ channel structure; the hollandite-romanechite family with a $(2 \times n)$ channel structure;

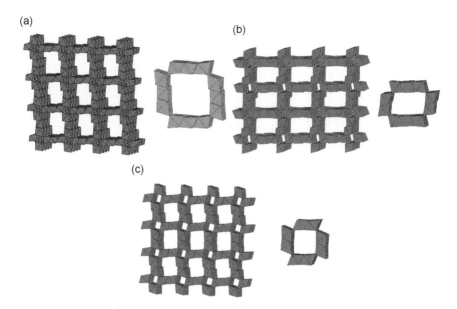

Figure 3.1 Structure and pore of manganese oxide octahedral molecular sieves (OMS): (a) todorokite (3×3) channel; (b) romanechite (2×3) channel; and (c) hollandite (2×2) channel

and the todorokite family with a $(3 \times n)$ channel structure. The structures contain infinite chains of edge-shared MnO_6 octahedra structural units, and the numbers 1, 2, 3 and n correspond to the numbers of octahedra in the unit chain width. The chains are linked by corner sharing to form a one-dimensional channel structural network (Figure 3.1). Microporous properties of hollandite (2×2), romanechite (2×3), RUB-7 (2×4) and todorokite (3×3) have been characterised by using the gas adsorption technique.[17–19] Names, formulae, channel structures and pore sizes of the ordered microporous manganese oxides together with a related Mo-V oxide are summarised in Table 3.1.

Pore sizes estimated by the nitrogen and argon adsorption techniques were reasonable, because the size corresponds to the short width of the pore and the longest diagonal length of the pores. Furthermore, the molecular sieve property of todorokite, which adsorbed cyclohexane but not 1,3,5-triisopropylbenzene, has been reported.[20]

Because of the mixed-valent manganese framework, valences of manganese are $(+2, +3,$ and $+4)$ or $(+3$ and $+4)$, a small number of guest cations usually being required for charge balance in the channel of the manganese oxide. These cations do not completely block the pores.

Table 3.1 Names, formulas, channel structures and pore sizes of manganese oxide and Mo-V oxide OMS

Name	Formula	Channel structure	Calc. pore sizes (nm)	Obs. pore size (nm)	Ref.
Hollandite	$KMn_8O_{16} \cdot nH_2O$	2×2	$0.46 \times 0.46 \times 0.65^a$	0.58^b	[17]
Romanechite	$Na_xMn_5O_{10} \cdot nH_2O$	2×3	$0.46 \times 0.69 \times 0.83^a$	$0.75^c, 0.68^b$	[18, 19]
RUB-7	$Na_2Mn_3O_6 \cdot nH_2O$	2×4	$0.46 \times 0.92 \times 1.03^a$	0.76^b	[19]
Todorokite	$Mg_xMn_6O_{12} \cdot nH_2O$	3×3	$0.69 \times 0.69 \times 0.98^a$	$0.7{-}0.8^b$	[17]
Orthorhombic MoVO	$Mo_{30}V_{10}O_{112}$	Seven-membered ring	0.4	0.4^d	[22]

^{a}Channel width (long) $\times$ channel width (short) $\times$ diagonal length.
^{b}Obtained from argon isotherm (at 87 K).
^{c}Obtained from nitrogen isotherm (at 77 K).
^{d}Obtained using the molecular probe method.

Hollandite (2×2) was prepared by reaction aqueous of $KMnO_4$ and $MnSO_4$ at pH $\sim$1 and refluxing at 100 °C.[17] Other manganese oxide OMS were prepared from a layered manganese oxide (Na-Birnessite, Na_4 $Mn_{14}O_{27} \cdot 9H_2O$, $2 \times \infty$) under hydrothermal reaction conditions with different pH values of the solutions.[17–19,21]

Another example, crystalline Mo-V mixed metal oxide, has recently been reported.[22] The Mo-V mixed metal oxide ($Mo_{30}V_{10}O_{112}$) has a layered orthorhombic structure with a slab composed of six- and seven-membered rings of corner-sharing MO_6 octahedra and pentagonal $(M)M_5O_{21}$ units with a pentagonal bipyramid and five edge-sharing octahedra. The layered six- and seven-membered rings form channel structures (Figure 3.2). Aperture diameters of the six- and seven-membered rings estimated by X-ray structure analysis were about 0.25–0.28 and about 0.33–0.37 nm, respectively.

This orthorhombic Mo-V oxide was prepared by hydrothermal reaction of $(NH_4)_6Mo_7O_{24}$ and $VOSO_4$. Similar to manganese oxide OMS, cations ($[NH_4]^+$) occupy both channels. To form an empty channel, $[NH_4]^+$ cations are removed by calcination. Ordered microporosity was first confirmed by a single uptake of N_2 and Ar adsorption at very low P/P_0 (type I behaviour), and the pore diameter was estimated by the molecular probe technique to be ca 0.4 nm (Figure 3.3). This value is close to the aperture diameter of the seven-membered channel. The micropore volume calculated by the Dubinin–Astakhov (DA) equation for smaller gases is about 0.025 $cm^3 g^{-1}$, which is the calculated pore volume of a seven-membered channel. These results indicate that this seven-membered ring channel is a micropore.

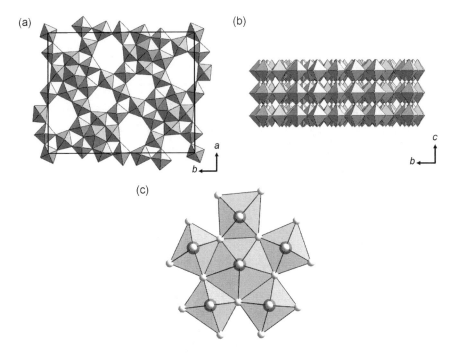

Figure 3.2   Polyhedral representation of the orthorhombic $Mo_{30}V_{10}O_{112}$: (a) ab-plane; (b) bc-plane; and (c) pentagonal unit $[Mo_6O_{21}]^{n-}$

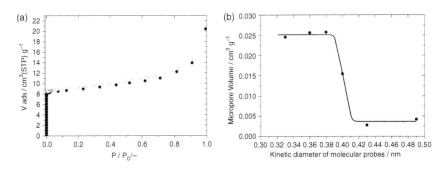

Figure 3.3   (a) Adsorption and desorption isotherm for nitrogen at 77 K and (b) micropore volume estimated by the Dubinin–Astakhov (DA) method plotted against the kinetic diameter of carbon dioxide (0.33 nm), Kr (0.36 nm), methane (0.38 nm), ethane (0.4 nm), n-butane (0.43 nm), and n-hexane (0.49 nm) of orthorhombic Mo-V oxide

The orthorhombic Mo-V oxide forms rod-like crystals with lengths of up to several tens of micrometres (Figure 3.4). The cross-section is rhombic in shape with submicrometre thickness. The microporous channel grows along the long axis of the rod-like crystals.

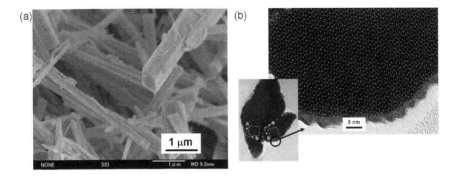

Figure 3.4 (a) SEM image and (b) TEM image viewed along the [001] direction of orthorhombic Mo-V oxide

An important feature of both manganese oxides and the Mo-V oxide is that they are redox active. Therefore, applications as catalysts, conductive materials, and electrode materials have been investigated. For example, both the manganese oxides[14–16] and the Mo-V oxide[23–26] have excellent catalytic activity for selective oxidation of organic molecules. However, catalytic activity characteristic of ordered porosity has not yet been reported, because pores are so small that only very small organic molecules can enter the pores.

Both microporous manganese oxides[17–21] and the Mo-V oxide[22,23,27,28] were prepared by heating aqueous solutions of the metal precursors. In order to produce pure samples, conditions (pH, temperature, reaction time, and concentration of metal precursor) must be carefully controlled. Further investigation is needed to demonstrate the importance of these bifunctional (redox and microporosity) materials for understanding the formation mechanism of these microporous materials and for the development of a new strategy to form other microporous transition metal oxides.

3.4 MESOPOROUS TRANSITION METAL OXIDE MATERIALS

Ordered mesoporous crystalline metal oxides have been synthesised using template methods, which are generally divided into the 'soft template method' and the 'hard template method', depending on the nature of the templates.

3.4.1   Soft Template Method

In the early 1990, a strategy to prepare ordered mesoporous silica was reported.[4–7] By using an assembly of organic surfactants as templates, ordered hexagonal or cubic organic-silica hybrid materials were synthesised. After removing the templates, ordered hexagonal or cubic mesoporous silica was obtained (Figure 3.5).[29,30]

It was difficult to produce ordered mesoporous crystalline transition metal oxides because of the necessity of heat treatment for removing templates and crystallisation of wall transition metal oxides. Under calcination conditions, the mesoporous structure of transition metal oxides is less stable than that of silica. Redox reaction and sintering during the calcination destroy the mesoporous structure.

The first mesoporous crystalline transition metal oxide was reported in 1997 for manganese oxides using cetyltrimethylammonium bromide (CTAB) as a template.[31] A general method for producing mesoporous crystalline transition metal oxides by using a triblock copolymer nonionic surfactant as a template was reported in 1998.[32,33] Metal precursors, metal chlorides in most cases, were mixed with a poly(alkylene oxide) block copoymer such as $HO(CH_2CH_2O)_{20}(CH_2CH(CH_3)O)_{70}(CH_2\ CH_2O)_{20}H$

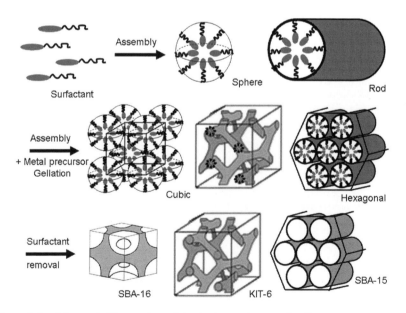

Figure 3.5   Schematic illustration of the 'soft template method'

[designated $(EO)_{20}(PO)_{70}(EO)_{20}$, BASF, Pluronic P-123], $HO(CH_2$-$CH_2O)_{106}(CH_2CH(CH_3)O)_{70}(CH_2$ $CH_2O)_{106}H$ (designated EO_{106}-$PO_{70}EO_{106}$, Pluronic P-127), and $HO(CH_2CH_2O)_{75}(CH_2CH$-$(CH_2CH_3)O)_{45}H$ (designated $EO_{57}BO_{45}$, Dow Chemical) (where EO is ethylene oxide, PO is propylene oxide and BO is 1-butene oxide) in alcohol solution. The resulting sol solution was gelled at a temperature, during which the metal precursor is hydrolysed and polymerised into a metal oxide network. Hexagonal and cubic mesophases are preferred when $(EO)_{20}(PO)_{70}(EO)_{20}$ and $EO_{57}BO_{45}$ are used as the template, respectively. Calcination of the solid produced results in removal of the template, and the amorphous metal oxide network is crystallised into crystalline metal oxide. The wall thickness of materials prepared by using a triblock copolymer is larger than that of materials prepared by using other ionic surfactants, and this method is suitable for preparing a variety of mesoporous crystalline transition metal oxides. Optimisation of the general methods (surfactants, additives, and removal of templates) have been reported.[29,30,34]

The disadvantage of this approach is that the resulting oxides are not highly crystallised because heat treatment cannot be conducted at a temperature above about 400 °C, where structural collapse occurs. This soft method has not been applied for ordered mesoporous crystalline late transition metal (such as Cu, Co, Ni and Fe) oxides. As a rare case, preparation of mesoporous nickel or iron oxide was reported. However, nickel oxide has an amorphous wall and iron oxide (crystalline γ-Fe_2O_3) has disordered wormhole mesopores.[64,65]

Recently, new crystallisation methods for mesoporous amorphous oxides using combined assembly by the soft and hard (CASH) chemistry method[45] and two-step wall reinforcing method[34–37] have been reported.

3.4.2 Hard Template Method

The other method for preparing ordered mesoporous materials is the so-called 'hard template method' using hard mesoporous silica or replicated carbon templates. The metal precursors are filled into hard templates. In this method, heat treatment can be performed at a higher temperature without structural collapse and highly crystallised materials can be obtained[29,30,34,66] (Figure 3.6).

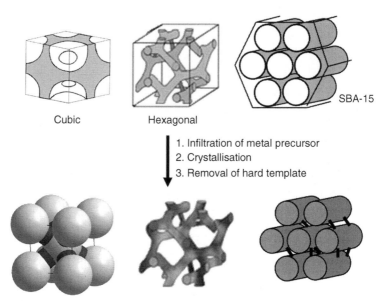

Figure 3.6 Schematic illustration of the 'hard template method'

The steps involved in this method are as follows: (1)The template materials are highly ordered two-dimensional hexagonal mesoporous silica such as SBA-15 and three-dimensional caged mesoporous silica. (2) Metal precursors are introduced by capillary force, chelation force, and hydrophilic affinity. (3) A thermal treatment is employed to crystallise the guest materials and to form the mesoporous pattern. (4) The template silica is etched by HF or NaOH to obtain self-standing ordered metal oxide guest frameworks with mesostructure.

In hexagonal mesoporous SBA-15, straight mesopores are connected three-dimensionally by some microporous bridges. Therefore, a three-dimensional assembly of nanowires with bridges can be obtained from the SBA-15 template.

Infiltration of a sufficient amount of metal precursors in the mesopores is necessary. The simplest way is to mix mesoporous silica in an ethanol solution of metal salts and then to evaporate the ethanol. The precursor molecules might enter the pores during the evaporation of ethanol by capillary action. The 'two solvent' impregnation method, in which a suspension of mesoporous silica in dry hexane is mixed with an aqueous solution of metal nitrate and then dried, was applied to complete the infiltration.[68,75,76] Modification of the mesopore surface to accelerate infiltration was reported. If the melting point of the metal precursor is lower than the decomposition temperature (55 °C and 75 °C for $Co(NO_3)_2 \cdot 6H_2O$), the neat metal precursor can infiltrate into the mesoporous silica by capillary action.[88]

Metal precursors can be calcined at a temperature higher than 400 °C, which is the temperature limit for the soft template method. The high temperature changes amorphous materials into crystalline materials. Porous single crystals, in which pores are in a single crystal, are frequently observed.

A limitation of the hard template method is that the resulting materials must be stable in HF or NaOH solution and that the precursors must not react with the silica template at a high temperature. For example, in the formation of lithium containing transition metal oxides, it is necessary to first form the transition metal oxide as a mesoporous solid to prevent reaction of the alkali metal Li with the silica template and then to react the mesoporous transition metal solid with a lithium source, such as LiOH.

These methods, structural characterisation, and applications are summarised in Table 3.2.

3.4.3 Mesoporous Oxides of Group 4 Elements (Ti, Zr)

Crystalline TiO_2 is most attractive because it has excellent performance in photocatalytic reactions and energy conversion materials.[38] Ordered crystalline mesoporous TiO_2 materials have attracted much attention because of the increased surface area of mesoporous materials.

The first ordered mesoporous crystalline TiO_2 (anatase) was obtained using a nonionic surfactant template. Depending on the type of template [$(EO)_{20}(PO)_{70}(EO)_{20}$ or $EO_{57}BO_{45}$], a hexagonal or cubic mesoporous structure was obtained.[32,33] Wall thickness (*ca* 5 nm) was greater than that when an ionic surfactant was used (maximum of a few nanometres) and the wall is therefore sufficiently stable for crystallisation.[39] Crystallite size (*ca* 3 nm) was smaller than the wall thickness, indicating that the mesoporous structure was built up from aggregation of small nanocrystallites. Thin films of mesoporous TiO_2 were obtained by controlling gelation and solvent evaporation conditions.[40–42] Control of the amount of water and humidity during the gelation is one of the most important factors for obtaining ordered mesoporous TiO_2. Furthermore, by increasing the amount of template, the mesopore structure changes from cubic to hexagonal.

Table 3.2 Precursors, templates, solidifications, template removal methods, crystal phase, and properties of ordered mesoporous crystalline materials

Metal	Precursor solution	Template[a]	Solidification gelation	Template removal	Crystal phase[b]	Properties and application[c]	Ref.
Ti	$TiCl_4$ + template in EtOH	$(EO)_{20}(PO)_{70}(EO)_{20}$	40 °C in air	Calc. (400 °C)	TiO_2 (anatase)	Pore: Hexagonal (XRD) Wall thickness: ~5 nm (TEM) Crystallite size: ~3 nm (PXRD) Crystallite size: ~3 nm (TEM) Surface area. 200 m² g⁻¹ (BET) Pore diameter: 6.5 nm (BJH) Porosity: ~50 %	[32,33]
Ti	$TiCl_4$ + template in EtOH	$EO_{57}BO_{45}$	40 °C in air	Calc. (400 °C)	TiO_2 (anatase)	Pore: Cubic ($Im\overline{3}m$) (XRD) Wall thickness: ~5 nm (TEM) Crystallite size: ~3 nm (PXRD) Crystallite size: ~3 nm (TEM) Surface area. 200 m² g⁻¹ (BET) Pore diameter: 6.8 nm (BJH) Porosity: ~50 %	[32,33]
Ti	$TiCl_4$ + template + EtOH + H_2O (1:0.005:40: 10 mol)	$(EO)_{106}(PO)_{70}(EO)_{106}$	25–35 °C under controlled relative humidity	Calc. (400 °C)	TiO_2 (anatase) (pdf 21-1272)	Pore: cubic ($Im\overline{3}m$) (SAXS) Wall thickness: ~10 nm (TEM)	[40–42]
Ti	$TiCl_4$ + template + EtOH + H_2O (1:0.09:40: 10 mol)	$(EO)_{20}(PO)_{70}(EO)_{20}$	25–35 °C under controlled relative humidity	Calc. (400 °C)	TiO_2 (anatase) (pdf 21-1272)	Pore: hexagonal ($Im\overline{3}m$) (SAXS) Wall thickness: ~10 nm (TEM)	[41]

	Precursor	Template	Condition	Calcination	Product	Characterization	Ref.
Ti	$TiCl_4$ (4.2 g) + template (1.7 g) + EtOH + HCl (12.1 M) ($Ti:H_2O$ = 1:6 mol)	$(EO)_{20}(PO)_{70}(EO)_{20}$	8 °C	Calc. (400 °C)	TiO_2 (anatase)	Pore: cubic ($Im\bar{3}m$) (SAXS)	[43,44]
Ti	$TiCl_4$ + $Ti(OEt)_4$ + template in THF (Ti:template = 1:2 mol)	PI-b-PEO (M_w: 27 220 g mol^{-1}, 16.7 wt% PEO)	Heated (up to 130 °C) in air	1. 700 °C in argon 2. 450 °C in air	TiO_2 (anatase) (pdf 21-1272)	Pore: hexagonal (SAXS); Wall thickness: 8 nm (TEM); Crystallite size: 10 nm (PXRD); Surface area. 89 m^2 g^{-1} (BET); Pore diameter: 2.3 nm (BJH)	[45]
Zr	$ZrCl_4$ + template in EtOH	$(EO)_{20}(PO)_{70}(EO)_{20}$	40 °C in air	Calc. (400 °C)	ZrO_2 (tetragonal)	Pore: Hexagonal (XRD); Wall thickness: ~6.5 nm (TEM); Crystallite size: ~2 nm (PXRD); Crystallite size: ~2 nm (TEM); Surface area. 150 m^2 g^{-1} (BET); Pore diameter: 5.8 nm (BJH); Porosity: ~45 %	[32,33]
Zr	$ZrCl_4$ + template in EtOH	$EO_{57}BO_{45}$	40 °C in air	Calc. (400 °C)	TiO_2 (anatase)	Pore: Cubic ($Im\bar{3}m$) (XRD); Wall thickness: ~6.5 nm (TEM); Crystallite size: ~2 nm (PXRD); Crystallite size: ~2 nm (TEM); Surface area. 150 m^2 g^{-1} (BET); Pore diameter: 6.2 nm (BJH); Porosity: ~45 %	[32,33]
Zr	$ZrCl_4$ + template + EtOH + H_2O (1:0.05:40: 20 mol)	$(EO)_{106}(PO)_{70}(EO)_{106}$	25–35 °C under controlled relative humidity	Calc. (400 °C)	ZrO_2 (tetragonal) (pdf 50-1089)	Pore: hexagonal ($P6m$) (SAXS); Crystallite size:4 nm (PXRD)	[40]

(continued overleaf)

Table 3.2 (*continued*)

Metal	Precursor solution	Template[a]	Solidification gelation	Template removal	Crystal phase[b]	Properties and application[c]	Ref.
Zr, Y	ZrCl$_4$ + YCl$_3$ + template + H$_2$O in EtOH (Zr:Y:template: H$_2$O:EtOH = 4:1:2.5:100:200)	(EO)$_{106}$(PO)$_{70}$ (EO)$_{106}$	20–23 °C under controlled relative humidity (50 %)	Calc. (400 °C)	YSZ (pdf 82-1245)	Pore: cubic($Im\bar{3}m$) (SAXS) Crystallite size: 7–9 nm (PXRD)	[47]
Zr, Ce	ZrCl$_4$ + CeCl$_3$ + template + H$_2$O in EtOH (Zr:Y:template: H$_2$O:EtOH = 4:1:2.5:100:200)	(EO)$_{106}$(PO)$_{70}$ (EO)$_{106}$	20–23 °C under controlled relative humidity (50 %)	Calc. (400 °C)	Ceria-zirconia (tetragonal)	Pore: cubic($Im\bar{3}m$) (SAXS) Crystallite size: 7–9 nm (PXRD)	[47]
Nb	NbCl$_5$ + template in EtOH	(EO)$_{20}$(PO)$_{70}$ (EO)$_{20}$	40 °C in air	Calc. (400 °C)	Nb$_2$O$_5$ (nucleation just started)	Pore: Hexagonal (XRD) Wall thickness: ~4 nm (TEM) Crystallite size: <1 nm (PXRD) Surface area. 196 m^2 g^{-1} (BET) Pore diameter: 5 nm (BJH) Porosity: ~50 %	[32,33]
Nb	NbCl$_5$ + template in CHCl$_3$ (Ti:template = 1:1.45 mol)	PI-*b*-PEO (M_w: 33 500 g mol^{-1}, 23 wt% PEO)	Heated (up to 130 °C) in air	1. 700 °C in argon 2. 450 °C in air	Nb$_2$O$_5$ (pdf 28-0317)	Pore: hexagonal (SAXS) Wall thickness: 6 nm (TEM) Crystallite size: 12 nm (PXRD) Surface area. 54 m^2 g^{-1} (BET) Pore diameter: 35 nm (BJH)	[45]
Nb	NbCl$_5$ + template in EtOH	PE-*co*-PB-*b*-PE	Dip coating under controlled relative humidity (15–20 %)	Calc. (600 °C)	Nb$_2$O$_5$ JCPFF (30-873)		[48]

Ta	TaCl$_5$ + template in EtOH	(EO)$_{20}$(PO)$_{70}$(EO)$_{20}$	40 °C in air	Calc. (400 °C)	Ta$_2$O$_5$ (nucleation just started)	Pore: Hexagonal (XRD) Wall thickness: ~4 nm (TEM) Crystallite size: <1 nm (PXRD) Surface area. 165 m^2 g^{-1} (BET) Pore diameter: 3.5 nm (BJH) Porosity: ~50 %	[32,33]
Ta	TaCl$_5$ (6 mmol) + template (1 g) + H$_2$O (6 mmol) in EtOH (10 g)	(EO)$_{20}$(PO)$_{70}$(EO)$_{20}$	40 °C in air	1. Calc. (500 °C) in air 2. Treated with bis(trimethylsiloxy)methylsilane (BTMS) at 70 °C 3. Calc. (860 °C) in air 4. Washed with NaOH (1.5 M) solution at 70 °C	Ta$_2$O$_5$ (orthorhombic)	Pore: Hexagonal (XRD) Surface area: 109 m^2 g^{-1} (BET) Pore diameter: 3.7 nm (BJH)	[51,53, 54]
Nb, Ta	NbCl$_5$ + TaCl$_5$ (Metal: 5.5 mmol) + template (1 g) + H$_2$O (18 mmol) in EtOH (10 g)	(EO)$_{20}$(PO)$_{70}$(EO)$_{20}$	40 °C in air	1. Calc. (450 °C) in air 2. Treated with furfuryl alcohol vapour at 200 °C 3. Calc. (550 °C) in vacuo 4. Calc. (650 °C) in He 5. Calc. (500 °C) in air	(NbTa)$_2$O$_5$	Pore: Hexagonal (XRD)	[57,58]

(continued overleaf)

Table 3.2 (continued)

Metal	Precursor solution	Template[a]	Solidification gelation	Template removal	Crystal phase[b]	Properties and application[c]	Ref.
Cr	Cr(NO$_3$)$_3$·3H$_2$O + template + ethylene glycol (EG) in PrOH (Cr:template:EG:PrOH = 134:1:1080:1115)	(EO)$_{106}$(PO)$_{70}$(EO)$_{106}$	1. 40–50 °C 2. 150 °C in air	1. Extracted with EtOH 2. Calc. (400 °C)	Cr$_2$O$_3$ (rhombohedral)	Pore: cubic ($Im\bar{3}m$) (SAXS) Wall thickness: 13.3 nm (TEM) Crystallite size: <1 nm (PXRD) Surface area. 96 m^2 g^{-1} (BET) Pore diameter: 7.9 nm (BJH) Catalysis	[62]
Cr	Kr$_2$Cr$_2$O$_7$ (10 wt%) + template (1 g) in H$_2$O (pH 1.5 with HCl)	SBA-15 (amine modified)	Calc. (>350 °C)	Washed with 10 % HF	Cr$_2$O$_3$ (rhombohedral)	Pore: Nanowires (TEM) Wall thickness: 13 nm (TEM) Surface area. 58 m^2 g^{-1} (BET) Pore diameter: 3.4 nm (BJH)	[67]
Cr	Cr(NO$_3$)$_3$ (8.0 g) in H$_2$O (3.7) + template (1 g) in hexane	KIT-6	Calc. (550 °C)	Washed with 10 % HF	Cr$_2$O$_3$ (pdf 01-1294)	Wall thickness: 6 nm (TEM) Surface area. 86 m^2 g^{-1} (BET) Pore diameter: 3.1 + 12.3 nm (BJH)	[68]
Cr	Cr(NO$_3$)$_3$ hydrate (1.25 mmol) + template (0.15 g) in EtOH (5 g)	SBA-15 (microwave-digested)	Calc. (550–650 °C)	Washed with 10 % HF or 2 M NaOH	Cr$_2$O$_3$	Pore: Nanowires (TEM) Surface area. 65 m^2 g^{-1} (BET) Pore diameter: 3.6 nm (BJH)	[69]
Cr	Cr(NO$_3$)$_3$ nonhydrate (0.6 g) + template (0.15 g) in EtOH (6.5 g)	SBA-15	Calc. (500 °C)	Washed with 10 % HF	Cr$_2$O$_3$ (rhombohedral)	Pore: Nanowires (TEM)	[70]
Cr	Cr(NO$_3$)$_3$ nonhydrate (0.6 g) + template (0.15 g) in EtOH (6.5 g)	KIT-6	Calc. (500 °C)	Washed with 10 % HF	Cr$_2$O$_3$ (rhombohedral)	Pore: Bicontinuous channel network (TEM)	[70]

Element	Precursor	Template	Treatment	Calcination/Wash	Product (pdf)	Characterization	Ref.
Mo	$MoCl_5$ + template in EtOH	PE-co-PB-b-PE	Dip coating under controlled relative humidity (15–20 %)	Calc. (450 °C)	MoO_3 (pdf 5-508)		[48]
W	WCl_6 + template in EtOH	$(EO)_{20}(PO)_{70}(EO)_{20}$	40 °C in air	Calc. (400 °C)	WO_3 (pdf 20-1324)	Pore: Hexagonal (XRD) Wall thickness: ~5 nm (TEM) Crystallite size: ~3 nm (PXRD) Surface area. 125 m² g⁻¹ (BET) Pore diameter: 5 nm (BJH) Porosity: ~48 %	[32,33]
W	WCl_6 (0.6 g) + template (20–25 wt%) in EtOH (6 ml)	PE-co-PB-b-PE	Dip coating under controlled relative humidity (5–10 %)	Calc. (550 °C)	WO_3 (pdf 43-1035)	Pore: Cubic (XRD, TEM) Wall thickness: ~5 nm (TEM) Crystallite size: 10–15 nm (PXRD) Pore diameter: 14 nm (TEM) **Electrochromic material**	[63]
W	$H_3PW_{12}O_{40}$	SBA-15 (amine modified)	Calc. (600 °C)	Washed with 10 % HF	WO_3 (orthorhombic) (pdf 20-1324)	Pore: Nanowires (TEM)	[72]
W	$H_3PW_{12}O_{40}$ (1.2 g) + template (0.4 g) in EtOH (1 ml)	KIT-6	Calc. (500 °C)	Washed with 2 M HF	WO_3 (monoclinic) (pdf 20-1324)	Crystallite size: 6.2 nm (PXRD) Surface area. 86 m² g⁻¹ (BET)	[73]

(continued overleaf)

Table 3.2 (continued)

Metal	Precursor solution	Template[a]	Solidification gelation	Template removal	Crystal phase[b]	Properties and application[c]	Ref.
Mn	Layered Mn(OH)$_2$ generated from MnCl$_2$ + NaOH was mixed with template in H$_2$O	CTAB				$Mn_2O_3 + Mn_3O_4$	Pore: Hexagonal (at high CTAB conc.) and cubic (at low CTAB conc.) (XRD) Pore size: 3.0 nm (BJH) Wall thickness: 1.7 nm **Catalyst, conductive material**	[31]
Mn	Mn(NO$_3$)$_3$ hydrate (1.25 mmol) + template (0.15 g) in EtOH (5 g)	SBA-15 (microwave-digested)	Calc. (550–650 °C)	Washed with 10 % HF or 2 M NaOH	$Mn_2O_3 + MnO_2 + Mn_3O_4$	Pore: Nanowires (TEM) Surface area. 103 m^2 g^{-1} (BET) Pore diameter: 3.4 nm (BJH)	[69]
Mn	Mn(NO$_3$)$_2$ (6 g) + template (1 g) in EtOH (25 ml)	KIT-6	Calc. (350 °C)	Washed with 2 M NaOH	β-MnO$_2$ (rutile structure) (pdf 24-735)	Wall thickness: 7.5 nm (TEM) Crystallite size: 6–26 nm (PXRD) Surface area: 64 m^2 g^{-1} (BET) Pore diameter: 4.9, 18.2 nm (BJH) **Lithium battery**	[74]
Mn	Mn(NO$_3$)$_2$ (7.5 g) + in H$_2$O (5 ml) + template (5 g) in hexane (200 ml)	KIT-6	Calc. (400 °C)	Washed with 2 M NaOH	β-MnO$_2$ (rutile structure) (pdf 24-735)	Wall thickness: 7.5 nm (TEM) Surface area. 127 m^2 g^{-1} (BET) Pore diameter: 3.7 nm (BJH) **Lithium battery**	[75]

Element	Precursor	Template	Calcination	Washing	Product	Characterization	Ref.
Mn, Li	$Mn(NO_3)_2$ (7.5 g) + in H_2O (5 ml) + template (5 g) in hexane (200 ml)	KIT-6	Calc. (600 °C)	1. Washed with 2 M NaOH 2. Calc. [280 °C in H_2/Ar (5/95)] 3. Mixed with LiOH 4. Calc. (350 °C)	$Li_{1.12}Mn_{1.88}O_4$	Wall thickness: 7 nm (TEM) Surface area: 90 m² g⁻¹ (BET) **Lithium battery**	[76]
Fe	$Fe(NO_3)_3$ hydrate (1.25 mmol) + template (0.15 g) in EtOH (5 g)	SBA-15 (microwave-digested)	Calc. (550–650 °C)	Washed with 10 % HF or 2 M NaOH	Fe_2O_3	Pore: Nanowires (TEM) Surface area: 137 m² g⁻¹ (BET) Pore diameter: 4.0 nm (BJH)	[69]
Fe	$Fe(NO_3)_3 \cdot 9H_2O$ (1.5 g) + template (1 g) in EtOH (20 ml)	KIT-6	Calc. (600 °C)	Washed with 2 M NaOH	α-Fe_2O_3	Wall thickness: ~7 nm (TEM) Surface area: 139 m² g⁻¹ (BET) Pore diameter: 3.9 nm (BJH) Crystallite size: 6 nm (PXRD)	[81,82, 91]
Co	$Co(NO_3)_2$ hydrate (1.25 mmol) + template (0.15 g) in EtOH (5 g)	SBA-15 (microwave-digested)	Calc. (550–650 °C)	Washed with 10 % HF or 2 M NaOH	Co_3O_4	Pore: Nanowires (TEM) Surface area: 82 m² g⁻¹ (BET) Pore diameter: 3.4 nm (BJH)	[69]
Co	$Co(NO_3)_2$ hydrate (1.25 mmol) + template (0.15 g) in EtOH (5 g)	SBA-16 (microwave-digested)	Calc. (550–650 °C)	Washed with 10 % HF or 2 M NaOH	Co_3O_4	Pore: Nanosphere (TEM) Surface area: 92 m² g⁻¹ (BET) Pore diameter: 6.5 nm (BJH)	[69,83]
Co	$Co(NO_3)_2 \cdot 6H_2O$ (0.6 g) + template (0.15 g) in EtOH (6.5 g)	SBA-15	Calc. (500 °C)	Washed with 10 % HF	Co_3O_4 (pdf 78-1970)	Pore: Nanowires (TEM)	[70]

(continued overleaf)

Table 3.2 (continued)

Metal	Precursor solution	Template[a]	Solidification gelation	Template removal	Crystal phase[b]	Properties and application[c]	Ref.
Co	Co(NO$_3$)$_2$·6H$_2$O (0.6 g) + template (0.15 g) in EtOH (6.5 g)	KIT-6	Calc. (500 °C)	Washed with 10 % HF	Co$_3$O$_4$ (pdf 78-1970)	Pore: Bicontinuous channel network (TEM)	[70]
Co	Co(NO$_3$)$_2$ (0.8 M) + template (0.5 g) in EtOH (5 ml)	Vinyl-functionalised cubic ($Ia3d$) mesoporous silica	Calc. (450 °C)	Washed with 2 M NaOH	Co$_3$O$_4$ (pdf 42-1467)	Wall thickness: 3 nm; Surface area: 122 m^2 g^{-1} (BET); Pore diameter: 3.8 nm (BJH); Crystallite size: 7.8 nm (PXRD)	[84]
Co	Co(NO$_3$)$_2$·6H$_2$O (1 g) + template (2 g) in EtOH (20 ml)	SBA-15	Calc. (500 °C)	Washed with 10 % HF	Co$_3$O$_4$ (cubic spinel)	Pore: Nanowires ($P6mm$) (TEM); Wall thickness: 7–8 nm (TEM); Surface area: 90 m^2 g^{-1} (BET); Pore diameter: 4 nm (BJH); Lithium battery electrode	[85]
Co	Co(NO$_3$)$_2$·6H$_2$O (1 g) + template (2 g) in EtOH (20 ml)	KIT-6	Calc. (500 °C)	Washed with 10 % HF	Co$_3$O$_4$ (cubic spinel)	Pore: Bicontinuous channel network ($Ia\overline{3}d$) (TEM); Wall thickness: 7–8 nm (TEM); Surface area: 130 m^2 g^{-1} (BET); Pore diameter: 4 nm (BJH); Lithium battery electrode	[85]
Co	Co(NO$_3$)$_2$·6H$_2$O (0.8 M) + template (0.5 g) in EtOH (5 ml)	KIT-6 with different pore size	Calc. (450 °C)	Washed with 2 M NaOH	Co$_3$O$_4$	Surface area: 70-153 m^2 g^{-1} (BET); Pore diameter: 3–6 nm (BJH)	[86,87]

Co	Co(NO₃)₂ (1 mmol) + template (0.15 g) was ground for a few minutes	SBA-16	Calc. (500 °C)	Washed with 10 % HF	Co_3O_4	Surface area. $122\,m^2\,g^{-1}$ (BET) Pore diameter: 6.6 nm (BJH)	[88]
Co	Co(NO₃)₂ (1 mmol) + template (0.15 g) was ground for a few minutes	FDU-12	Calc. (500 °C)	Washed with 10 % HF	Co_3O_4	Surface area. $151\,m^2\,g^{-1}$ (BET) Pore diameter: 6.1 nm (BJH)	[88]
Co, Fe	1. Co(NO₃)₂·6H₂O (0.8 M) + template (0.5 g) in EtOH (5 ml) and heated at 200 °C 2. Co(NO₃)₂·6H₂O + Fe(NO₃)₃·9H₂O (Co:Fe = 1:2) + template (0.5 g) in EtOH (5 ml)	KIT-6	Calc. (450 °C)	Washed with 2 M NaOH	$Co_3O_4 + CoFe_2O_4$		[87]
Co, Li	Co(NO₃)₂ (1 g) + template (2 g) in hexane (20 ml)	SBA-15	Calc. (500 °C)	1. Washed with aqueous HF 2. Mixed with LiOH·H₂O (0.6 g) in EtOH (20 ml) 3. Calc. (400 °C)	$Li_{0.95}CoO_2$	Pore: Nanowire (TEM) Wall thickness: 10 nm (TEM) Surface area. $70\,m^2\,g^{-1}$ (BET) Pore diameter: 3.8 nm (BJH)	[89]

(continued overleaf)

Table 3.2 (*continued*)

Metal	Precursor solution	Template[a]	Solidification gelation	Template removal	Crystal phase[b]	Properties and application[c]	Ref.
Co, Li	$Co(NO_3)_2$ (1 g) + template (2 g) in hexane (20 ml)	KIT-6	Calc. (500 °C)	1. Washed with aqueous HF 2. Mixed with $LiOH \cdot H_2O$ (0.6 g) in EtOH (20 ml) 3. Calc. (400 °C)	$Li_{0.95}CoO_2$	Pore: Cubic ($Ia\overline{3}d$) (TEM) Wall thickness: 8 nm (TEM) Surface area: 92 $m^2 g^{-1}$ (BET) Pore diameter: 3.7 nm (BJH)	[89]
Co, La	$La(NO_3)_3$ (4 mmol) + $Co(NO_3)_2$ (4 mmol) + citric acid (8 mmol) template (1 g) in EtOH (15 ml) + H_2O (5 ml)	Mesoporous vinylsilica ($Ia3d$)	Calc. (700 °C)	Washed with 2 M NaOH	$LaCoO_3$ (pdf 48-0123)	Pore: Cubic ($Ia\overline{3}d$) (SAXS + TEM) Surface area: 97 $m^2 g^{-1}$ (BET) Pore diameter: 6 nm (BJH) **Methane oxidation catalyst**	[90]
Ni	$Ni(NO_3)_2$ hydrate (1-.25 mmol) + template (0.15 g) in EtOH (5 g)	SBA-15 (microwave-digested)	Calc. (550–650 °C)	Washed with 10 % HF or 2 M NaOH	NiO	Pore: Nanowires (TEM) Surface area: 56 $m^2 g^{-1}$ (BET) Pore diameter: 3.5 nm (BJH)	[69]
Ni	$Ni(NO_3)_2 \cdot 6H2O$ (4 g) + template (0.15 g) in EtOH (150 ml)	KIT-6	Calc. (550 °C)	Washed with 2 M NaOH	NiO	Pore: Cubic (TEM) Surface area: 82–109 $m^2 g^{-1}$ (BET) Pore diameter: 3.3 + 11 nm (BJH)	[93]
Cu	$Cu(NO_3)_2$ (0.4 M) + template (2 g) in EtOH (20 ml)	CMK-3	Calc. (400 °C in N_2)	Calc. (550 °C in air)	CuO (pdf 48-1548)	Pore: Hexagonal (SXAS) Wall thickness: 3.7 nm (TEM) Surface area: 149 $m^2 g^{-1}$ (BET) Pore diameter: 5.5 nm (BJH)	[94]

| Ce | 1. CeCl$_3$ (2 g) + template (0.5 g) in H$_2$O (1 g) 2. Heated at 100 °C 3. Reaction with NH$_3$ gas | SBA-15 | Calc. (700 °C) | Washed with 2 M NaOH | CeO$_2$ | Pore: Nanowires (TEM) | [95] |
| Ce | 1. CeCl$_3$ (2 g) + template (0.5 g) in H$_2$O (1 g) 2. Heated at 100 °C 3. Reaction with NH$_3$ gas | Cubic silica (Ia$\bar{3}$d) | Calc. (700 °C) | Washed with 2 M NaOH | CeO$_2$ | Surface area. 198 m^2 g^{-1} (BET) Pore diameter: 3.5 nm (BJH) | [95] |

[a] (EO)$_{20}$(PO)$_{70}$(EO)$_{20}$: HO(CH$_2$CH$_2$O)$_{20}$(CH$_2$CHCH$_3$O)$_{70}$(CH$_2$CH$_2$O)$_{20}$H (BASF, Pluronic P-123); EO$_{57}$BO$_{45}$: HO(CH$_2$CH$_2$O)$_{75}$(CH$_2$CH(CH$_2$-CH$_3$)O)$_{45}$H (Dow Chemical); (EO)$_{106}$(PO)$_{70}$(EO)$_{106}$: HO(CH$_2$CH$_2$O)$_{106}$(CH$_2$CHCH$_3$O)$_{70}$(CH$_2$CH$_2$O)$_{106}$H (BASF, Pluronic P-127); PI-b-PEO: poly(iso-prene-$block$-ethylene oxide); PE-co-PB-b-PE: poly(ethylene-co-butylene)-$block$-poly(ethylene oxide) (KLE); SBA-15: hexagonal ($P6mm$) mesoporous silica, KIT-6: cubic (Ia$\bar{3}$d) mesoporous silica; CTAB: cetyltrimethylammonium bromide; SBA-16: cubic ($Im\bar{3}m$) mesoporous silica; FDU-12 cubic ($Fm\bar{3}m$) mesoporous silica; CMK-3: mesoporous carbon prepared by SBA-15 as a template.
[b] pdf: The Powder Diffraction File, Joint Committee on Powder Diffraction Standards (JCPDS), International Centre for Diffraction Data.
[c] Crystallite sizes were estimated from X-ray diffraction broadening using the Scherrer equation and/or TEM. Surface areas were calculated by the Brunauer–Emmett–Teller (BET) method. Pore sizes were calculated by the Barrett–Joyner–Halenda (BJH) method.

The above-mentioned ordered mesoporous TiO_2 (anatase) was prepared by calcination at a temperature below 400 °C. The temperature is crucial; higher temperature destroys the mesopore structure due to crystal growth and lower temperature is not sufficient to obtain crytalline TiO_2.[43,44,46] A new technique to overcome this problem using olefin containing a diblock copolymer, poly(isoprene-*block*-ethylene oxide) (PI-*b*-PEO), as a template has been reported.[45] By heating PI-*b*-PEO in an inert gas, PEO is easily decomposed on heating, whereas the more thermally stable PI, containing two sp^2 carbons per monomer unit, is converted to a sturdy amorphous carbon material. This *in situ* prepared carbon is sufficient to act as a rigid support to the mesostructured oxide walls, preventing collapse at the temperature required to obtain highly crystallised materials. The produced carbon materials can be easily removed by calcination in the presence of oxygen without destroying mesoporous TiO_2. This method combines the so-called 'soft' organic template and 'hard' *in situ* generated carbon template, and the authors referred to this as the CASH chemistry method.

Ordered mesoporous ZrO_2 was prepared by employing a similar method using nonionic surfactant templates.[32,33,40] Similar to TiO_2, the ZrO_2 wall collapses by increasing the heating temperature due to crystal growth. By adding yttrium (Y) or cerium (Ce), more stable ordered cubic mesoporous yttria-stabilised zirconia (YSZ) or ceria-zirconia could be obtained in thin film form, and they were stable at up to 700 °C.[47]

3.4.4   Mesoporous Oxides of Group 5 Elements (Nb, Ta)

Ordered mesoporous Nb_2O_5 was first prepared by using the nonionic surfactant $(EO)_{20}(PO)_{70}(EO)_{20}$ templating method.[32,33] $NbCl_5$ was hydrolysed in the presence of $(EO)_{20}(PO)_{70}(EO)_{20}$ in ethanol. After gelation, the template was removed by calcination at 400 °C in air. The produced material had an ordered mesoporous structure, though crystallinity was low.

Similar to TiO_2, it is generally difficult to achieve both ordered mesoporous structure and crystallinity. In a thin film, mesoporous crystalline Nb_2O_5 was obtained.[48] The thin film was formed on an indium-doped tin oxide (ITO) electrode, and an application of this mesoporous Nb_2O_5-coated electrode as a matrix in assembling proteins has been reported.[92] Recently, ordered hexagonal mesoporous crystalline Nb_2O_5 has been prepared by using the CASH chemistry method.[45]

Ordered mesoporous Ta_2O_5 was first prepared by using the nonionic surfactant $(EO)_{20}(PO)_{70}(EO)_{20}$ templating method.[32,33] The crystallinity of the produced mesoporous materials was low. Several optimisations of production of mesoporous tantalum (Ta) oxide and magnesium (Mg)-containing Mg-Ta oxides have been reported. However, these mesoporous compounds have amorphous walls.[49–52] Although the walls of these oxides were amorphous, both tantalum oxide and Mg-Ta mixed oxides exhibited higher photochemical activity for overall water decomposition than did crystallised nonporous Ta_2O_5 and $MgTa_2O_6$, respectively.[49,52] The advantage of the mesoporous structure originates from the very thin walls, the distance for the photoexcited electrons and holes to reach the surface being much shorter than that in nonporous samples. Crystallisation of the wall was achieved by reinforcement of the wall. Amorphous mesoporous tantalum oxide was reacted with bis(trimethylsiloxy)methylsilane ($Me_3SiOSiHMeOSiMe_3$; BTMS) or 1,1,1,3,4,4,4-heptamethyltrisiloxane, and the silicon species on the wall was condensed to amorphous SiO_2 to reinforce the wall.[53,54] The SiO_2-covered mesoporous tantalum oxide was successfully crystallised without collapse of the mesopore structure. By washing with alkaline solution, the amorphous SiO_2 covering was removed and mesoporous crystalline Ta_2O_5 was produced. The mesoporous crystalline Ta_2O_5 exhibits better photocatalytic activity than that of an amorphous one. Generally, high crystallinity, *i.e.* lower density of lattice defects to reduce recombination of the photoexcited electrons and holes intended to participate in the catalytic reaction, is suitable for photocatalyst design.[55,56] Therefore, crystalline mesoporous oxides have attracted much attention as photochemical catalysts.

Hexagonal mesoporous crystalline Nb-Ta mixed oxide was also prepared by using the wall reinforcing method.[58] Hexagonal mesoporous amorphous Nb-Ta mixed oxide was prepared by using the nonionic surfactant templating method.[57] Furfuryl alcohol was incorporated into the mesoporous amorphous Nb-Ta oxide and was carbonised in the pores. In the presence of the amorphous carbon, the amorphous Nb-Ta oxide was crystallised to $(NbTa)_2O_5$ under an inert atmosphere without collapse of the mesoporous structure. The amorphous carbon can be easily removed by calcination in the presence of oxygen to produce hexagonal mesoporous $(NbTa)_2O_5$. This mesoporous $(NbTa)_2O_5$ shows an interesting feature. In other mesoporous materials, crystallite sizes are equal to or smaller than wall thickness and aggregates of small crystallites from the mesoporous structure. Therefore, crystallite sizes should not exceed wall size in order to maintain the mesoporous structure. In mesoporous $(NbTa)_2O_5$, a crystallite is larger than a mesopore, and mesopores are present in a crystal, which has been also reported as unordered wormhole mesoporous $(NbTa)_2O_5$ and $Zr_6Nb_2O_{17}$.[59–61]

3.4.5   Mesoporous Oxides of Group 6 Elements (Cr, Mo, W)

Cubic ordered mesoporous crystalline Cr_2O_3 was prepared using a non-ionic surfactant template. A combination of $Cr(NO_3)_3$ and ethylene glycol (EG) in propanol (PrOH) solution was essential to produce the desired ordered mesoporous Cr_2O_3.[62] The prepared mesoporous Cr_2O_3 exhibits high efficiency as a catalyst for decomposition of volatile organic compounds (VOCs).

Well-ordered mesoporous crystalline Cr_2O_3 was prepared by using a variety of mesoporous silica templates (SBA-15, KIT-6).[67–70] Meso-pores are in a Cr_2O_3 single crystal and the whole particle is a three-dimensional porous single crystal, with the crystal structure of the walls of the mesopores being uniformly oriented. Thus prepared mesoporous Cr_2O_3 single crystal shows a greater capacity and better sustained rever-sibility than does bulk Cr_2O_3.[71]

Mesoporous crystalline MoO_3 and WO_3 were obtained in thin films.[48,63] Ordered mesoporous WO_3 was first produced using the soft template method.[32,33,63] By changing formation conditions, cubic and hexagonal mesoporous structures were obtained. The hard template method using SBA-15 or KIT-6 produced ordered mesoporous crystalline WO_3 materials. With SBA-15 as a template, a porous single crystal was formed,[72] while polycrystalline porous WO_3 materials were formed with a KIT-6 template,[73] although reaction conditions were similar.

3.4.6   Mesoporous Oxides of Group 7 Elements (Mn)

The first mesoporous crystalline manganese oxide was prepared by oxi-dation of a Mn(OH) nanosheet in the presence of an ionic organic surfactant (CTAB).[31]

Recently, synthesis of ordered mesoporous crystalline β-MnO_2 using KIT-6 as a hard template has been reported.[74,75] Because of the large surface area and accessibility to the crystalline wall with the ordered mesoporous structure, mesoporous β-MnO_2 facilitates fast transport of electrolytes with lithium ions. Furthermore, a thin pore wall reduces the lithium ion diffusion path in the crystal. Therefore, ordered β-MnO_2 is a promising cathode material for rechargeable lithium-ion batteries. Ordered mesoporous spinel-type $Li_{1.12}Mn_{1.88}O_4$ was synthesised by the two-step method. First, ordered mesoporous Mn_2O_3 was prepared using KIT-6 as a hard template. Then the produced Mn_2O_3 was reduced to Mn_3O_4 and further converted to spinel-type $Li_{1.12}Mn_{1.88}O_4$ by solid-state reaction without collapse of the ordered mesoporous structure.[76]

In contrast to Cr_2O_3 materials, manganese oxide mesopores are constructed by assembly of nanocrystallites, and crystallite sizes are smaller than or equal to the wall thickness.

3.4.7   Mesoporous Oxides of Elements of Groups 8–11 (Fe, Co, Ni, Cu)

Mesoporous crystalline late transition metal oxides were prepared by hard templating methods. Mesoporous iron oxides have been prepared using soft templating methods. However, such methods invariably lead to mesoporous materials with amorphous walls.[65,77–80] Ordered mesoporous crystalline α-Fe_2O_3 was prepared by using the hard templating method with KIT-6.[81,82] The α-Fe_2O_3 mesopores have a near-single crystal-like arrangement of atoms within the walls. The α-Fe_2O_3 could be reduced to Fe_3O_4 under an H_2/Ar atmosphere and then oxidised to γ-Fe_2O_3 in the presence of air while preserving the ordered mesostructure and crystalline walls throughout.[91] Such solid-state transformation demonstrates the stability of the mesoporous structure to phase transition.

Ordered mesoporous Co_3O_4 was prepared by using hard templates such as SBA-15, FDU-12, KIT-6 and CMK-4.[69,70,83–88] Both mesoporous single crystal Co_3O_4,[69,88] in which mesopores are in a single crystal, and polycrystallite Co_3O_4,[84] with nanocrystallite-constructed mesopore wall, were reported. Because the melting point (55 °C) of the cobalt precursor ($Co(NO_3)_2 \cdot 6H_2O$) is lower than the decomposition temperature (75 °C), simple infiltration of $Co(NO_3)_2 \cdot 6H_2O$ into the mesoporous silica template was possible by mixing both solid $Co(NO_3)_2 \cdot 6H_2O$ and the template and then heating above the melting point of $Co(NO_3)_2 \cdot 6H_2O$. Liquid $Co(NO_3)_2 \cdot 6H_2O$ was incorporated into the mesopores by capillary action.[88]

Mesoporous Co_3O_4 showed superior capacity at high rates for lithium battery applications compared with bulk materials.[85] Mesoporous

Co_3O_4 showed good catalytic performance for CO oxidation, similar to that of the best reported catalysts.[86]

Mesoporous mixed metal oxides such as $CoFe_2O_4/Co_3O_4$ composite,[87] $LiCoO_2$,[89] and $LaCoO_3$[90] were obtained by using similar hard templating methods. Mesoporous crystalline $LiCoO_2$ was prepared from mesoporous Co_3O_4 by solid-state reaction with LiOH without collapse of mesoporous structure. The starting mesoporous Co_3O_4 and produced $LiCoO_2$ were mesoporous single crystals in which pores are in a single crystal, indicating that the single crystal structure is maintained during the lithium insertion reaction.[89] Mesoporous perovskite-type $LaCoO_3$ has also been prepared.[90] Mixing of citric acid as a chelating agent and water is necessary to form pure $LaCoO_3$ perovskite phase. Large surface area ($97\,m^2\,g^{-1}$) was achieved by introducing mesopores in $LaCoO_3$.

Ordered mesoporous crystalline NiO was prepared by using the hard template method.[69,93] Mesoporous crystalline NiO with a bimodal pore size distribution can be produced.[93]

Ordered mesoporous crystalline CuO was produced by using a 'hard' carbon template (CMK-3) prepared by using an SBA-15 template. This two-step templating method produced well-ordered hexagonal (similar to SBA-15) mesoporous CuO material.[94]

3.4.8   Mesoporous Oxides of Lanthanide Elements (Ce)

Ordered mesoporous nanocrystalline CeO_2 was obtained by using silica hard templates.[95] $CeCl_3$ was incorporated into the mesoporous silica and then converted to cerium hydroxides by reacting with ammonia vapour. Calcination of cerium silica composite and removal of the silica using NaOH solution produced mesoporous CeO_2.

3.5   MACROPOROUS MATERIALS

Ordered macroporous materials with pore sizes of more than 50 nm appeared in the late 1990s with the development of a method using colloidal crystals of monodisperse spheres as a new template.[8,9] The walls of macroporous materials are larger than those of mesoporous materials, and a number of well-ordered macroporous crystalline transition metal oxides have been prepared.[96–101] The preparation method includes the following steps: (1) a colloidal crystal template is prepared by ordering a monodisperese sphere, *e.g.* PS, PMMA or SiO_2, into a face-centered close-packed array (opal structure); (2) interstices in the colloidal crystal are then filled with a liquid metal precursor solution, either neat or in a solvent, which solidify in voids, resulting in an intermediate composite structure; and (3) an ordered foam is produced after removing the template by calcination or extraction. The ordered (inverse opal) structures synthesised using this method consist of a skeleton surrounding uniform close-packed macropores (Figure 3.7). The macropores are three-dimensionally interconnected through windows that form as a result of the contact between the template spheres prior to infiltration of the metal precursor solution.

In order to produce a three-dimensionally ordered macroporous structure, the metal precursor solution should fulfil the following criteria: (1) Sufficient metal should be present in the voids to form a macroporous wall, and the metal concentration should therefore be high. (2) Reactivity of the metal precursor should be mild so that it can infiltrate the voids. If the metal precursor reacts with a functional group on the surface of the template or moisture in the air before it infiltrates the voids, an ordered porous structure cannot be obtained. (3) The metal precursor should be solidified in the voids before the template is removed and the produced

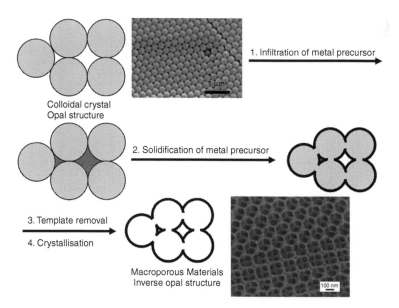

Figure 3.7 Colloidal crystal templating method to produce three-dimensionally ordered macroporous materials

solid should be stable during removal of the template. Polymer spheres, PS and PMMA, are removed by calcination or extraction with organic solvents. Silica spheres are removed by extraction with HF solution or NaOH solution. The solid walls should be stable under these conditions. (4) If the solid walls are amorphous, the amorphous walls should be crystallised without collapse of the macroporous structure.

Inverse opal structures have been classified into three structures, the so-called 'residual volume structure', 'shell structure' and 'skeleton structure'. The residual volume structure is a perfect inverse opal structure, which can be produced if the whole space among the opal spheres is completely filled by the product materials. If the space is incompletely filled, the surface of the sphere template is covered by the product materials, and a shell structure is generated. Most amorphous compounds tend to form a shell structure. On the other hand, crystalline compounds tend to form a skeleton structure.

The skeleton structure consists of strut-like bonds and vertices, with the struts connecting two kinds of vertices. The two vertices are replicas of the former octahedral and tetrahedral voids of the opal structure (Figure 3.8). These struts and vertices form a CaF_2 lattice, in which the

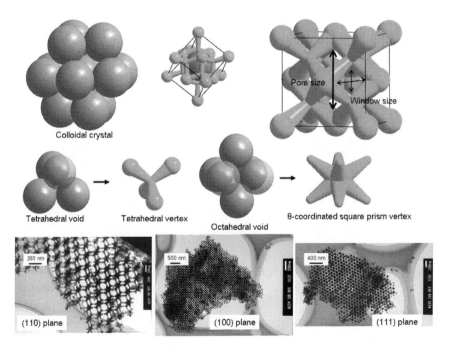

Figure 3.8 Structure of inverse opals with skeleton structure

8-coordinated square prism calcium (former octahedral voids of the opal) vertex is larger than the tetrahedral fluorine (former tetrahedral voids of the opal) vertex. The views towards the (111), (100) and (110) planes of the skeleton structure preset the hexagonal, square and lozenge arrangements, respectively.

3.5.1 Macroporous Monometal Oxides

The most successful metal precursors are alkoxides of Ti and Zr, because their reactivity is easily controllable. They are liquids at room temperature and can be used without a solvent, and the metal concentration is high. After infiltration into the voids of templates, they can be solidified *via* a sol-gel transformation. Therefore, well-ordered macroporous crystalline TiO_2 and ZrO_2 can often be prepared.

On the other hand, most of the other transition metal alkoxides react so quickly that the reaction cannot be controlled and they cannot be infiltrated into the voids of the template. Furthermore, obtaining alkoxide precursors of transition metals and lanthanide metals is difficult and expensive. Commercially available common salts of these transition metals are often not suitable for starting materials due to their low melting points. Common transition metal salts melt at a temperature at which the template polymer decomposes and therefore do not form an ordered macroporous structure. For such metals, two-step or *in situ* solidification before removal of the polymer template is necessary. Solidification of the transition metal salts (nitrate or acetate) by reacting oxalate[138,145,147] in the void to form oxalate salts before removal of the polymer template is often a good method (two-step solidification method) for producing ordered macroporous materials. These solids can be converted to metal oxides without melting. Solidification can be performed also *in situ* by heating (*in situ* method). The metal salts (nitrates or acetates) can infiltrate the voids with EG,[139] lactic acid,[148,149] citric acid[150,151] or ethylenediaminetetraacetic acid (EDTA).[152] By heating the template with the infiltrated materials, the metal salts react with the EG, citric acid or EDTA to form solid materials such as oxalate salts, citric acid salts or EDTA salts, respectively, which can be converted to the desired metal oxide without melting.

Table 3.3 summarises the precursors, templates, solidification methods, template removal methods, crystal phases and structural characterisation of ordered macroporous crystalline monotransition metal oxides.

Table 3.3 Precursors, templates, solidifications, template removal methods, crystal phase and properties of ordered macroporous crystalline monometal oxides

Metal	Precursor solution	Template[a]	Solidification	Template removal	Crystal Phase[b]	Properties[c] and application	Ref.
Ti	Ti(OEt)$_4$	PS (470 nm)	Dry	Calc. (575 °C) in air flow	TiO$_2$ (anatase)	Surface area: 38 m^2 g^{-1} (BET) Wall thickness: 8–13 nm (TEM) Crystallite size: 26–30 nm (PXRD), 13 nm (BET), 9–13 nm (TEM)	[8]
Ti	Ti(OPr)$_4$ (100–20 vol%) in ethanol	PS (180, 1460 nm)	Air dry	Calc. (450 °C)	TiO$_2$ (anatase)		[9]
Ti	Ti(OBu)$_4$ (40 vol%) in alcohol	PS (203 nm)	Heating at 160 °C	Calc. (450 °C) in air	TiO$_2$	Pore diameter: 125 nm (SEM)	[102]
Ti	Ti(OEt)$_4$	PS (421, 697 nm)	Air dry	Extraction (THF-acetone)	Amorphous	Surface area: 157 m^2 g^{-1} (BET) Wall thickness: 37-58 nm (TEM)	[103]
Ti	Ti(OEt)$_4$	PS (421, 697 nm)	Air dry	Calc. (575 °C)	TiO$_2$ (anatase) (pdf 21-1272)	Surface area: 50 m^2 g^{-1} (BET) Wall thickness: 24-36 nm (TEM) Crystallite size: 20-35 nm (TEM)	[103]

	Precursor	Template	Drying	Calcination	Product	Properties	Ref.
Ti	Ti(OEt)₄	PS (421, 697 nm)	Air dry	Calc. (1000 °C)	TiO_2 (anatase + rutile)	Surface area: 18 m² g⁻¹ (BET) Wall thickness: 19–67 nm (TEM) Crystallite size: 18–99 nm (TEM)	[103]
Ti	Ti(OEt)₄	PS (400 nm)	Air dry	Calc. (575 °C) in air	TiO_2	Pore diameter: 330 nm (SEM)	[104]
Ti	Ti(OEt)₄ in EtOH	PS (778, 3190 nm)	Air dry	Calc. (600–800 °C) in air	TiO_2	Pore shrinkage: 25–30 % (SEM)	[105]
Ti	Ti(OiPr)₄ in iPrOH	PS (640 nm) coated with polyelectrolyte multilayer	Air dry	Calc. (500 °C)		Pore diameter: 414–567 nm (SEM) Wall thickness: 13–42 nm s (SEM)	[106]
Ti	Ti(OEt)₄ in H₂O and EtOH	SiO₂ (300, 351 nm)	Reflux under N₂	Extraction with HF	TiO_2 (1025 °C)		[107]
Ti	Ti(OPr)₄ (20–100 vol%), Ti(OiPr)₄, or Ti(OEt)₄ in EtOH	PS (180–1460 nm)	Air dry	Calc. 450 °C	TiO_2 (anatase)		[108]
Ti	Ti(OEt)₄ and EtOH (1:50–1:100 vol ratio)	PS	Heated at 40–50 °C	Calc (540 °C) in air	TiO_2		[109]
Ti	Ti(OiPr)₄ (80 vol%) in EtOH	PS (270 nm)	Air dry	Calc. (300–700 °C)	TiO_2 (anatase)		[110, 111]
Ti	Ti(OPr)₄ (5 ml), HCl (1 ml), EtOH (5 ml) and H₂O (2 ml)	PMMA (310, 375, 425 nm)	Air dry	Calc. (400 °C)	TiO_2 (anatase)	Pore diameter: 250–310 nm (SEM) Crystallite size: 10 nm (PXRD)	[112]

(continued overleaf)

Table 3.3 (*continued*)

Metal	Precursor solution[a]	Template[a]	Solidification	Template removal	Crystal Phase[b]	Properties[c] and application	Ref.
Ti	Ti(OiPr)$_4$, EtOH, diethanolamine, and H$_2$O (1:40:0.6:3.3 vol ratio)	SiO$_2$ (309 nm)	Air dry	Extraction with NaOH (20 %) solution at 60 °C	TiO$_2$ (anatase) + Ti$_5$O$_9$ (520 °C)		[113]
Ti	TiCl$_3$ (50 mM) aqueous solution (pH 2.4)	Carboxylate-modified PS (243 nm)	Anodic oxidative deposition	Calc. (450 °C)	TiO$_2$ (anatase)	Pore shrinkage: 30 % (SEM) **Lithium battery**	[114]
Ti	Ti(OiPr)$_4$ in iPrOH	Carboxylate-modified PS (243 nm)	Air dry	Calc. (575 °C)	TiO$_2$ (anatase)	Pore shrinkage: 30 % (SEM) Crystallite size: 10–20 nm (TEM, PXRD) **Lithium battery**	[114]
Ti	Ti(OiPr)$_4$ (50 vol%) in iPrOH	PMMA (800 nm)	Air dry	Calc. (600 °C) in air flow	TiO$_2$	Surface area: 49 m^2 g^{-1} (BET) Pore diameter: 630 nm (SEM) Wall thickness: 60 nm (SEM) Crystallite size: 26 nm (PXRD)	[115]
Ti	Ti(OBu)$_4$ and EtOH (3:4)	PS-carboxyl groups on surface	Air dry	Calc. (700 °C)	TiO$_2$ (anatase) 500 °C TiO$_2$ (rutile) 700 °C		[116]

	Precursor	Template	Processing	Calcination	Product	Properties	Ref.
Ti	Ti(OiPr)$_4$ (1.2 w/v%) and HNO$_3$ (0.12 %) in EtOH	Carboxylate-modified PS (150, 243 nm)	1. Air dry to produce TiO$_2$ seed 2. (NH$_4$)$_2$TiF$_6$ (0.2M) + boric acid (0.25 M) at 51 °C (pH 3)	Calc. (400 °C)	TiO$_2$ (anatase)		[117]
Ti	Ti(OiPr)$_4$ + EtOH (1:1)	PS (100, 400 nm)	Heated at 80 °C	Calc. (450 °C) in air	TiO$_2$ (anatase)	Pore diameter: 250 nm (SEM) Crystallite size: 10 nm **Solar cell**	[118]
Ti	Ti(OBu)$_4$	PS (130, 150, 180, 210, 300, 380 nm)	Air dry	Calc. (450 °C)	TiO$_2$ (anatase)	Pore shrinkage:: 20 % (SEM) Crystallite size: 6–12 nm (TEM) **Photocatalyst**	[119, 120]
Ti	Ti(OiPr)$_4$	Latex (245 nm)	Heated at 50 °C	Calc. (450 °C)	TiO$_2$ (anatase)	**Solar cell**	[121]
Ti	Ti(OiPr)$_4$ in PrOH (1:4 or 1:8 molar ratio)	PS and PS copolymerised with 2-hydroxyethyl methacrylate, acrylic acid, methacrylic acid, or itaconic acid (288–486 nm)	Air dry	Calc. (500 °C)	TiO$_2$ (anatase)	Surface area: 28–79 m^2 g^{-1} (BET) Pore diameter: 160–360 nm (SEM) Wall thickness : 60 nm (SEM) Crystallite size: 4–25 nm (PXRD)	[122]

(continued overleaf)

Table 3.3 (continued)

Metal	Precursor solution	Template[a]	Solidification	Template removal	Crystal Phase[b]	Properties[c] and application	Ref.
Ti	Ti(OEt)4 (2 ml) + CF$_3$CO$_2$H (1.6 ml) + HCl (12 M, 0.4 ml) + ethanol or ethylene glycol	PS (330, 430, 500, 740 nm)	Air dry	Calc. (450 °C)	TiO$_2$ (anatase)		[123]
Ti	TiCl$_4$ in H$_2$O	PS (1000 nm)	Air dry	Extraction with toluene	TiO$_2$ (amorphous)		[124]
Ti	Ti(OBu)$_4$	PS (1000 nm)	Air dry	Calc. (575 °C)	TiO$_2$ (anatase)	Pore diameter: 800 nm (SEM)	[124]
Ti	(Ammonium lactate) titanium dihydroxide (50 %) in H$_2$O	PS (340 nm)	Air dry	Calc. (450 °C)	TiO$_2$ (anatase)	Crystallite size: 7 nm	[125]
Ti	(Ammonium lactate) titanium dihydroxide in H$_2$O	PS (350 nm)	Air dry	Calc. (450 °C) in O$_2$ flow	TiO$_2$ (anatase)	Pore diameter: 300 nm (SEM) Crystallite size: <10 nm **Solar cell**	[126]
Ti	Atomic layer deposition using TiCl$_4$ and H$_2$O	SiO$_2$		Extraction with HF (2 %)			[127]
Ti	TiO$_2$ dispersion (34 nm, 55 wt%) in H$_2$O (pH 8.5)	PS (560 nm)	Air dry	Calc. (450 °C) in air	TiO$_2$		[128]
Ti	TiO$_2$ nanopowder (30 nm, 5 wt%) + polyethylenimine (1 wt%) in H$_2$O (pH 8)	PMMA (350, 800, 1300 nm)	Air dry	Calc. (500 °C) in air	TiO$_2$ (1100 °C)		[129]

	Precursor	Template	Deposition/Drying	Calcination	Product	Properties	Ref.
Ti	TiO$_2$ nanoparticles (15 nm, 10 %) suspension	PS (329 nm)	Air dry	Calc. (500 °C)	TiO$_2$		[130]
Ti	Ti(SO$_4$)$_2$ (0.02 M) + KNO$_3$ (0.1 M) + H$_2$O$_2$ (0.5 M) in EtOH–H$_2$O (2:8 vol%)	PS (300 nm)	Cathodic electrodeposition at −1.8 V (*vs* SCE)	Extraction with toluene	TiO$_2$ (anatase) 450 °C	Pore diameter: 260 nm (SEM)	[131]
Ti	TiCl$_3$ (0.1 M) (~10% in 20~30 wt% HCl)	PS (215 nm)	Electrodeposited at 0 V (*vs* SCE)	Extraction with toluene at 60 °C	TiO$_2$		[132]
Ti	Ti(OiPr)$_4$ (0.15 wt%) + HNO$_3$ (0.015 %) in EtOH	Carboxylate-modified PS (243 nm)	1. Air dry 2. (NH$_4$)$_2$TiF$_6$ (0.1–0.3 M) + H$_3$BO$_3$ (0.2–0.5 M) in H$_2$O	Calc. (400 °C)	TiO$_2$ (anatase)	Solar cell	[133]
Ti	(NH$_4$)$_2$TiF$_6$ + boric acid (pH 3) in H$_2$O	Carboxy-modified PS (291 nm)	Air dry	1. Calc. (700 °C) under Ar 2. Calc. (450 °C) in air	TiO$_2$ (anatase) (pdf 89-4921)		[134]
Zr	Zr(OnPr)$_4$	PS (470 nm)	Dry	Calc. (575 °C) in air flow	ZrO$_2$ (baddeleyite)	Surface area: 38 m^2 g^{-1} (BET) Wall thickness: 24–36 nm (TEM) Crystallite size: 34 nm (PXRD), 35 nm (BET), 20–35 nm (TEM)	[8]

(continued overleaf)

Table 3.3 (*continued*)

Metal	Precursor solution	Template[a]	Solidification	Template removal	Crystal Phase[b]	Properties[c] and application	Ref.
Zr	$Zr(OnPr)_4$ (70 %) in nPrOH	PS (421, 697 nm)	Air dry	Calc. (575 °C)	ZrO_2 (baddelyeytite) (pdf 37-1484)	Surface area: $9\,m^2\,g^{-1}$ (BET) Wall thickness: 9–13 nm (TEM) Crystallite size: 8–13 nm (TEM)	[103]
Zr	$Zr(OiPr)_4$	PS (1000 nm)	Air dry	Calc. (575 °C)	ZrO_2 (monoclinic)		[124]
Zr	$ZrO(NO_3)_2$ (~1.7 M) in 1 M HNO_3 + H_2SO_4 + MeOH	PMMA (300–500 nm)	Air dry	Calc. (450–700 °C) in air flow	ZrO_2 (~600 °C) Tetragonal (pdf 42-1164) ZrO_2 (~600 °C) Tetragonal (pdf 42-1164) + monoclinic (pdf 37-1484)	Surface area: $5–123\,m^2\,g^{-1}$ (BET) Pore shrinkage: ~35 % (SEM) Crystallite size: ~7 nm (PXRD)	[135]
Zr	$Zr(OAc)_4$ in dilute acetic acid (6 ml) and MeOH (6 ml)	PMMA (310, 375, 425 nm)	Air dry	Calc. (450 °C)	ZrO_2 (tetragonal)	Pore diameter: 310–425 nm (SEM) Crystallite size: 1–2 nm (PXRD)	[112]
Zr	ZrO_2 nanopowder (30 nm, 5 wt%) + polyethylenimine (1 wt%) in H_2O (pH 8)	PMMA (350, 800, 1300 nm)	Air dry	Calc. (500 °C) in air	ZrO_2 (850–1100 °C)		[129]

	Precursor	Drying	Treatment	Product	Characterization	Ref.
V	VO(OiPr)$_3$ + iPrOH (1:8 vol ratio)	PS (1000 nm)	Air dry	Extraction with toluene	V$_2$O$_5$	Pore diameter: 800 nm (SEM) Wall thickness: 150 nm (SEM) **Lithium battery**	[136]
V	Protonated [VO$_3$]$^-$ (0.5 M) in H$_2$O mixed with MeOH (H$_2$O/MeOH = 2/1)	PMMA (495 nm)	Air dry	Calc. (400 °C) in N$_2$/air	V$_2$O$_5$ (shcherbinaite) (pdf 41-1426)	Surface area: 30–37 m^2 g^{-1} (BET) Pore diameter: 385 nm (SEM) Crystallite size: 33 nm (PXRD)	[137]
Nb	NH$_4$NbF$_6$ + boric acid (pH 3) in H$_2$O	Carboxy-modified PS (291 nm)	Air dry	1. Calc. (700 °C) under Ar 2. Calc. (450 °C) in air	Nb$_2$O$_5$ (orthorhombic) (pdf 27-1003)	Crystallite size: 10 nm (PXRD)	[134]
Cr	Cr(oxalate) prepared by CrO$_3$ + oxalic acid in EtOH (5 ml) and H$_2$O (5 ml)	PS (760 nm)	Dry	Calc. (450 °C) in air flow	Cr$_2$O$_3$ (eskolaite) (pdf 38-1479)	Surface area: 42 m^2 g^{-1} (BET) Pore diameter: 465 nm (SEM) Wall thickness: 45 nm (SEM) Crystallite size: 30 nm (PXRD)	[138]

(continued overleaf)

Table 3.3 (*continued*)

Metal	Precursor solution	Template[a]	Solidification	Template removal	Crystal Phase[b]	Properties[c] and application	Ref.
Cr	$Cr(NO_3)_3$ (Metal: 2 M) in ethylene glycol and MeOH (MeOH 40 vol%)	PMMA (416 nm)	Heating	Calc. (400 °C)	Cr_2O_3	Crystallite size: 39 nm (PXRD)	[139]
W	$W(V)(OEt)_5$	PS (421, 697 nm)	Air dry	Calc. (450 °C)	WO_3, (pdf 43-1035)	Wall thickness: 14–32 nm (TEM) Crystallite size: 8–68 nm (TEM)	[103]
W	WCl_6 (0.5 M) + HCl (1.5 M) in MeOH	PMMA	Air dry	Calc. (400 °C) in N_2/air	WO_3 (orthorhombic) (pdf 20-1324)	Pore diameter: 269 nm (SEM) Wall thickness: 24 nm (SEM) Crystallite size: 5 nm (PXRD)	[140]
W	WCl_6 (0.5 M) + HCl (1.5 M) in MeOH	PMMA	Air dry	Calc (400 °C) in N_2	WO_3 (cubic) (pdf 46-1096)		[140]
W	Tungsten peroxy ethyl ester in EtOH	PS (300, 760 nm)	Heated at 80 °C	Calc. (460 °C)	WO_3 (hexagonal)	Pore diameter: 283 and 702 nm (SEM) Crystallite size: 26 nm (PXRD) **Photonic device**	[141]
W	Tungsten peroxy ethyl ester in EtOH	PS (890 nm)	Heated at 110 °C	Calc. (500 °C)	WO_3	Pore diameter: 850 nm (SEM)	[142]
W	$(NH_4)_6W_{12}O_{40}$ (50 wt% as WO_3) in H_2O with MeOH (2:1 vol)	PMMA (180, 260 and 460 nm)	Air dry	Calc. (500–600 °C)	WO_3 (monoclinic) (pdf 20-1324)	Surface area: 7–21 $m^2 g^{-1}$ (BET) Pore diameter: 166–444 nm (TEM) **Photocatalysis**	[143]

	Precursor	Template	Method	Treatment	Product	Properties	Ref.
W	W species (0.25 M), prepared by reaction of W powder with H_2O_2, in H_2O/iPrOH (70/30)	PS (267 nm)	Electrodeposition at −0.9V vs Ag/AgCl	Extraction with toluene	WO_3	Electrochromism	[144]
Mn	$Mn(OAc)_2 \cdot 4H_2O$ (2 g) in EtOH (10 ml)	PS (660 nm)	Oxalic acid (~30 g) in EtOH (100 ml)	Calc (450 °C) in air flow	Mn_2O_3 (Bixbyite-C) (pdf 41-1442)	Surface area: 20 $m^2\,g^{-1}$ (BET) Pore diameter: 380 nm (SEM) Wall thickness: 70 nm (SEM) Crystallite size: 20 nm (PXRD)	[138]
Mn	$Mn(NO_3)_2$ (Metal: 2 M) in ethylene glycol and MeOH (MeOH 40 vol%)	PMMA (291 nm)	Heating	Calc. (400 °C)	Mn_2O_3 + Mn_3O_4		[139]
Fe	$Fe(OEt)_3$ (25 wt%) in EtOH	PS (421, 697 nm)	Air dry	Calc. (450 °C)	Fe_2O_3 (hematite, pdf 33-0664)	Wall thickness: 22–52 nm (TEM) Crystallite size: 22–120 nm (TEM)	[103]
Fe	$FeC_2O_4 \cdot 2H_2O$ in EtOH (5 ml) and H_2O_2/H_2O (5 ml)	PS (760 nm)	Dry	Calc. (450 °C) in air flow	Fe_2O_3 (hematite) (pdf 33-0664)	Surface area: 39 $m^2\,g^{-1}$ (BET) Pore diameter: 550 nm (SEM) Wall thickness: 60 nm (SEM) Crystallite size: 23 nm (PXRD) Changeable to 3DOM Fe (4)	[138, 145]

(continued overleaf)

Table 3.3 (*continued*)

Metal	Precursor solution	Template[a]	Solidification	Template removal	Crystal Phase[b]	Properties[c] and application	Ref.
Fe	Fe(NO$_3$)$_3$·9H$_2$O (Metal: 2 M) in ethylene glycol and MeOH (MeOH 40 vol%)	PMMA (291 nm)	Heating	Calc. (400 °C)	Fe$_2$O$_3$, (pdf 33-0664)	Surface area: 57 m^2 g^{-1} (BET) Pore diameter: 170 nm (TEM) Crystallite size: 19 nm (PXRD)	[139]
Fe	(NH$_4$)$_2$Fe(SO$_4$)$_2$ + AcOK in H$_2$O	SiO$_2$ (300 nm)	Potentiostatic deposition at −0.5 V and −0.6 V *vs* Ag/AgCl (Constant current density 0.05 mA cm^{-2})	Extraction with 2% HF	Fe$_2$O$_3$	Pore diameter: 400 nm (SEM) Wall thickness: 80 nm (SEM) Crystallite size: 4–25 nm (PXRD)	[146]
Co	Co(OAc)$_2$·4H$_2$O (2 g) in AcOH (10ml)	PS (640 nm)	Oxalic acid (~30 g) in EtOH (100 ml)	Calc. (400 °C) in air flow	Co$_3$O$_4$ (pdf 42-1467)	Surface area: 25 m^2 g^{-1} (BET) Pore diameter: 500 nm (SEM) Wall thickness: 55 nm (SEM) Crystallite size: 35 nm (PXRD) **Changeable to 3DOM Co (4**	[138]

Ni	$Ni(OAc)_2 \cdot 4H_2O$ (2 g) in AcOH (10 ml)	PS (660 nm)	Oxalic acid (~30 g) in EtOH (100 ml)	Calc. (360 °C) in air flow	NiO (bunsenite) (pdf 04-0835)	Surface area: 63 m² g⁻¹ (BET) Pore diameter: 470 nm (SEM) Wall thickness: 60 nm (SEM) Crystallite size: 10 nm (PXRD) **Changeable to 3DOM Ni (4)**	[138, 145,147]
Cu	$CuSO_4$ (0.4 M) + lactic acid (3 M) in alkaline solution (pH 12)	PS (350 nm)	Cathodic electrodeposition at −0.5 V (vs SCE) at 45 °C	Extraction with THF	Cu_2O (pdf 78-2076)	**Photocathodic material**	[148, 149]
La	$La(NO_3)_3 \cdot 6H_2O$ (1.90 g) + citric acid (1.05 g) in EtOH (5 ml)	PS (676 nm)	Heating	Calc. (600 °C)	$La_2O(CO_3)_2$	Pore diameter: 353 nm (SEM)	[150]
La	$La(NO_3)_3 \cdot 6H_2O$ (1.90 g) + citric acid (1.05 g) in EtOH (5 ml)	PS (600 nm)	Heating (60–70 °C)	Calc. (800 °C)	La_2O_3	Pore diameter: 286 nm (SEM)	[150]
Ce	$Ce(NO_3)_3 \cdot 6H_2O$ (2.17 g) + citric acid (1.05 g) in EtOH (5 ml)	PS (676 nm)	Heating (60–70 °C)	Calc. (400 °C)	CeO_2	Pore diameter: 390 nm (SEM)	[150]
Ce	$Ce(NO_3)_3 \cdot 6H_2O$ (2.17 g) + citric acid (1.05 g) in EtOH (5 ml)	PS (676 nm)	Heating (60–70 °C)	Calc. (800 °C)	CeO_2	Pore diameter: 377 nm (SEM)	[150]

(continued overleaf)

Table 3.3 (*continued*)

Metal	Precursor solution	Template[a]	Solidification	Template removal	Crystal Phase[b]	Properties[c] and application	Ref.
Ce	Ce(NO$_3$)$_3$·6H$_2$O (2.17 g) + citric acid (1.05 g) in EtOH	PMMA (325 nm)	Heating	Calc. (400 °C)	CeO$_2$ (cubic)	Surface area: 51 m^2 g^{-1} (BET) Pore diameter: 240 nm (SEM) Crystallite size: 10 nm (PXRD)	[151]
Nd	Nd-EDTA in H$_2$O	PS (726, 833 nm)	Air dry	Calc. (600–720 °C)	Nd$_2$O$_2$CO$_3$ (640 °C) Nd$_2$O$_3$ (hexagonal) (720 °C)		[152]
Sm	Sm-EDTA in H$_2$O	PS (726, 833 nm)	Air dry	Calc. (600–720 °C)	Sm$_2$O$_3$ (cubic)		[152]
Eu	Eu-EDTA in H$_2$O	PS (726, 833 nm)	Air dry	Calc. (600–720 °C)	Eu$_2$O$_3$ (cubic)		[152]

[a]PMMA: poly(methyl methacrylate); PS: polystyrene.
[b]pdf: The Powder Diffration File, Joint Committee on Powder Diffraction Standards (JCPDS), International Centre for Diffraction Data.
[c]Crystallite sizes were estimated from X-ray diffraction broadening using the Scherer equation and/or TEM. Surface areas were calculated by the Brunauer–Emmett–Teller (BET) method. Pore sizes and wall thickness were estimated using SEM or TEM.

3.5.2   Macroporous Oxides of Group 4 Elements (Ti, Zr)

Three-dimensionally ordered macroporous crystalline TiO_2 could be obtained by using titanium alkoxides as a metal source. Various titanium alkoxides [$Ti(OEt)_4$, $Ti(O^iPr)_4$, $Ti(OBu)_4$] can be used in alcoholic solutions or neat.[8,9,102–124] After the titanium alkoxides have infiltrated the voids of the template, they can be solidified by drying in air (sometimes heated at ca 50 °C) by a sol–gel process. Another titanium source, ammonium lactate titanium dihydroxide, can be used.[125,126] A gaseous Ti source, $TiCl_4$, can be used. Gaseous $TiCl_4$ was deposited on a SiO_2 template, and then $TiCl_4$ was hydrolysed with vaporous H_2O. This two-step deposition of TiO_2 on the template was repeated to deposit TiO_2 in the voids of the template.[127]

A suspension of TiO_2 nanoparticles can be used as a Ti source. Solid TiO_2 nanoparticles can be packed into the voids of the template, and solidification is not necessary.[128–130] Electrochemical deposition of TiO_2 in the voids of templates is possible if the colloidal crystal templates are deposited on electrodes.[131,132] Deposition of TiO_2 by reacting $(NH_4)_2TiF_6$ and H_3BO_3 in the voids of the template is possible if the surface of the template PS has been modified with carboxylic acid.[114,133,134]

By calcination of the solidified amorphous TiO_2 materials at ca 450 °C anatase-phase TiO_2 is formed, and the anatase-phase TiO_2 can be transferred to rutile phase TiO_2 by further calcination.

Three-dimensionally ordered macroporous crystalline ZrO_2 could be obtained by using zirconium alkoxides as a metal source. Various zirconium alkoxides [$Zr(OnPr)_4$ or $Zr(OiPr)_4$] can be used in alcoholic solutions or neat.[8,103,135] After the zirconium alkoxides have infiltrated the voids of the template, they can be solidified by drying in air by a sol–gel process. Other zirconium salts, $ZrO(NO_3)_2$ and $Zr(OAc)_4$, can be used.[112,135]

A suspension of ZrO_2 nanoparticles suspension can be used as a Zr source. Solid ZrO_2 nanoparticles can be packed into the voids of the template, and solidification is not necessary.[129]

Tetragonal or monoclinic (baddelyeytite) ZrO_2 was produced depending on the preparation method.

3.5.3   Macroporous Oxides of Group 5 Elements (V, Nb)

Three-dimensionally ordered macroporous crystalline V_2O_5 could be obtained by using vanadium alkoxide $VO(O^iPr)_3$ as a metal source,

which can be solidified by a sol-gel process.[136] Protonated metavanadate ($H[VO_3]$) prepared by protonation of sodium metavanadate can be
also used as a metal source.[137]

Niobium oxide can be deposited into the voids of a PS template by
reacting NH_4NbF_6 and H_3BO_3 if the PS surface has been modified with
carboxylic acid [134]. The carboxylic acid is needed to increase affinity of
the niobium oxide with the PS and to promote selective deposition of
niobium oxide on the surface of the PS template. Well-ordered crystalline Nb_2O_5 could be obtained by two-step calcination in Ar and air. By
the first calcination (700 °C) in Ar, the PS was carbonised and the
deposited niobium oxide was crystallised. In this step, the remaining
carbon prevents collapse of the three-dimensional Nb_2O_5 framework
due to crystal growth. The carbon materials can be removed by
the second calcination at 450 °C in air. The temperature of the second
calcination is lower than that of the first calcination, and three-
dimensionally ordered macroporous crystalline orthorhombic Nb_2O_5
was produced.

3.5.4   Macroporous Oxides of Group 6 Elements (Cr, W)

Two methods for preparation of ordered macroporous crystalline
Cr_2O_3 have been reported. Chromium oxalate is soluble in H_2O–
EtOH solution, and the chromium oxalate solution infiltrates the
voids of the templates. Solvents are removed by heating, and solid
chromium oxalate is deposited in the voids and can be converted to
Cr_2O_3 without melting.[138] Another method is an *in situ* method.
$Cr(NO_3)_3$ is dissolved in EG–MeOH mixed solution, and the solution
infiltrates the voids of the templates. By heating, EG reacts with
$Cr(NO_3)_3$ to form chromium oxalate derivatives in the voids by nitrate
oxidation, and the chromium oxalate derivatives can be converted to
Cr_2O_3 without melting.[139]

Tungsten alkoxide, $W^{(V)}(OEt)_5$, is commercially available and is
soluble in alcoholic solution. Tungsten alkoxide solution infiltrates
the voids and is solidified in the voids by a sol-gel method as described
by Holland *et al.*[103] Ethanol solution of WCl_6 can infiltrate the voids of
the templates, and W species in the voids can be solidified by a sol-gel
method.[140] The resultant WO_3 materials have either a poorly ordered
structure[103] or low pore fraction.[140] Well-ordered macroporous
WO_3 materials can be obtained using other W starting compounds

such as tungsten peroxy ethyl ester[141,142] and ammonium metatungstate [$(NH_4)_6W_{12}O_{40}$].[143] $(NH_4)_6W_{12}O_{40}$ was dissolved in methanol–water mixed solvent and infiltrated the voids of a PMMA template. Evaporation of solvents by heating results in deposition of $(NH_4)_6W_{12}O_{40}$ solid in the voids, and the $(NH_4)_6W_{12}O_{40}$ solid can be converted to WO_3 by calcination.[143]

WO_3 can be electrochemically deposited in the voids of the template if the template has been attached to an electrode.[144]

3.5.5 Macroporous Oxides of Elements of Groups 7–11 (Mn, Fe, Co, Ni, Cu)

Alkoxides of late transition metals are not easily available. Some chemical companies produce alkoxides of late transition metals, but they are expensive and too reactive to infiltrate into the template. They react quickly on the surface of the template, and solids that are produced prevent complete infiltration. An example using iron alkoxide, $Fe(OEt)_3$, has been reported, but in this case, some nonporous bulk material existed in the sample.[103]

Well-ordered macroporous Fe_2O_3 can be prepared by using iron oxalate.[138,145] Iron oxalate is soluble in ethanol–water mixed solvents, which can infiltrate the voids of templates. Iron oxalate can be converted to Fe_2O_3 without melting to form well-ordered macroporous material. If the oxalate salts are not soluble in alcoholic solution, oxalate or oxalate derivatives can be produced in the voids by a two-step method or an *in situ* method. In the two-step method, nitrate or acetate salts of Mn, Co or Ni infiltrate the voids and then react with oxalic acid to form oxalate salts in the voids. These oxalate salts can be converted to oxide by calcination.[138,145,147] In the *in situ* method, nitrate of Mn or Fe is dissolved in EG–methanol solution, and the solution infiltrates the voids of templates. By heating the solution, EG is oxidised by nitrate to form metal oxalate derivatives. The metal oxalate derivatives can be converted to metal oxides.[139]

Fe_2O_3[146] or Cu_2O[148,149] can be deposited in the voids of the template by an electrochemical method. A colloidal crystal template of SiO_2 or PS is deposited on electrodes, and the electrodes are electrolysed in alkaline aqueous solution of $(NH_4)_2Fe(SO_4)_2$ or $CuSO_4$ to form oxides. After removal of the template by extraction, ordered macroporous materials can be obtained.

3.5.6 Macroporous Oxides of Lanthanide Elements (La, Ce, Nd, Sm, Eu)

Like alkoxides of late transition metals, lanthanide alkoxides are not suitable for production of ordered macroporous lanthanide oxides. Common lanthanide nitrates are also not suitable. Addition of citric acid or EDTA is necessary to produce ordered macroporous $La_2O(CO_3)_2$, La_2O_3, CeO_2, $Nd_2O_2CO_3$, Nd_2O_3, Sm_2O_3, and Eu_2O_3 materials.[151,152]

3.5.7 Macroporous Multi-Component Metal Oxides

If mixed-metal alkoxide precursors can be prepared and their reactivity can be controlled, ordered macroporous mixed metal oxides can be produced by a sol-gel method using a colloidal crystal template. Alkoxides of Ti and Zr with other metals are easily prepared by mixing titanium alkoxide and zirconium alkoxide with alkoxides of other metals or metal salts, and the formed mixed metal alkoxides can infiltrate the voids of templates. Therefore, production of mixed metal oxides with Ti and Zr is straightforward.

In the case of other transition metals, their alkoxides are not suitable as metal sources. Furthermore, a two-step method in which deposited metal salts are solidified by a reaction with oxalate or a base is not suitable for the preparation of ordered macroporous mixed metal oxides. Each metal has a different reactivity with oxalic acid or a base, and the produced oxalate or metal hydroxide has different solubilities in the reacting media, which causes a mixed metal oxide with an undesired metal ratio.[137,178–180] On the other hand, *in situ* methods, in which an additive such as EG, citric acid or EDTA is present with mixed metals, ensure the chemical homogeneity of the products and are suitable methods for producing ordered macroporous mixed metal oxides with a desired ratio. Synthesis methods, structural characterisation and applications of macroporous mixed metal oxides are summarised in Table 3.4.

Mixed metal oxides with Ti or Zr are produced by using alkoxides of mixed metals. Alkoxides of mixed metals with Ti or Zr are produced by mixing titanium alkoxide or zirconium alkoxide with other metal salts or metal alkoxide, and various ordered macroporous $BaTiO_3$,[153–155] $CsAlTiO_4$,[156] $Li_4Ti_5O_{12}$,[157,158] $La_4Sr_8Ti_{11}Mn_{0.5}Ga_{0.5}O_{38-\delta}$,[159] TiO_2+CeO_2,[160] $PbTiO_3$,[124] $PbZr_{0.5}Ti_{0.5}O_3$,[124] $Pb_{0.91}La_{0.09}(Zr_{0.65}Ti_{0.35})_{0.9775}O_3$,[161,162] $Ti_{1-x}Ta_xO_{2+x/2}$[163] and $Y_{0.043}Zr_{0.957}O_2$[103] can be produced.

Table 3.4 Precursors, templates, solidifications, template removal methods, crystal phase and properties of ordered macroporous crystalline multicomponent metal oxides

Metal	Precursor solution	Template[a]	Solidification	Template removal	Crystal Phase[b]	Properties[c] and application	Ref.
Ti, Ba	Ba(OAc)$_2$ (1 mmol) in AcOH (5 ml) + Ti(OBu)$_4$ in EtOH (1:2 vol)	PS (726 nm)	Dry in vacuum desiccator	Calc. (700 °C)			[153]
Ti, Ba	Ba(OEt)$_2$ (1.4 g) in 6 ml MeOH/ MeOCH$_2$CH$_2$OH (60/40 vol) + 1.9 g Ti(OiPr)$_4$	Polymer (350–410 nm)	Air dry	Calc. (600, 750 °C)	BaTiO$_3$ (600 °C)	Pore diameter: 280–330 nm (SEM) Wall thickness: ~0 nm (SEM) Crystallite size: ~10 nm (600 °C), ~55 nm (750 °C) (PXRD and TEM)	[154]
Ti, Ba	BaTi(C$_2$H$_2$O$_2$)$_3$- 4C$_2$H$_6$O$_2$·H$_2$O in ethylene glycol	Polymer beads (380 nm)	Air dry	Calc. (650 °C) in air flow	BaTiO$_3$	Pore diameter: 230 nm (TEM) Wall thickness: 33–55 nm (TEM)	[155]
Cs. Al, Ti	Alkoxide prepared by mixing Cs$_2$CO$_3$, Ti(OPr)$_4$, and Al(OiBu)$_3$ in AcOH and EtOH	PS (1000 nm)	Air dry	Calc. (500–900 °C) under O$_2$ flow	CsAlTiO$_4$	Surface area: 24–49 m^2 g^{-1} (BET) Pore diameter: 500–800 nm (SEM) Wall thickness: 10–20 nm (SEM) Crystallite size: 16–25 nm (PXRD)	[156]
Li, Ti	Li(OAc) (1 M) in H$_2$O + H$_2$TiO(C$_2$O$_4$) (1 M) in H$_2$O (4:5 vol) + MeOH (20 vol%)	PS (300 nm) PMMA (300 nm)	Air dry	Calc. (675 °C) in air	Li$_4$Ti$_5$O$_{12}$ + small amount of TiO$_2$ (rutile and anatase) and Li$_2$TiO$_3$	**Lithium battery**	[157]

(continued overleaf)

Table 3.4 (*continued*)

Metal	Precursor solution	Template[a]	Solidification	Template removal	Crystal Phase[b]	Properties[c] and application	Ref.
Li, Ti	LiOAc + Ti(OiPr)$_4$ + AcOH + iPrOH (4:5:60:100 mol ratio)	PS (1000 nm)	80 °C in air	Calc. (450, 600, 700, 800 °C) in air	Li$_4$Ti$_5$O$_{12}$ + TiO$_2$ (anatase) (450 °C) Li4Ti5O$_{12}$ + TiO$_2$ (anatase and rutile) (600 °C) Li$_4$Ti$_5$O$_{12}$ + TiO$_2$ (rutile) (700 °C) Li$_4$Ti$_5$O$_{12}$ (800 °C)	Surface area: 8 m^2 g^{-1} (800 °C) (BET) Pore diameter: 800 nm (800 °C) (SEM) Porosity: 64 % (Hg porosimetry) **Lithium battery**	[158]
La, Sr, Ti, Mn, Ga	Ln$_2$O$_3$ + Sr(NO$_3$)$_2$, Mn(NO$_3$)$_3$·6H$_2$O + Ti(OiPr)$_4$ in diluted HNO$_3$ in H$_2$O + citric acid (citric acid/metal, M, molar ratio = 0.3–0.5) (pH = 3 adjusted by HNO$_3$), concentration (1 g oxide in 25 ml)	PMMA (400 nm)	Air dry	Calc. (650 °C) in air	La$_4$Sr$_8$Ti$_{11}$Mn$_{0.5}$Ga$_{0.5}$O$_{38-\delta}$	Crystallite size: 22 nm (PXRD)	[159]

Ce, Ti	$Ce(NO_3)_3 \cdot 6H_2O$ and $Ti(OnBu)_4$ (4.148 g) with 5 ml EtOH (Ce:Ti molar ratio, 0.05, 0.07, 0.10, 0.20)	PS (~300 nm)	Air dry	Calc. (750 °C)	TiO_2 (anatase) + TiO_2 (rutile) + CeO_2 (cerianite)		[160]
Pb, Ti	0.1 M $Pb(OAc)$ + 0.1 M $Ti(OBu)_4$ in MeOH heated at 60 °C and H_2O	PS (1000 nm)	Air dry	Extraction with toluene	$PbTiO_3$ (tetragonal) (600 °C)	No shrinkage	[124]
Pb, Ti, Zr	0.1 M $Pb(OAc)$ + 0.1 M $Ti(OBu)_4$ + 0.1 M $Zr(OiPr)_4$ in MeOH heated at 60 °C and H_2O	PS (1000 nm)	Air dry	Extraction with toluene	$PbZr_{0.5}Ti_{0.5}O_3$ (amorphous) (600 °C)		[124]
Pb, La, Zr, Ti	Alkoxide precursor prepared from $Zr(OPr)_4$ (70 wt%) in nPrOH, $Ti(OBu)_4$, $Pb(OAc)_2 \cdot 3H_2O$, $La(NO_3)_3 \cdot 6H_2O$ in AcOH, H_2O, and MeOH	PS (400 nm)	Air dry	Calc. (450 °C)	$Pb_{0.91}La_{0.09}$ $(Zr_{0.65}Ti_{0.35})_{0.9775}O_3$ (600–750 °C)	Pore diameter: 280 nm (SEM)	[161, 162]
Ti, Ta	$TaCl_5$ and $Ti(OiPr)_4$ in EtOH (10:1:190, 19.5:0.5:195 or 18.5:1.5:185 molar ratio)	PS (400 nm)	Air dry	Calc. (500 °C)	$Ti_{1-x}Ta_xO_{2+x/2}$ ($x = 0.025, 0.05$ or 0.075) (anatase) (pdf 21-1272)	Surface area: 76 m² g⁻¹ (BET) Pore diameter: 290–310 nm (SEM) Wall thickness: 50–80 nm (SEM) Crystallite size: 10–12 nm (TEM) **Photocatalyst**	[163]

(continued overleaf)

Table 3.4 (*continued*)

Metal	Precursor solution	Template[a]	Solidification	Template removal	Crystal Phase[b]	Properties[c] and application	Ref.
Zr, Y	Y(OiPr)$_3$ in Zr(OnPr)$_4$ (70 %) in nPrOH	PS (421, 697 nm)	Air dry	Calc. (1000 °C)	Y$_{0.043}$Zr$_{0.957}$O$_2$, cubic YSZ, (pdf 30-1468) (ED)	Wall thickness: 8–41 nm (TEM) Crystallite size: 6–29 nm (TEM)	[103]
Ce, Zr	Nanoparticle-sol of Zr-Ce oxide in H$_2$O	PS (1000 nm)		Calc. (550 °C)	Ce$_{0.5}$Zr$_{0.5}$O$_2$, tetragonal (a = 3.6654, c = 5.263 Å)	Surface area: 14 m^2 g^{-1} (BET) Crystallite size: 20 nm (PXRD) **Oxidation catalyst**	[164]
Zr, Y	Y$_2$O$_3$ + ZrO(NO$_3$) in diluted HNO$_3$ in H$_2$O + citric acid (citric acid/M = 0.3–0.5) (pH = 3 adjusted by HNO$_3$, concentration (1 g oxide in 25 ml)	PMMA (400 nm)	Air dry	Calc. (650, 800, 1000 °C) in air	Zr$_{0.8}$Y$_{0.2}$O$_{2-δ}$	Surface area: 21 (650 °C), 19 (800 °C), 14 (1000 °C) m^2 g^{-1} (BET) Pore diameter : 175-225 nm (SEM) Wall thickness : 25-50 nm (SEM) Crystallite size : 12 (650 °C) nm (PXRD), 10 nm (650 °C) (TEM)	[159]
Ni, Y, Zr	Y$_2$O$_3$ + ZrO(NO$_3$)$_2$ + Ni(NO$_3$)$_2$·6H$_2$O in diluted HNO$_3$ in H$_2$O + Citric acid (citric acid/M = 0.3–0.5) (pH = 3 adjusted by HNO$_3$, concentration (1 g oxide in 25 ml)	PMMA(400 nm)	Air dry	Calc. (650 °C) in air	NiO-Zr$_{0.8}$Y$_{0.2}$O$_{2-δ}$	Crystallite size: 11 nm (Zr$_{0.8}$Y$_{0.2}$ Zr$_{2-d}$) and 18 nm (NiO) (PXRD)	[159]

V, P	1. H_3PO_3 in EtOH 2. $VO(COH(Me)_2)_3$ in ethanol (EtOH/V, wt/wt = 1)	PS (~400 nm)	Air dry	Calc. (450 °C) in air	$VOPO_4 \cdot 2H_2O$	Surface area: $64\,m^2\,g^{-1}$ (BET)	[165, 166]
V, P	1. H_3PO_3 in EtOH 2. $VO(COH(Me)_2)_3$ in Ethanol (EtOH/V, wt/wt = 3)	PS (~400 nm)	Air dry	Extraction (THF/ acetone)	$VOPO_4 \cdot 2H_2O$	Surface area: $50\,m^2\,g^{-1}$ (BET)	[165, 166]
V, P	1. H_3PO_3 in EtOH 2.) V_2O_5 in ethanol (EtOH/V, wt/wt = 5)	PS (~400 nm)	Air dry	Calc. (450 °C) in air	$VOPO_4 \cdot 2H_2O$	Surface area: $41\,m^2\,g^{-1}$ (BET)	[165, 166]
V, P	1. H_3PO_3 in EtOH 2. $VO(COH(Me)_2)_3$ in ethanol (EtOH/V, wt/wt = 10)	PS (~400 nm)	Air dry	Extraction (THF/ acetone)	$VOPO_4 \cdot 2H_2O$ and b-$VOHPO_4 \cdot 2H_2O$	Surface area: $75\,m^2\,g^{-1}$ (BET)	[165, 166]
V, P	1. H_3PO_4 in EtOH 2.) V_2O_5 in ethanol (EtOH/V, wt/wt = 10)	PS (~400 nm)	Air dry	Calc. (450 °C) in air	$(VO)_2P_2O_7$	Surface area: $44\,m^2\,g^{-1}$ (BET)	[165, 166]
Li, Nb	$LiNb(OEt)_6$ (1.5 M) in EtOH	PS (640 nm) coated with polyelectrolyte multilayer	Air dry	Calc. (500 °C under N_2 then under air)	$LiNbO_3$		[167]
Zn, Cr	$Zn(NO_3)_3 \cdot 6H_2O$, $Cr(NO_3)_3 \cdot 9H_2O$ (Metal: 2 M) in ethylene glycol and MeOH (MeOH 40 vol%)	PMMA (291 nm)	Heating	Calc. (600 °C)	$ZnCr_2O_4$ (pdf 22-1107)	Crystallite size: 13 nm (PXRD)	[139]

(continued overleaf)

Table 3.4 (continued)

Metal	Precursor solution	Template[a]	Solidification	Template removal	Crystal Phase[b]	Properties[c] and application	Ref.
Zn, Cr	$ZnCl_2$ and $AlCl_3$ (1 M, Zn:Al = 2:1) in $EtOH-H_2O$ (50:50)	PS (820 nm)	NaOH (2 M) solution	Extracted with toluene or THF-acetone	$Zn_{2.0}Cr(OH)_6(CO_3)_{0.42}Cl_{0.32} \cdot nH_2O$	Surface area: $72\,m^2\,g^{-1}$ (BET); Pore diameter: 620 nm (SEM); Wall thickness: 31 nm (SEM)	[168]
La, Mo	La_2O_3 in diluted HNO_3 in H_2O + MoO_3 dissolved in NH_3-H_2O + citric acid (citric acid/M = 0.3–0.5) (pH = 3 adjusted by HNO_3), concentration (1 g oxide in 25 ml)	PMMA (400 nm)	Air dry	Calc. (550 °C) in air	$La_2Mo_2O_9$	Crystallite size: 60 nm (PXRD)	[159]
La, Ca, Mn	La-Ca-Mn alkoxide solution in EtOH	PMMA (400 nm)	Dry	Calc. (800 °C) in O_2 flow	$La_{0.7}Ca_{0.3}MnO_3$ pseudo cubic (a = 3.841 Å)	Surface area: $24\,m^2\,g^{-1}$ (BET); Pore diameter: 340 nm (SEM); Wall thickness: 60 nm (SEM); Crystallite size: 10 nm (PXRD)	[169, 170]
Li, Mn	$LiNO_3$ (1.3 M) + $Mn(NO_3)_2$ (2.6 M) in EtOH	PS-*co*-methacrylic acid (380 nm)	Air dry	Calc. (500–700 °C)	$LiMn_2O_4$	Surface area: $24\,m^2\,g^{-1}$ (BET); **Lithium battery**	[171]
La, Mn	$La(NO_3)_3 \cdot 6H_2O$, $Mn(NO_3)_2 \cdot 9H_2O$ (Metal: 2 M) in ethylene glycol and MeOH (MeOH 40 vol%)	PMMA (291 nm)	Heating	Calc. (700 °C)	$LaMnO_3$ (pdf 32-0484)	Crystallite size: 28 nm (PXRD)	[139]

Elements	Precursor solution	Template	Drying	Calcination	Product	Characterization	Ref.
La, Sr, Mn	La_2O_3 + $Sr(NO_3)_2$, $Mn(NO_3)_2 \cdot 6H_2O$ in diluted HNO_3 in H_2O + citric acid (citric acid/M = 0.3–0.5) (pH = 3 adjusted by HNO_3), concentration (1 g oxide in 25 ml)	PMMA (400 nm)	Air dry	Calc. (600 °C) in air	$La_{0.8}Sr_{0.2}MnO_{3-d}$	Crystallite size: 22 nm (PXRD)	[159]
La, Ca, Sr, Mn	La-Ca-Sr-Mn alkoxide solution in EtOH	PMMA (400 nm)	Air dry	Calc. (800 °C)	$La_{0.7}Ca_{0.3-x}Sr_xMnO_3$ (pseudo cubic)	Wall thickness: 40–60 nm (SEM)	[172]
La, Sr, Fe	$La(NO_3)_3 \cdot 6H_2O$, $Sr(NO_3)_2$, $Fe(NO_3)_3 \cdot 9H_2O$ (Metal: 2 M) in ethylene glycol and MeOH (MeOH 40 vol%)	PS (161 nm)	Heating	Calc. (500–600 °C)	$La_{1-x}Sr_xFeO_3$ (x = 0–0.2)	Surface area: 24–49 $m^2\ g^{-1}$ (BET) Pore diameter: 87–94 nm (SEM) Wall thickness: 13–20 nm (SEM) Crystallite size: 16–26 nm (PXRD)	[173]
La, Fe	$La(NO_3)_3 \cdot 6H_2O$, $Fe(NO_3)_3 \cdot 9H_2O$ (Metal: 2 M) in ethylene glycol and MeOH (MeOH 40 vol%)	PMMA (291 nm)	Heating	Calc. (600 °C)	$LaFeO_3$ (pdf 37-1493)	Pore diameter: 198 nm (TEM) Crystallite size: 25 nm (PXRD)	[174]
Zn, Fe	$Zn(NO_3)_2 \cdot 6H_2O$, $Fe(NO_3)_3 \cdot 9H_2O$ (Metal: 2 M) in ethylene glycol and MeOH (MeOH 40 vol%)	PMMA (291 nm)	Heating	Calc. (600 °C)	$ZnFe_2O_4$ (pdf 22-1012)	Pore diameter: 183 nm (TEM) Crystallite size: 23 nm (PXRD)	[174, 175]

(continued overleaf)

Table 3.4 (continued)

Metal	Precursor solution	Template[a]	Solidification	Template removal	Crystal Phase[b]	Properties[c] and application	Ref.
Ni, Fe	$Ni(NO_3)_2 \cdot 6H_2O$, $Fe(NO_3)_3 \cdot 9H_2O$ (Metal: 2 M) in ethylene glycol and MeOH (MeOH 40 vol%)	PMMA (291 nm)	Heating	Calc. (600 °C)	$NiFe_2O_4$ (pdf 10-0325)	Pore diameter: 169 nm (TEM) Crystallite size: 17 nm (PXRD)	[174, 175]
Co, Fe	$Co(NO_3)_3 \cdot 6H_2O$, $Fe(NO_3)_3 \cdot 9H_2O$ (Metal: 2 M) in ethylene glycol and MeOH (MeOH 40 vol%)	PMMA (291 nm)	Heating	Calc. (600 °C)	$CoFe_2O_4$ (pdf 3-0864)	Crystallite size: 18 nm (PXRD)	[175]
Ni, Zn, Fe	$Zn(NO_3)_2 \cdot 6H_2O$, $Ni(NO_3)_2 \cdot 6H_2O$, $Fe(NO_3)_3 \cdot 9H_2O$ (Metal: 2 M) in ethylene glycol and MeOH (MeOH 40 vol%)	PMMA (291 nm)	Heating	Calc. (600 °C)	$Zn_xNi_{1-x}Fe_2O_4$ ($x = 0.2$–0.8)	Crystallite size: 19–22 nm (PXRD)	[175]
La, Sr, Fe	$La_2O_3 + Sr(NO_3)_2$, $Fe(NO_3)_3 \cdot 9H_2O$ in diluted HNO_3 in H_2O + citric acid (citric acid/M = 0.3–0.5) (pH = 3 adjusted by HNO_3), concentration (1 g oxide in 25 ml)	PMMA (400 nm)	Air dry	Calc. (650 °C) in air	$La_{0.7}Sr_{0.3}FeO_{3-d}$	Crystallite size: 20 nm (PXRD)	[159]

Mg, Al, Fe	MgCl$_2$, FeCl$_3$, and AlCl$_3$ (1 M, Mg:Fe:Al = 4:1:1) in EtOH–H$_2$O (50:50)	PS (820 nm)	NaOH (2 M) solution	Extracted with toluene or THF–acetone	Mg$_{1.85}$Al$_{0.5}$Fe$_{0.43}$(OH)$_{5.7}$(CO$_3$)$_{0.51}$·nH$_2$O	Pore diameter: 535 nm (SEM) Wall thickness: 41 nm (SEM)	[168]
Sr, Sm, Co	Metal nitrates	PS (200 nm)	Calc. (800 °C) in air	Sr$_{0.5}$Sm$_{0.5}$CoO$_3$	**SOFC cathode**	[176]
Li, Co	LiOAc·2H$_2$O (0.53–0.90 M) + Co(OAc)$_2$·4H$_2$O (0.53 M) + polyethylene glycol (av. M_n = 1000) + H$_2$PtCl$_6$·H$_2$O in MeOH	PMMA (530 nm)	Oxalic acid (saturated) in EtOH	Pt doped LiCoO$_2$ + Co$_3$O$_4$ (400–800 °C)	Crystallite size: 13–89 nm (500–800 °C) (PXRD) **Lithium battery**	[177]
Al, Co	CoCl$_2$ and AlCl$_3$ (1 M, Co:Al = 2:1) in EtOH–H$_2$O (50:50)	PS (820 nm)	NaOH (2 M) solution	Extracted with toluene or THF–acetone	Co$_{1.9}$Al(OH)$_{5.8}$(CO$_3$)$_{0.5}$·nH$_2$O	Surface area: 57 m^2 g^{-1} (BET) Pore diameter: 430 nm (SEM) Wall thickness: 42 nm (SEM)	[168]
Mg, Al, Co	MgCl$_2$, CoCl$_2$, and AlCl$_3$ (1 M, Mg:Co:Al = 1:1:1) in EtOH–H$_2$O (50:50)	PS (820 nm)	NaOH (2 M) solution	Extracted with toluene or THF–acetone	Mg$_{0.95}$Co$_{0.9}$Al(OH)$_{5.7}$(CO$_3$)$_{0.5}$·nH$_2$O	Pore diameter: 550 nm (SEM) Wall thickness: 46 nm (SEM)	[168]
Li, Ni	LiOAc (1.6 g) and Ni(OAc)$_2$·4H$_2$O (4 g) in AcOH (20ml)	PMMA (495 nm)	Oxalic acid (ca 2.4 g) in EtOH (20 ml)	Calc. (700 °C) in air flow	LiNiO$_2$		[137]

(continued overleaf)

Table 3.4 (continued)

Metal	Precursor solution	Template[a]	Solidification	Template removal	Crystal Phase[b]	Properties[c] and application	Ref.
Al, Ni	$NiCl_2$ and $AlCl_3$ (1 M, Ni:Al = 2:1) in $EtOH$–H_2O (50:50)	PS (820 nm)	NaOH (2 M) solution	Extracted with toluene or THF–acetone	$Ni_{1.95}Al(OH)_{5.9}$ $(CO_3)_{0.52} \cdot nH_2O$	Surface area: 63 $m^2\ g^{-1}$ (BET) Pore diameter: 420 nm (SEM) Wall thickness: 35 nm (SEM)	[168]
Mg, Al, Ni	$MgCl_2$, $NiCl_2$, and $AlCl_3$ (1 M, Mg:Ni;Al = 1:1:1) in $EtOH$–H_2O (50:50)	PS (820 nm)	NaOH (2 M) solution	Extracted with toluene or THF–acetone	$Mg_{0.9}Ni_{1.1}$ $Al(OH)_{6.4}$ $(CO_3)_{0.5} \cdot nH_2O$	Pore diameter: 520 nm (SEM) Wall thickness: 52 nm (SEM)	[168]
La, Al	$La(NO_3)_3 \cdot 6H_2O$, $Al(NO_3)_3 \cdot 9H_2O$ (Metal: 2 M) in ethylene glycol and MeOH (MeOH 40 vol%)	PMMA (291 nm)	Heating	Calc. (800 °C)	$LaAlO_3$, (pdf 31-0022)	Crystallite size: 37 nm (PXRD)	[139]
Ce, Gd	Ln_2O_3 in diluted HNO_3 in H_2O + citric acid (citric acid/M = 0.3–0.5) (pH = 3 adjusted by HNO_3), Concentration (1 g oxide in 25 ml)	PMMA (400 nm)	Air dry	Calc. (450 °C) in air	$Ce_{0.8}Gd_{0.2}O_{2\text{-}d}$	Crystallite size: 9 nm (PXRD)	[159]

[a]PMMA: poly(methyl methacrylate); PS: polystyrene.
[b]pdf: The Powder Diffraction File, Joint Committee on Powder Diffraction Standards (JCPDS), International Centre for Diffraction Data.
[c]Crystallite sizes were estimated from X-ray diffraction broadening using the Scherrer equation and/or TEM. Surface areas were calculated by the Brunauer–Emmett–Teller (BET) method. Pore sizes and wall thickness were estimated by using SEM or TEM.

Nanoparticles of $Ce_{0.5}Zr_{0.5}O_2$ can be deposited into the voids of templates, and ordered macroporous $Ce_{0.5}Zr_{0.5}O_2$ can be produced after calcination.[164]

If nitrates of Zr and Y are used, addition of citric acid is necessary for production of ordered macroporous $Zr_{0.8}Y_{0.2}O_{2-\delta}$.[159]

Ordered macroporous mixed metal oxides of V or Nb are produced by using their alkoxide precursors. Macroporous vanadyl phosphate can be produced by using various V- and P-containing alkoxides.[165,166] Ordered macroporous $LiNbO_3$ can be produced using its mixed metal alkoxide $LiNb(OEt)_6$.[167]

Ordered macroporous spinel-type $ZnCr_2O_4$ is produced by using an *in situ* method with EG.[139] EG–methanol solution of $Zn(NO_3)_2$ and $Cr(NO_3)_3$ infiltrates the voids of the template. EG is oxidised by nitrates to form a Zn-Cr oxalate derivative in the voids, and further calcination results in removal of the template and formation of the desired ordered macroporous spinel-type $ZnCr_2O_4$. Ordered macroporous layered double hydroxide $Zn_2Cr(OH)_6(CO_3)_{0.42}Cl_{0.32}$ is produced by reacting $ZnCl_2$ and $AlCl_3$ with a base in the voids of template. Ethanol–water mixed solution of $ZnCl_2$ and $AlCl_3$ infiltrates the voids of the template. After the solvent has been removed, the deposited $ZnCl_2$ and $AlCl_3$ react with NaOH to produce $Zn_2Cr(OH)_6(CO_3)_{0.42}Cl_{0.32}$ in the voids. After the template has been removed by extraction, ordered macroporous $Zn_2Cr(OH)_6(CO_3)_{0.42}Cl_{0.32}$ can be produced.[168]

Ordered macroporous $La_2Mo_2O_9$ can be produced using a La and Mo aqueous solution containing citric acid.[159]

An alkoxide precursor of a La-Ca-Mn mixture can be produced by repeated heating of metal salts in the presence of 2-methoxyethanol and HNO_3. By using this alkoxide solution in EtOH, well-ordered macroporous perovskite-type $La_{0.7}Ca_{0.3}MnO_3$ or $La_{0.7}Ca_{0.3-x}Sr_xMnO_3$ can be prepared.[169,170,172] Addition of EG or citric acid to a mixture of mixed nitrates solution is an alternative method for preparing ordered macroporous perovskite-type $LaMnO_3$[139] or $La_{0.8}Sr_{0.2}MnO_3$.[159] If a polymer template is modified with carboxylic acid, an ethanol solution of a mixture of the nitrates [$Li(NO_3)$ and $Mn(NO_3)_2$] without any additive can be used as a metal precursor to form ordered macroporous $LiMn_2O_4$.[171] The carboxylic acid of the template may play the same role as that of citric acid.

Well-ordered macroporous spinel-type and perovskite-type mixed iron oxide, $La_{1-x}Sr_xFeO_3$ and MFe_2O_4 (M = Ni, Zn, Co, $Ni_{1-x}Zn_x$) can be prepared using a mixed metal nitrates solution in mixed solvent of EG and methanol.[139,174,175] Ordered macroporous layered double hydroxide $Mg_2Al_{0.5}Fe_{0.43}(OH)_{5.7}(CO_3)_{0.51}$ is produced by reacting $MgCl_2$, $FeCl_3$ and $AlCl_3$ with a base in the voids of the template. Ethanol–water mixed solution of $MgCl_2$, $FeCl_3$ and $AlCl_3$ infiltrates the voids of the template. After the solvent has been removed, the deposited $MgCl_2$, $FeCl_3$ and $AlCl_3$ react with NaOH to produce $Mg_2Al_{0.5}Fe_{0.43}(OH)_{5.7}(CO_3)_{0.51}$ in the voids. After the template has been removed by extraction, ordered macroporous $Mg_2Al_{0.5}$-$Fe_{0.43}(OH)_{5.7}(CO_3)_{0.51}$ can be produced.[168]

Ordered macroporous $Sm_{0.5}Sr_{0.5}CoO_3$ is prepared from mixed nitrate salts.[176] This is the only example of no additive or template modification being needed. Methanol solution of LiOAc, H_2PtCl_6 and $Co(OAc)_2$ infiltrates the voids of templates. After evaporation of methanol, mixed metal oxalate salts are deposited in the voids by treatment with saturated oxalic acid solution. After calcination, a Pt-doped macroporous $LiCoO_2$ and Co_3O_4 mixture is produced.[177] Ethanol–water mixed solution of $CoCl_2$ and $AlCl_3$ or $MgCl_2$, $CoCl_2$ and $AlCl_3$ infiltrates the voids of a template. After the solvent has been removed, the deposited chloride salts react with NaOH to produce $Co_{1.9}Al(OH)_{5.8}(CO_3)_{0.5}$ or $Mg_{0.95}Co_{0.9}Al(OH)_{5.7}(CO_3)_{0.5}$ in the voids. After the template has been removed by extraction, ordered macroporous $Co_{1.9}Al(OH)_{5.8}(CO_3)_{0.5}$ or $Mg_{0.95}Co_{0.9}Al$ $(OH)_{5.7}(CO_3)_{0.5}$ can be produced.[168]

An acetic acid solution of LiOAc and $Ni(OAc)_2$ infiltrates the voids of templates. After evaporation of methanol, mixed metal oxalate salts are deposited in the voids by treatment with saturated oxalic acid solution. After calcination, macroporous $LiNiO_2$ is produced.[137] Ethanol–water mixed solution of $NiCl_2$ and $AlCl_3$ or $MgCl_2$, $NiCl_2$ and $AlCl_3$ infiltrates the voids of the template. After the solvent has been removed, the deposited chloride salts react with NaOH to produce $Ni_{1.95}Al$-$(OH)_{5.9}(CO_3)_{0.52}$ or $Mg_{0.9}Ni_{1.1}Al$ $(OH)_{6.4}(CO_3)_{0.5}$ in the voids. After the template has been removed by extraction, ordered macroporous $Ni_{1.95}Al(OH)_{5.9}(CO_3)_{0.52}$ or $Mg_{0.9}Ni_{1.1}Al$ $(OH)_{6.4}(CO_3)_{0.5}$ can be produced.[168]

Ordered macroporous perovskite-type $LaAlO_3$ is produced by using an *in situ* method with EG.[139] EG–methanol solution of $La(NO_3)_3$ and $Al(NO_3)_3$ infiltrates the voids of the template. EG is oxidised by nitrates to form an La-Al oxalate derivative in the voids, and further calcination results in removal of the template and formation of the desired ordered macroporous perovskite-type $LaAlO_3$.[139]

Mixed lanthanide nitrate solution with citric acid as an additive infiltrates the voids of templates. The lanthanide nitrates react with citric acid to form mixed lanthanide citrate in the voids. Further calcination results in removal of the polymer template and conversion of the mixed lanthanide citrate to mixed lanthanide oxide ($Ce_{0.8}Gd_{0.2}O_{2-\delta}$) without collapse of the macroporous structure.[159]

3.5.8 Two-Step Templating Method

There are a few examples of initially produced macroporous (inverse opal strucuture) materials being used as templates (Table 3.5 and Figure 3.9). After metal salts are deposited on the surfaces of marcroporous (inverse opal strucuture) materials, the macroporous template is removed. Replication of the inverse opal structure results again in the production of an opal structure into which hollow spheres are packed. Macroporous carbon materials are prepared using a colloidal crystal template. Ethanol solution of titanium alkoxide [$Ti(OBu)_4$] infiltrates the voids of the macroporous carbon materials. After the alkoxide has been solidified by a sol-gel method, the carbon template is removed by calcination to form macroporous TiO_2.[181] Using the same template, macroporous $Nd_2O_2CO_3$ and Nd_2O_3 can be prepared.[181] If the macroporous carbon template has been covered by polyelectrolytes, TiO_2 can be deposited on the surface by hydrothermal reaction.[182]

3.5.9 Applications

The synthesis of ordered macroporous crystalline materials has been attracting much attention. Walls of macroporous materials are larger than those of mesoporous materials, and this macroporosity can be introduced into a wide variety of transition metal oxides. Potential applications of these materials include photonic materials, catalysts and electrode materials. The ordering scale is close to the wavelength of light, and interest has therefore been shown in photonic materials. In some cases, introduction of macroporosity increases the surface area, and these materials show better catalytic performance than that of nonporous materials. Similar to mesoporous materials, macropores are favoured for diffusion of reactants compared with nonporous materials and many applications, such as in a Li battery electrode, have been reported.

Table 3.5 Two-step templating method for metal oxides

Metal	Precursor solution	Template[a]	Solidification	Template removal	Crystal Phase	Properties and application	Ref.
Ti	Ti(OBu)$_4$ in ethanol (1:1 vol)	3DOM carbon	Air dry	Calc. (600 °C)	TiO$_2$ (anatase)		[181]
Ti	Titanium(IV) bis(ammonium lactate) dihydroxide (0.05–0.2 M) in H$_2$O	3DOM carbon treated with polyelectrolytes	Hydrothermal treatment at 140–240 °C)	Not removed	TiO$_2$ (anatase) 140 °C TiO$_2$ (anatase + rutile) 190 °C		[182]
Nd	Nb-EDTA (1 mmol) in acetic acid– NH$_3$–H$_2$O (5 ml)	3DOM carbon	Air dry	Calc. (720 °C)	Nd$_2$O$_2$CO$_3$ (600 °C) Nd$_2$O$_3$ (700 °C)		[181]

[a]3DOM: three-dimensionally ordered macroporous.

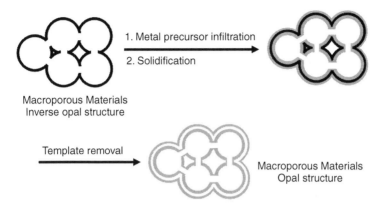

Figure 3.9 Two-step preparation of macroporous materials

3.6 CONCLUSION

We have summarised methods for the synthesis of ordered porous (micro, meso and macro) crystalline transition metal oxides. Crystalline transition metal oxides have unique redox, magnetic and electronic properties due to an incomplete d subshell. Properties such as size selectivity, facile and homogeneous access of reactants to the surface and photonic properties are added by introduction of ordered pores. The design and syntheses of ordered porous crystalline transition metal oxides remain areas of intensive research.

REFERENCES

[1] F. Schüth, K.S.W. Sing and J. Weitkamp (Eds), *Handbook of Porous Solids*, Wiley-VCH, Weinheim, 2002.

[2] K.S.W. Sing, D.H. Everett, R.A.W. Haul, L. Moscou, R.A. Pierotti, J. Rouquérol and T. Siemieniewska, *Pure Appl. Chem.*, **57**, 603 (1985).

[3] S.M. Auerbach, K.A. Carrado and P.K. Dutta (Eds), *Handbook of Zeolite Science and Technology*, Marcel Dekker, New York, 2003.

[4] T. Yanagisawa, T. Shimizu, K. Kuroda and C. Kato, *Bull. Chem. Soc. Jpn.*, **63**, 988 (1990).

[5] C.T. Kresge, M.E. Leonowicz, W.J. Roth, J.C. Vartuli and J.S. Beck, *Nature*, **359**, 710 (1992).

[6] J.S. Beck, J.C. Vartuli, W.J. Roth, M.E. Leonowicz, C.T. Kresge, K.D. Schmitt, C.T.-W. Chu, D.H. Olson, E.W. Sheppard, S.B. McCullen, J.B. Higgins and J.L. Schlenker, *J. Am. Chem. Soc.*, **114**, 10834 (1992).

[7] S. Inagaki, Y. Fukushima and K. Kuroda, *J. Chem. Soc., Chem. Commun.*, 680 (1993)

[8] B.T. Holland, C.F. Blanford and A. Stein, *Science*, 281, 538 (1998).

[9] J.E.G.J. Wijnhoven and W.L. Vos, *Science*, 281, 802 (1998).

[10] A.D. McNaught and A. Wilkinson, *IUPAC. Compendium of Chemical Terminology, The Gold Book*, 2nd Edn, Blackwell Scientific Publications, Oxford, 1997.

[11] S. Natarajan and S. Mandal, *Angew. Chem. Int. Ed.*, 47, 4798 (2008).

[12] G. Férey, C. Mellot-Draznieks and T. Loiseau, *Solid State Sci.*, 5, 79 (2003).

[13] A.K. Cheetham, G. Férey and T. Loiseau, *Angew. Chem. Int. Ed.*, 38, 3268 (1999).

[14] S.L. Suib, *J. Mater. Chem.*, 18, 1623 (2008).

[15] S.L. Suib, *Acc. Chem. Res.*, 41, 479 (2008).

[16] Q. Feng, H. Kanoh and K. Ooi, *J. Mater. Chem.*, 9, 319 (1999).

[17] C.-L. O'Young, R.A. Sawicki and S.L. Suib, *Miroporous Mater.*, 11, 1 (1997).

[18] X. Shen, Y. Ding, J. Liu, K. Laubernds, R.P. Zerger, M. Polverejan, Y.-C. Son, M. Aindow and S.L. Suib, *Chem. Mater.*, 16, 5327 (2004).

[19] X.-F. Shen, Y.-S. Ding, J. Liu, J. Cai, K. Laubernds, R.P. Zerger, A. Vasiliev, M. Aindow and S.L. Suib, *Adv. Mater.*, 17, 805 (2005).

[20] Y.F. Shen, R.P. Zerger, R. DeGuzman, S.L. Suib, L. McCurdy, D.I. Potter and C.L. O'Young, *Science*, 260, 511 (1993).

[21] G.-G. Xia, W. Tong, E.N. Tolentino, N.-G. Duan, S.L. Brock, J.-Y. Wang, S.L. Suib and T. Ressler, *Chem. Mater.*, 13, 1585 (2001).

[22] M. Sadakane, K. Kodato, T. Kuranishi, Y. Nodasaka, K. Sugawara, N. Sakaguchi, T. Nagai, Y. Matsui and W. Ueda, *Angew. Chem. Int. Ed.*, 47, 2493 (2008).

[23] M. Sadakane, N. Watanabe, T. Katou, Y. Nodasaka and W. Ueda, *Angew. Chem. Int. Ed.*, 119, 1493 (2007).

[24] N. Watanabe and W. Ueda, *Ind. Eng. Chem. Res.*, 45, 607 (2006).

[25] F. Wang and W. Ueda, *Appl. Catal., A*, 346, 155 (2008).

[26] F. Wang and W. Ueda, *Chem. Lett.*, 37, 184 (2008).

[27] T. Katou, D. Vitry and W. Ueda, *Chem. Lett.*, 32, 1028 (2003).

[28] M. Sadakane, K. Yamagata, K. Kodato, K. Endo, K. Toriumi, Y. Ozawa, T. Ozeki, T. Nagai, Y. Matsui, N. Sakaguchi, W.D. Pyrz, D.J. Buttrey, D.A. Bolm, T. Vogt and W. Ueda, *Angew. Chem. Int. Ed.*, 48, 3782 (2009).

[29] F. Schüth, *Angew. Chem. Int. Ed.*, 42, 3604 (2003).

[30] A. Taguchi and F. Schüth, *Microporous Mesoporous Mater.*, 77, 1 (2005).

[31] Z.-R. Tian, W. Tong, J.-Y. Wang, N.-G. Duan, V.V. Krishnan and S.L. Suib, *Science*, 276, 926 (1997).

[32] P. Yang, D. Zhao, D.I. Margolese, B.F. Chmelka and G.D. Stucky, *Nature*, 396, 152 (1998).

[33] P. Yang, D. Zhao, D.I. Margolese, B.F. Chmelka and G.D. Stucky, *Chem. Mater.*, 11, 2813 (1999).

[34] J.N. Kondo and K. Domen, *Chem. Mater.*, 20, 835 (2008).

[35] T. Katou, B. Lee, D. Lu, J.N. Kondo, M. Hara and K. Domen, *Angew. Chem. Int. Ed.*, 42, 2382 (2003).

[36] N. Shirokura, K. Nakajima, A. Nakabayashi, D. Lu, M. Hara, K. Domen, T. Tatsumi and J.N. Kondo, *Chem. Commun.*, 2188 (2006).

[37] Y. Noda, B. Lee, K. Domen and J.N. Kondo, *Chem. Mater.*, 20, 5361 (2008).

[38] X. Chen and S.S. Mao, *Chem. Rev.*, 107, 2891 (2007).

[39] Z. Peng, Z. Shi and M. Liu, *Chem. Commun.*, 2125 (2000).

[40] E.L. Crepaldi, G.J. de A.A. Soler-Illia, D. Grosso and C. Sanchez, *New J. Chem.*, **27**, 9 (2003).

[41] E.L. Crepaldi, G.J. de A.A. Soler-Illia, D. Grosso, F. Cagnol, F. Ribot and C. Sanchez, *J. Am. Chem. Soc.*, **125**, 9770 (2003).

[42] D. Grosso, G.J. de A.A. Soler-Illia, D.L. Crepaldi, B. Charleux and C. Sanchez, *Adv. Funct. Mater.*, **13**, 37 (2003).

[43] P.C.A. Alberius, K.L. Frindell, R.C. Hayward, E.J. Kramer, G.D. Stucky and B.F.Chmelka, *Chem. Mater.*, **14**, 3284 (2002).

[44] B.L. Kirsch, E.K. Richman, A.E. Riley and S.H. Tolbert, *J. Phys. Chem. B*, **108**, 12698 (2004).

[45] J. Lee, M.C. Orilall, S.C. Warren, M. Kamperman, F.J. Disalvo and U. Wiesner, *Nat. Mater.*, **7**, 222 (2008).

[46] J.N. Kondo, T. Yamashita, K. Nakajima, D. Lu, M. Hara and K. Domen, *J. Mater. Chem.*, **15**, 2035 (2005).

[47] E.L. Crepaldi, G.J. de A.A. Soler-Illia, A. Bouchara, D. Grosso, D. Durand and C. Sanchez, *Angew. Chem. Int. Ed.*, **115**, 361 (2003).

[48] T. Brezesinski, M. Groenewolt, N. Pinna, H. Amenitsch, M. Antonietti and B.M. Smarsly, *Adv. Mater.*, **18**, 1827 (2006).

[49] Y. Takahara, J.N. Kondo, T. Takata, D. Lu and K. Domen, *Chem. Mater.*, **13**, 1194 (2001).

[50] M. Uchida, J.N. Kondo, D. Lu and K. Domen, *Chem. Lett.*, **31**, 498 (2002).

[51] K. Nakajima, M. Hara, K. Domen and J.N. Kondo, *Chem. Lett.*, **34**, 394 (2005).

[52] J.N. Kondo, M. Uchida, K. Nakajima, L. Daling, M. Hara and K. Domen, *Chem. Mater.*, **16**, 4303 (2004).

[53] N. Shirokura, K. Nakajima, A. Nakabayashi, D. Lu, M. Hara, K. Domen, T. Tatsumi and J.N. Kondo, *Chem. Commun.*, 2188 (2006).

[54] Y. Noda, B. Lee, K. Domen and J.N. Kondo, *Chem. Mater.*, **20**, 5361 (2008).

[55] S. Ikeda, N. Sugiyama, S. Murakami, H. Kominami, Y. Kera, H. Noguchi, K. Uosaki, T. Torimoto and B. Ohtani, *Phys. Chem. Chem. Phys.*, **5**, 778 (2003).

[56] B. Ohtani and S.-I. Nishimoto, *J. Phys. Chem.*, **97**, 920 (1993).

[57] T. Katou, D. Lu, J.N. Kondo and K. Domen, *J. Mater. Chem.*, **12**, 1480 (2002).

[58] T. Katou, B. Lee, D. Lu, J.N. Kondo, M. Hara and K. Domen, *Angew. Chem. Int. Ed.*, **42**, 2382 (2003).

[59] B. Lee, D. Lu, J.N. Kondo and K. Domen, *Chem. Commun.*, 2118 (2001).

[60] B. Lee, T. Yamashita, D. Lu, J.N. Kondo and K. Domen, *Chem. Mater.*, **14**, 867 (2002).

[61] D. Lu, B. Lee, J.N. Kondo and K. Domen, *Microporous Mesoporous Mater.*, **75**, 203 (2004).

[62] A.K. Sinha and K. Suzuki, *Angew. Chem. Int. Ed.*, **44**, 271 (2005).

[63] T. Brezesinski, D.F. Rohlfing, S. Sallard, M. Antonietti and B.M. Smarsly, *Small*, **2**, 1203 (2006).

[64] S. Banerjee, A. Santhanam, A. Dhathathreyan and P.M. Rao, *Langmuir*, **19**, 5522 (2003).

[65] D.N. Srivastava, N. Perkas, A. Gedanken and I. Felner, *J. Phys. Chem. B*, **106**, 1878 (2002).

[66] A.-H. Lu and F. Schüth, *Adv. Mater.*, **18**, 1793 (2006).

[67] K. Zhu, B. Yue, W. Zhou and H. He, *Chem. Commun.*, 98 (2003).

[68] K. Jiao, B. Zhan, B. Yue, Y. Ren, S. Liu, S. Yan, C. Dickinson, W. Zhou and H. He, *Chem. Commun.*, 5618 (2005).

[69] B. Tian, X. Liu, H. Yang, S. Xie, C. Yu, B. Tu and D. Zhao, *Adv. Mater.*, **15**, 1370 (2003).

[70] C. Dickinson, W. Zhou, R.P. Hodgkins, Y. Shi, D. Zhao and H. He, *Chem. Mater.*, **18**, 3088 (2006).

[71] L. Dupont, S. Laruelle, S. Grugeon, C. Dickinson, W. Zhou and J.-M. Tarascon, *J. Power Sources*, **175**, 502 (2008).

[72] B. Yue, H. Tang, Z. Kong, K. Zhu, C. Dickinson, W. Zhou and H. He, *Chem. Phys. Lett.*, **407**, 83 (2005).

[73] X. Cui, H. Zhang, X. Dong, H. Chen, L. Zhang, L. Guo and J. Shi, *J. Mater. Chem.*, **18**, 3575 (2008).

[74] J.-Y. Luo, J.-J. Zhang and Y.-Y. Xia, *Chem. Mater.*, **18**, 5618 (2006).

[75] F. Jiao and P.G. Bruce, *Adv. Mater.*, **19**, 657 (2007).

[76] F. Jiao, J. Bao, A.H. Hill and P.G. Bruce, *Angew. Chem. Int. Ed.*, **47**, 9711 (2008).

[77] F. Jiao and P.G. Bruce, *Angew. Chem. Int. Ed.*, **43**, 5958 (2004).

[78] A. Lezau, M. Trudeau, G.M. Tsoi, L.E. Wenger and D. Antonelli, *J. Phys. Chem. B*, **108**, 5211 (2004).

[79] J.W. Long, M.S. Logan, C.P. Rhodes, E.E. Carpenter, R.M. Stroud and D.R. Rolison, *J. Am. Chem. Soc.*, **126**, 16879 (2004).

[80] A.S. Malik, M.J. Duncan and P.G. Bruce, *J. Mater. Chem.*, **13**, 2123 (2003).

[81] F. Jiao, A. Harrison, J.-C. Jumas, A.V. Chardwick, W. Kockelmann and P.G. Bruce, *J. Am. Chem. Soc.*, **128**, 5468 (2006).

[82] A.H. Hill, F. Jiao, P.G. Bruce, A. Harrison, W. Kockelmann and C. Ritter, *Chem. Mater.*, **20**, 4891 (2008).

[83] B. Tian, X. Liu, L.A. Solovyov, Z. Liu, H. Yang, Z. Zhang, S. Xie, F. Zhang, B. Tu., C. Yu, O. Terasaki and D. Zhao, *J. Am. Chem. Soc.*, **126**, 865 (2004).

[84] Y. Wang, C.-M. Yang, W. Schmidt, B. Spliethoff, E. Bill and F. Schüth, *Adv. Mater.*, **17**, 53 (2005).

[85] K.M. Shaju, F. Jiao, A. Débart and P.G. Bruce, *Phys. Chem. Chem. Phys.*, **9**, 1837 (2007).

[86] H. Tüysüz., M. Commotti and F. Schüth, *Chem. Commun.*, 4022 (2008).

[87] H. Tüysüz., C.W. Lehmann, H. Bongard, B. Tesche, R. Schmidt and F. Schüth, *J. Am. Chem. Soc.*, **130**, 11510 (2008).

[88] W. Yue, A.H. Hill, A. Harrison and W. Zhou, *Chem. Commun.*, 2518 (2007).

[89] F. Jiao, K.M. Shaju and P.G. Bruce, *Angew. Chem. Int. Ed.*, **44**, 6550 (2005).

[90] Y. Wang, J. Ren, Y. Wang, F. Zhang, X. Liu, Y. Guo and G. Lu, *J. Phys. Chem. C*, **112**, 15293 (2008).

[91] F. Jiao, J.-C. Jumas, M. Womes, A.V. Chadwick, A. Harrison and P.G. Bruce, *J. Am. Chem. Soc.*, **128**, 12905 (2006).

[92] X. Xu, B. Tian, J. Kong, S. Zhang, B. Liu and D. Zhao, *Adv. Mater.*, **15**, 1932 (2003).

[93] F. Jiao, A.H. Hill, A. Harrison, A. Berko, A.V. Chadwick and P.G. Bruce, *J. Am. Chem. Soc.*, **130**, 5262 (2008).

[94] X. Lai, X. Li, W. Geng, J. Tu, J. Li and S. Qui, *Angew. Chem. Int. Ed.*, **46**, 738 (2007).

[95] S.C. Laha and R. Ryoo, *Chem. Commun.*, 2138 (2003).

[96] R.C. Schroden and A. Stein, in *Colloids and Colloid Assemblies in 3D Ordered Macroporous Material*, F. Caruso (Ed.), Wiley-VCH Verlag GmbH and Co., KGaA, Weinheim, 2004, p. 465.

[97] A. Stein and R.C. Schroden, *Curr. Opin. Solid State Mater. Sci.*, **5**, 553 (2001).
[98] A. Stein, *Microporous Mesopo'rous Mater.*, **44–45**, 227 (2001).
[99] C.F. Blanford, H. Yan, R.C. Schroden, M. Al-Daous and A. Stein, *Adv. Mater.*, **13**, 401 (2001).
[100] M.L.K. Hoa, M. Lu and Y. Zhang, *Adv. Colloid Interface Sci.*, **121**, 9 (2006).
[101] A. Stein, F. Li and N.R. Denny, *Chem. Mater.*, **20**, 649 (2008).
[102] J.S. Yin and Z.L. Wang, *Adv. Mater.*, **11**, 469 (1999).
[103] B.T. Holland, C.F. Blanford, T. Do and A. Stein, *Chem. Mater.*, **11**, 795 (1999).
[104] N.P. Johnson, D.W. McComb, A. Richel, B.M. Treble and R.M. DeLaRue, *Synth. Met.*, **116**, 469 (2001).
[105] M. Lanata, M. Cherchi, A. Zappettini, S.M. Pietralunga and M. Martinelli, *Opt. Mater.*, **17**, 11 (2001).
[106] D. Wang, R.A. Caruso and F. Caruso, *Chem. Mater*, **13**, 364 (2001).
[107] M.E. Turner, T.J. Trentler and V.L. Colvin, *Adv. Mater.*, **13**, 180 (2001).
[108] J.E.G.J. Wijnhoven, L. Bechger and W.L. Vos, *Chem. Mater.*, **13**, 4486 (2001).
[109] P. Ni, B. Cheng and D. Zhang, *Appl. Phys. Lett.*, **80**, 1879 (2002).
[110] W. Dong, H. Bongard, B. Tesche and F. Marlow, *Adv. Mater.*, **14**, 1457 (2002).
[111] W. Dong, H.J. Bongard and F. Marlow, *Chem. Mater.*, **15**, 568 (2003).
[112] R.C. Schroden, M. Al-Daous, C.F. Blanford and A. Stein, *Chem. Mater.*, **14**, 3305 (2002).
[113] S. Kuai, S. Badilescu, G. Bader, R. Brüning, X. Hu and V.-V. Truong, *Adv. Mater.*, **15**, 73 (2003).
[114] L. Kavan, M. Zukalová, M. Kalbáč and M. Graetzel, *J. Electrochem. Soc.*, **151**, A1301 (2004).
[115] K. Sasahara, T. Hyodo, Y. Shimizu and M. Egashira, *J. Eur. Ceram. Soc.*, **24**, 1961 (2004).
[116] H. Li, H. Wang, A. Chen, B. Meng and X. Li, *J. Mater. Chem.*, **15**, 2551 (2005).
[117] L.I. Halaoui, N.M. Abrams and T.E. Mallouk, *J. Phys. Chem. B*, **109**, 6334 (2005).
[118] C.L. Huisman, J. Schoonman and A. Goossens, *Solar Energy Mater. Sol. Cells*, **85**, 115 (2005).
[119] J.I.L. Chen, G. vonFreymann, S.Y. Choi, V. Kitaev and G.A. Ozin, *Adv. Mater.*, **18**, 1915 (2006).
[120] J.I.L. Chen, G. vonFreymann, V. Kitaev and G.A. Ozin, *J. Am. Chem. Soc.*, **129**, 1196 (2007).
[121] I. Rodriguez, F. Ramiro-Manzano, P. Atienzar, J.M. Martinez, F. Meseguer, H. Garcia and A. Corma, *J. Mater. Chem.*, **17**, 3205 (2007).
[122] M.C. Carbajo, E. Enciso and M.J. Torralvo, *Colloids Surf. A*, **293**, 72 (2007).
[123] J.W. Galusha, C.-K. Tsung, G.D. Stucky and M.H. Bartl, *Chem. Mater*, **20**, 4925 (2008).
[124] G. Gundiah and C.N.R. Rao, *Solid State Sci.*, **2**, 877 (2000).
[125] C. Dionigi, P. Greco, G. Ruani, M. Cavallini, F. Borgatti and F. Biscarini, *Chem. Mater.*, **20**, 7130 (2008).
[126] P.R. Somani, C. Dionigi, M. Murgia, D. Palles, P. Nozar and G. Ruani, *Solar Energy Mater. Sol. Cells*, **87**, 513 (2005).
[127] J.S. King, D.P. Gaillot, E. Graugnard and C.J. Summer, *Adv. Mater.*, **18**, 1063 (2006).

[128] G. Subramanian, V.N. Manoharan, J.D. Thorne and D.J. Pine, *Adv. Mater.*, **11**, 1261 (1999).

[129] F. Tang, H. Fudouzi, T. Uchikoshi and Y. Sakka, *J. Eur. Ceram. Soc.*, **24**, 341 (2004).

[130] H.-H. Chen, H. Suzuki, O. Sato and Z.-Z. Gu, *Appl. Phys. A*, **81**, 1127 (2005).

[131] H. Yan, Y. Yang, Z. Fu, B. Yang, L. Xia, Y. Xu, S. Fu and F. Li, *Chem. Lett.*, **35**, 864 (2006).

[132] C.-Y. Kuo and S.-Y. Lu, *J. Am. Ceram. Soc.*, **90**, 1956 (2007).

[133] S. Nishimura, N. Abrams, B.A. Lewis, L.I. Halaoui, T.E. Mallouk, K.D. Benkstein, J. vande Lagemaat and A.J. Frank, *J. Am. Chem. Soc.*, **125**, 6306 (2003).

[134] M.C. Orilall, N.M. Abrams, J. Lee, F.J. DiSalvo and U. Wiesner, *J. Am. Chem. Soc.*, **130**, 8882 (2008).

[135] M.A. Al-Daous and A. Stein, *Chem. Mater.*, **15**, 2638 (2003).

[136] J.S. Sakamoto and B. Dunn, *J. Mater. Chem.*, **12**, 2859 (2002).

[137] H. Yan, S. Sokolov, J.C. Lytle, A. Stein, F. Zhang and W.H. Smyrl, *J. Electrochem. Soc.*, **150**, A1102 (2003).

[138] H. Yan, C.F. Blanford, B.T. Holland, W.H. Smyrl and A. Stein, *Chem. Mater.*, **12**, 1134 (2000).

[139] M. Sadakane, T. Horiuchi, N. Kato, C. Takahashi and W. Ueda, *Chem. Mater.*, **19**, 5779 (2007).

[140] J.C. Lytle, N.R. Denny, R.T. Turgeon and A. Stein, *Adv. Mater.*, **19**, 3682 (2007).

[141] S.-L. Kuai, G. Bader and P.V. Ashrit, *Appl. Phys. Lett.*, **86**, 221110 (2005).

[142] A. Sinitskii, V. Abramova, T. Laptinskaya and Y.D. Tretyakov, *Phys. Lett. A*, **366**, 516 (2007).

[143] M. Sadakane, K. Sasaki, H. Kunioku, B. Ohtani, W. Ueda and R. Abe, *Chem. Commun.*, 6552 (2008).

[144] T. Sumida, Y. Wada, T. Kitamura and S. Yanagida, *Chem. Lett.*, **31**, 180 (2002).

[145] C.F. Blanford, C.B. Carter and A. Stein, *J. Mater. Sci.*, **43**, 3539 (2008).

[146] J.C. Kim, Y.N. Kim, N.H. Hur, W.S. Kim and Y.G. Kang, *Phys. Status Solidi B*, **241**, 1587 (2004).

[147] H. Yan, C.F. Blanford, B.T. Holland, M. Parent, W.H. Smyrl and A. Stein, *Adv. Mater.*, **11**, 1003 (1999).

[148] X. Li, Y. Jiang, Z. Shi and Z. Xu, *Chem. Mater.*, **19**, 5424 (2007).

[149] X. Li, F. Tao, Y. Jiang and Z. Xu, *J. Colloid Interface Sci.*, **308**, 460 (2007).

[150] Q.Z. Wu, W. Shen, J.F. Liao and Y.G. Li, *Mater. Lett.*, **58**, 2688 (2004).

[151] G.I.N. Waterhouse, J.B. Metson, H. Idriss and D. Sun-Waterhouse, *Chem. Mater.*, **20**, 1183 (2008).

[152] Y. Zhang, Z. Lei, J. Li and S. Lu, *New. J. Chem.*, **25**, 1118 (2001).

[153] Z. Lei, J. Li, Y. Zhang and S. Lu, *J. Mater. Chem.*, **10**, 2629 (2000).

[154] I. Soten, H. Miguez, S.M. Yang, S. Petrov, N. Coombs, N. Tetreault, N. Matsuura, H.E. Ruda and G.A. Ozin, *Adv. Funct. Mater.*, **12**, 71 (2002).

[155] P. Harkins, D. Eustace, J. Gallagher and D.W. McComb, *J. Mater. Chem.*, **12**, 1247 (2002).

[156] S. Madhavi, C. Ferraris and T. White, *J. Solid State Chem.*, **179**, 866 (2006).

[157] E.M. Sorensen, S.J. Barry, H.-K. Jung, J.R. Rondinelli, J.T. Vaughey and K.R. Poeppelmeier, *Chem. Mater.*, **18**, 482 (2006).

[158] S.-W. Woo, K. Dokko and K. Kanamura, *Electrochim. Acta*, **53**, 79 (2007).

[159] D. Marrero-López, J.C. Ruiz-Morales, J. Peña-Martinez, J. Canales-Vázquez and P. Núñez, *J. Solid State Chem.*, **181**, 685 (2008).

[160] Y. Jin, Y. Zhu, X. Yang, C. Wei and C. Li, *Mater. Chem. Phys.*, **106**, 209 (2007).

[161] B. Li, J. Zhou, Q. Li, L. Li and Z. Gui, *J. Am. Ceram. Soc.*, **86**, 867 (2003).

[162] B. Li, J. Zhou, L. Li, X.J. Wang, X.H. Liu and J. Zi, *Appl. Phys. Lett.*, **83**, 4704 (2003).

[163] C. Wang, A. Geng, Y. Guo, S. Jiang and X. Qu, *Mater. Lett.*, **60**, 2711 (2006).

[164] G.A. Umeda, W.C. Chueh, L. Noailles, S.M. Haile and B.S. Dunn, *Energy Environ. Sci.*, **1**, 484 (2008).

[165] M.A. Carreon and V.V. Guliants, *Chem. Commun.*, 1438 (2001).

[166] M.A. Carreon and V.V. Guliants, *Chem. Mater*, **14**, 2670 (2002).

[167] D. Wang and F. Caruso, *Adv. Mater.*, **15**, 205 (2003).

[168] E. Géraud, S. Rafqah, M. Sarakha, C. Forano, V. Prevot and F. Leroux, *Chem. Mater.*, **20**, 1116 (2008).

[169] E.O. Chi, Y.N. Kim, J.C. Kim and N.H. Hur, *Chem. Mater.*, **15**, 1929 (2003).

[170] Y.N. Kim, E.O. Chi, J.C. Kim, E.K. Lee and N.H. Hur, *Solid State Commun.*, **123**, 339 (2003).

[171] D. Tonti, M.J. Torralvo, E. Enciso, I. Sobrados and J. Sanz, *Chem. Mater.*, **20**, 4783 (2008).

[172] Y.N. Kim, S.J. Kim, E.K. Lee, E.O. Chi, N.H. Hur and C.S. Hong, *J. Mater. Chem.*, **14**, 1774 (2004).

[173] M. Sadakane, T. Asanuma, J. Kubo and W. Ueda, *Chem. Mater.*, **17**, 3546 (2005).

[174] M. Sadakane, C. Takahashi, N. Kato, T. Asanuma, H. Ogihara and W. Ueda, *Chem. Lett.*, **35**, 480 (2006).

[175] M. Sadakane, C. Takahashi, N. Kato, H. Ogihara, Y. Nodasaka, Y. Doi, Y. Hinatsu and W. Ueda, *Bull. Chem. Soc. Jpn.*, **80**, 677 (2007).

[176] F. Chen, C. Xia and M. Liu, *Chem. Lett.*, **30**, 1032 (2001).

[177] N.S. Ergang, J.C. Lytle, H. Yan and A. Stein, *J. Electrochem. Soc.*, **152**, A1989 (2005).

[178] H. Yan, C.F. Blanford, W.H. Smyrl and A. Stein, *Chem. Commun.*, 1477 (2000).

[179] S. Sokokov and A. Stein, *Mater. Lett.*, **57**, 3593 (2003).

[180] H. Yan, C.F. Blanford, J.C. Lytle, C.B. Carter, W.H. Smyrl and A. Stein, *Chem. Mater.*, **13**, 4314 (2001).

[181] Z. Lei, J. Li, Y. Ke, Y. Zhang, H. Zhang, F. Li and J. Xing, *J. Mater. Chem.*, **11**, 2930 (2001).

[182] Z. Wang, N.S. Ergang, M.A. Al-Daous and A. Stein, *Chem. Mater.*, **17**, 6805 (2005).

4

Templated Porous Carbon Materials: Recent Developments

Yongde Xia, Zhuxian Yang and Robert Mokaya
School of Chemistry, University of Nottingham, University Park, Nottingham, UK

4.1 INTRODUCTION

Porous materials contain voids as the majority phase, either with random character (disordered pore systems) or with high regularity (ordered pore systems).[1] Ordered porous materials are of great scientific and technological interest due to their abilities to interact with atoms, ions and molecules not only at their surfaces, but also throughout the bulk of the materials.[2] A classification of pore systems according to the IUPAC definition based on pore diameter is as follows: pores below 2 nm are called micropores and the typical representative materials of this class are the well-known zeolites. Pores that fall in the range 2–50 nm are called mesopores and the M41S family (MCMs and SBA series materials) are prominent examples of this class. Those above 50 nm are called macropores and amorphous aluminosilicate and controlled porous glass are typical representatives of this type of materials.

Porous Materials Edited by Duncan W. Bruce, Dermot O'Hare and Richard I. Walton
© 2011 John Wiley & Sons, Ltd.

Porous carbons, a class of nonoxide porous materials, are of great importance due to their applications in water and air purification, as gas hosts, templates,[3, 4] or components of electrodes.[5] The widespread use of porous carbons results from their remarkable properties, such as the hydrophobic nature of their surfaces, high specific surface areas, large pore volumes, chemical inertness, good thermal stability, good mechanical stability, easy handling and low cost of manufacture.[6] Most porous carbons are primarily microporous[7, 8] and the microporous nature is well-suited to many applications involving small molecules, such as molecular sieving, adsorption and catalysis. However, there are a number of other potential applications in which the presence of mesopores or even macropores would be preferable, for instance, adsorption of large, hydrophobic molecules such as vitamins, dyes and polymers, chromatographic separations and electrochemical double-layer capacitors.[7] Therefore, the development of mesoporous and macroporous, as well as microporous, carbon materials is of great importance not only from a fundamental research point of view, but also from the practical application point of view.

Porous carbons are usually obtained *via* carbonisation of precursors of natural or synthetic origin, followed by activation.[9] Various synthesis strategies have been explored to prepare porous carbons with controlled pore structure both at the micropore and mesopore level.[7] At the micropore level, molecular sieving carbons that possess uniform micropores may be prepared *via* carbonisation of appropriate carbon precursors. For example, Miura *et al.* illustrated the potential of targeted synthesis of molecular sieving carbons with a uniform pore size of *ca* 0.35 nm by changing the carbonisation temperature and the mixing ratio of coal, pitch, phenol and formaldehyde precursors.[10] Microporosity may also be controlled simply by carbonising ion-exchange resins at 500–900 °C, whereby spherical polystyrene-based resins with a sulfonic acid group are ion-exchanged with several kinds of cations including H^+, K^+, Na^+, Zn^{2+}, Cu^{2+}, Fe^{2+}, Ni^{2+} and Fe^{3+}. The molecular sieving carbons prepared from the resins with di- or trivalent cations have sharp pore size distribution varying between 0.38 nm and 0.45 nm depending on the type of cation.[11] Recently, microporous carbons with narrow pore size distribution in the range of 0.77–0.91 nm were obtained *via* carbonisation and activation of waste-derived biomass.[12] By adjusting the reaction parameters, different carbon materials with controlled morphology and texture were prepared through the carbonisation and activation of hydrolytic lignin from inexpensive biomass waste. The biomass-derived microporous carbons exhibit high surface areas up to $3100\,m^2\,g^{-1}$, large micropore volume up to 1.68 cm^3 g^{-1} and considerable gas uptake capability. In the mesopore regime, catalytic activation[13, 14] and carbonisation of polymer blends,[15] organic gels[16] and colloidal imprinting[17–19] have been proposed for the preparation of mesoporous carbons. Carbon materials made from the catalytic activation and carbonisation usually display disordered structures with wide pore size distribution.[13–16]

The template carbonisation route allows the preparation of carbon materials with a controlled architecture, particle size and relatively narrow pore size distribution.[20] Therefore, template carbonisation has attracted much attention for the preparation of ordered porous carbon materials. Generally, two types of template, classified as soft template and hard template, have been described as moulds to form porous materials. The hard template carbonisation approach, as shown in Figure 4.1,[21] usually involves the following steps: (a) the preparation of a porous template with controlled structure; (b) the introduction of a suitable carbon precursor into the template pores either by wet impregnation, chemical vapour deposition (CVD) or a combination of both methods; (c) the polymerisation and carbonisation of the organic-inorganic composite; and (d) the removal of the template. Following this template procedure, the carbon formed in the pores of the host template turns into a continuous carbon framework, while the space once occupied by the host template is thus transferred into the pores of the resulting carbon material. Hard templating, therefore, offers a rigid, nanocasting mould as a true template to replicate other materials. The hard templates must be thermally stable, chemically inert to carbon precursors and be amenable to removal to generate a pure carbon framework. Ordered inorganic porous solids are most commonly used as hard templates. It is worthwhile pointing out that the pore size of the resulting porous carbon materials is not always exactly the size of the pore wall thickness of the host templates since there is some shrinkage during carbonisation at elevated temperatures.

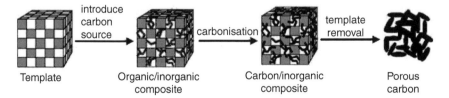

Template — Organic/inorganic composite — Carbon/inorganic composite — Porous carbon

introduce carbon source → carbonisation → template removal

Figure 4.1 Schematic illustration of hard template method for the preparation of porous carbon materials from porous inorganic templates.[21] Reprinted with permission from T. Kyotani, *Bull. Chem. Soc. Jpn.*, **79**, 1322. Copyright (2006) The Chemical Society of Japan

On the other hand, the soft template method involves cooperative assembly between the structure-directing agents (usually surfactants) and organic precursor species in solution. Therefore, the carbon structures obtained *via* soft templating are more flexible and their formation is dependent on temperature, type of solvent and ionic strength. However, there are currently only limited examples for the successful fabrication of porous carbon *via* the soft template method, which were reviewed recently by Wan *et al.*[22] Soft template and hard template routes have been classified as endotemplate and exotemplate, respectively.[23]

Historically, the hard template carbonisation method was first reported by Knox *et al.* in 1986.[24] They demonstrated the synthesis of graphitised porous carbons for liquid chromatography separation by impregnation of spherical porous silica gel with phenolic resin and subsequent carbonisation and removal of silica gel. Since the first reported use of template carbonisation, the method has been employed extensively to prepare ordered porous carbons.[20, 25-27] A variety of inorganic porous templates including zeolites,[27-30] various mesoporous silicas,[5, 20, 25, 31-36] colloid crystals,[37, 38] polystyrene[39] and aluminium oxide membrane have been explored to synthesise microporous, mesoporous and macroporous carbon materials. Various carbon precursors including sucrose,[25] furfuryl alcohol,[27, 28, 40-42] acrylonitrile,[27, 41] propylene,[28, 29] pyrene,[41] vinyl acetate[41] and acetonitrile[32, 43-45] have been applied to prepare porous carbons. Indeed, the past ten years have witnessed rapid advances in using template carbonisation to produce ordered porous carbon materials, ranging from microporous to mesoporous and macroporous carbons. The template carbonisation method has thus been regarded as one of the most effective approaches to prepare porous carbon materials with desirable physical and chemical properties. It has, therefore, opened up new opportunities in making novel porous materials for a wide range of applications.

Several reviews covering the synthesis, properties and applications of porous carbons, especially mesoporous carbon materials, can be found in the literature.[46-54] In this chapter, we summarise the recent developments in the synthesis and characterisation of templated porous carbon materials. Particular attention is paid to the synthesis of structurally ordered porous carbon materials with narrow pore size distribution *via* both hard and soft template methods. We especially emphasise those so-called breakthroughs in the preparation of porous carbon materials. The chapter is divided into three sections according to the pore size of carbon materials: we first consider the synthesis of microporous carbon materials using zeolites and clays as hard template, then summarise the preparation of mesoporous carbon materials *via* both hard template and self-assembly soft template methods, and finally we present the preparation of macroporous carbon materials from crystal colloid templates.

4.2 MICROPOROUS CARBON MATERIALS

4.2.1 Zeolites as Hard Template

Zeolites are highly crystalline aluminosilicate materials having uniform, subnanometre sized pores. The pore channel apertures are in the range of 0.3–1.5 nm[55] depending on the type of zeolite and its preparation history, e.g. calcination, leaching or various chemical treatments. Zeolites have been used widely in ion exchange, separation, catalysis, chemical sensing and host–guest chemistry, particularly as shape-selective catalysts owing to their uniform molecular-sized pores.[56] Since the wall thickness for zeolites is less than 1 nm, the periodic pore structures and well-defined internal nanospaces of zeolites provide opportunities to control the nanostructure and morphology of microporous carbon materials at the nanometre level. It is reasonable to expect that if such a nanospace in a zeolite is filled with carbon, the generated porous carbon should reflect the porosity of the original zeolite template. Zeolites have therefore been widely used as inorganic templates for the synthesis of microporous carbons with uniform pore sizes.[27–29, 41, 42, 57–61]

4.2.1.1 Disordered Carbon Replicas of Zeolite Templates

Kyotani et al. reported the preparation of high-surface-area porous carbons by carbonisation using zeolite Y as the template, with polyacrylonitrile, poly(furfuryl alcohol) and propylene as the carbon sources, respectively.[27] The resulting microporous carbons exhibited high surface areas of over 2000 m^2 g^{-1}. The morphology of the resulting microporous carbon particles was similar to that of the zeolite template particles, thus demonstrating that the carbonisation occurred inside the channels of the zeolite template rather than on the outside, which would generate irregular particles. However, the resulting carbon materials were devoid of any regular order stacking structures similar to zeolite Y. Enzel and Bein reported that polyacrylonitrile was carbonised inside the channels of zeolite NaY, sodium mordenite and silicalite to obtain microporous carbon.[57] Johnson et al. synthesised phenol-formaldehyde polymers inside various zeolites including zeolites Y, β, and L, and then carbonised the polymer/zeolite composites to obtain microporous carbon.[62] They found that the three-dimentional zeolite Y and β templates led to microporous carbon replicas, while the one-dimentional zeolite L gave a dense carbon with complete collapse of the replica occurring upon removal of the template. Rodriguez-Miraso et al. prepared porous carbons by chemical vapour infiltration of a wide-pore zeolite Y with propylene.[29] Meyers et al. reported the preparation of carbon materials with surface area from 2.1 to $947 \, m^2 \, g^{-1}$ at low temperature of 600 °C using zeolite Y, β, and ZSM-5 and montmorillonite clay K10 as templates with acrylonitrile, furfuryl alcohol, pyrene, and vinyl acetate as carbon precursors.[41] The use of an ammonium-exchanged form of zeolite Y as template and furfuryl alcohol as a carbon precursor to prepare microporous carbons by a simple impregnation method was reported by Su et al.[42] The resulting microporous carbon possesses high surface area and pore volume. However, the above-mentioned studies on zeolite-templated carbonisation result exclusively in microporous carbons with a significant proportion of mesopores,[27] and the ordered structure of the zeolite is not transferred faithfully to the resulting carbons.[27, 62] This could be due to low carbon precursor loading and/or spatial restrictions (within the zeolite template pores) during the precursor polymerisation[57] and limited subsequent carbon growth.[27] The incomplete carbon infiltration and filling of the zeolite pores implies that the resulting porous carbon is not a true replica of zeolite despite its morphology being similar to that of the zeolite template. The generation of mesopores results from the partial collapse of the carbon framework after the removal of the zeolite template by HF etching because the thin wall thickness of the carbon, derived from the small pores of the zeolite template, does not exhibit a sufficiently high mechanical strength to survive the removal of the template.

4.2.1.2   Ordered Carbon Replicas of Zeolite Templates

4.2.1.2.1   Zeolite Y as Hard Template
In order to improve the structural ordering of zeolite-templated carbons, Ma et al. have investigated systematically the synthesis of microporous carbons using zeolite Y as hard template. They used a two-step method to prepare an ordered, microporous carbon with high surface area, which retained the structural regularity of zeolite Y by filling as much carbon precursor as possible into the zeolite pores so as to prevent any subsequent partial collapse of the resulting carbon framework.[30] In the

two-step method, additional incorporation of carbon was achieved by a CVD process using propylene gas as carbon source after an initial carbonisation step involving heat-treatment of a zeolite/furfuryl alcohol composite at 700 °C. The carbon obtained after the removal of the zeolite template exhibited an ordered zeolite replica structure with long-range ordering, confirmed by the strong (111) reflection of zeolite Y at a 2θ angle of 6.26° in the X-ray diffraction (XRD) pattern, as shown in Figure 4.2. This was the first report on the successful replication of zeolite structures into microporous carbon material. It was claimed that the carbons obtained *via* this two-step method only contain microporosity.[30] Although ordered microporous carbon materials with a negative replica structure of zeolite Y were obtained by the two-step method, there was still an amorphous (002) peak at the 2θ angle of 23° in the XRD pattern, which demonstrated the partial collapse of the carbon framework in the zeolite channels, and/or the presence of graphitic/turbostratic domains deposited outside the zeolite pore system.

Ma *et al.* reported the synthesis of ordered microporous carbon having a rigid framework, but without the amorphous (002) peak, by heat treatment of the carbon/zeolite composite obtained by the two-step

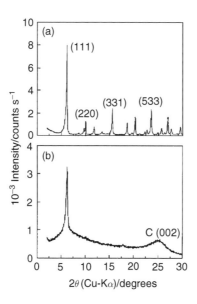

Figure 4.2 XRD patterns of (a) zeolite Y and (b) zeolite Y-templated carbon liberated from the carbon/zeolite Y composite.[30] Reprinted with permission from Z.X. Ma, T. Kyotani and A. Tomita, *Chem. Comm*, Preparation of a high surface area microporous carbon having the structural regularity of Y zeolite. Issue 23, 2365–2366. Copyright (2000) RSC Publishing

Figure 4.3 HRTEM image and corresponding diffraction pattern (inset) of zeolite Y-templated ordered microporous carbon.[63] Reprinted with permission from Z.X. Ma, T. Kyotani, Z. Liu, O. Terasaki and A. Tomita, *Chem. Mater.*, Very High Surface Area Microporous Carbon with a Three-Dimensional Nano-Array Structure: Synthesis and its Molecular Structure. **13**, 4413. Copyright (2001) American Chemical Society

method at 900 °C.[63] The carbon inside the channels seemed to be better carbonised and had a more rigid and stable structure with enhanced replication of zeolite-like ordering and exhibited high surface area up to $3600 \, m^2 \, g^{-1}$ and pore volume of $1.5 \, cm^3 \, g^{-1}$. Figure 4.3 shows the high-resolution transmission electron microscopy (HRTEM) image of the ordered microporous carbon obtained from zeolite Y.[63]

Kyotani *et al.* extended the two-step replication process to other zeolite templates including zeolite β, ZSM-5, mordenite and zeolite L, in order to make various ordered microporous carbon arrays.[28] The optimum conditions used to obtain carbons with the highest long-range ordering varied depending on the zeolite templates. When using a simple CVD method, unlike in the case of zeolite Y, some zeolites generated carbons with structural regularity of the corresponding zeolite template. The extent of such transferability, however, strongly depended on the kind of zeolite template employed. The order of the regularity in the resultant carbons was zeolite β>>zeolite L>mordenite> ZSM-5. The authors concluded that in order to obtain microporous carbons with high structural regularity, the zeolite templates should have a larger channel size (>0.6–0.7 nm) and, at the same time, the channel system should be three-dimensionally interconnected.[28] However, Hou *et al.* also prepared microporous carbons with zeolite structural ordering and BET surface areas exceeding $4000 \, m^2 \, g^{-1}$ *via* a simple CVD procedure without using the impregnation step.[64] It was claimed that low-temperature CVD followed by high temperature thermal treatment is essential to produce ordered porous carbon materials. Nitrogen-containing zeolite Y templated microporous carbon with good structural ordering has also been prepared by combining liquid impregnation of furfuryl alcohol and CVD of acetonitrile.[45]

Garsuch et al. reported on the preparation of microporous carbons that preserve the structural regularity of the zeolite Y templates.[60, 61] They found that the quality of the resulting ordered carbons strongly depends on heat treatment of the zeolite/carbon composite prior to etching.[60] In addition, the particle size of a zeolite template also affects the structural regularity of the resulting carbon materials. The particle size affects the diffusion of the carbon precursor into the zeolite pores: small particle size favours the formation of ordered microporous carbon materials.[61] Barata-Rodrigues et al. reported that the use of a zeolite template introduced controllable features into the resulting carbon, which could not be achieved by activated carbon preparation methods.[59] Additional porosity could be introduced by controlled carbonisation and demineralisation; and the porosity was also affected by incomplete filling of the template's pore network.[59] Ania et al. reported the synthesis of zeolite Y-templated microporous carbon material doped with nitrogen by a two-step, nanocasting process using acrylonitrile and propylene as precursors.[65] The resulting material inherited the ordered structure of the inorganic host and had a narrow pore-size distribution within the micropore range, and a large amount of heteroatoms including both oxygen and nitrogen. The templated carbon material displayed a large gravimetric capacitance of 340 F g^{-1} in aqueous media because of the combined electrochemical activity of the heteroatoms and the accessible porosity. This material can operate at 1.2 V in an aqueous medium with good cyclability (beyond 10 000 cycles) and is extremely promising for use in the development of high-energy-density supercapacitors.

4.2.1.2.2 Other Zeolites as Hard Template

Using a two-step method, Gaslain et al. reported the preparation of a porous carbon replica with a well resolved XRD pattern using wider pore channel zeolite EMC-2 as template.[66] The improved structural regularity of the replicated microporous carbons has been demonstrated by three, well-resolved XRD peaks corresponding to (100), (002) and (101) reflections of the EMC-2 zeolite, as shown in Figure 4.4. The resulting carbon materials display high surface area and pore volume without any significant contribution from mesoporosity, possibly due to enhanced

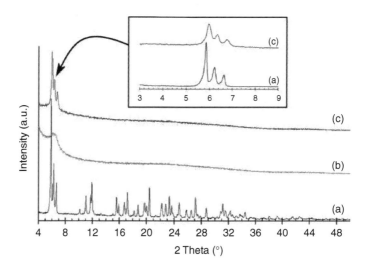

Figure 4.4 XRD patterns of (a) zeolite EMC-2, (b) carbon replica prepared by furfuryl alcohol infiltration and propylene CVD and (c) carbon replica prepared by furfuryl alcohol infiltration, propylene CVD and heat treatment at 900 °C under argon.[66] Reprinted with permission from F.O.M. Gaslain, J. Parmentier, V.P. Valtchev and J. Patarin, *Chem. Commun.*, 991. Copyright (2006) Royal Society of Chemistry

carbon precursor infiltration occasioned by the presence of a straight pore channel system in the zeolite EMC-2 template. Lei *et al.* reported a nanostructured porous carbon material synthesised by applying zeolite LTA as template and using CVD of methanol as the carbon source.[67] The resulting carbon materials have long-range periodic structure of a nanoscale curvature according to XRD and Raman spectroscopic studies. The use of zeolite MCM-22 and natural zeolites as templates with sucrose as carbon source to prepare porous carbon materials has been reported by Srinivasu *et al.* and Wang *et al.*, respectively.[68, 69]

Mokaya and co-workers has also extensively explored the use of zeolites as template to synthesise porous carbon materials.[44, 70–72] Using zeolite beta or silicalite-I as templates and acetonitrile as carbon source *via* a simple CVD method, hollow shells of porous nitrogen-doped carbon materials with high surface area can be produced.[44] The carbon materials generally retain the particle morphology of the zeolite templates. However, when the CVD is performed at temperatures ≥900 °C, hollow carbon shells that are hexagonal, cubic, or rectangular in shape are obtained as the predominant particle morphology, as indicated in Figure 4.5. Carbon materials prepared below 950 °C with zeolite β as template have high surface area up to $2270\,m^2\,g^{-1}$ and contain significant

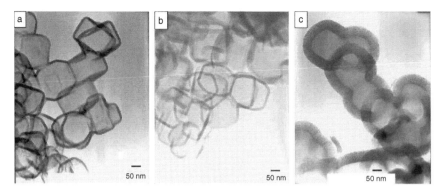

Figure 4.5 TEM images of carbon materials prepared *via* CVD method using silicalite-I as template at various CVD temperatures: (a) 900 °C; (b) 950 °C; and (c) 1000 °C.[44] Reprinted with permission from Z.X. Yang, Y.D. Xia and R. Mokaya, *Microporous Mesoporous Mater.*, Hollow shells of high surface area graphitic N-doped carbon composites nanocast using zeolite templates, **86**, 69, Copyright (2005) Elsevier

amounts of nongraphitic carbon that exhibits structural pore channel regularity replicated from the zeolite. Carbon materials templated by silicalite-I have lower surface area and do not exhibit structural pore channel ordering. It is thus possible, by choice of zeolite template and CVD conditions, to nanocast N-doped carbon materials that exhibit hollow-cored particle morphology, high surface area and zeolite-type pore channel ordering or hollow shells with significant levels of graphitisation.[44] In a later report, porous carbon materials were been prepared using zeolite 13X or zeolite Y as template and acetonitrile or ethylene as carbon source *via* CVD at 550–1000 °C.[70] Materials obtained from acetonitrile have high surface areas up to 1920 $m^2\,g^{-1}$, high pore volumes up to 1.4 $cm^3\,g^{-1}$, and exhibit some structural ordering replicated from the zeolite templates. Templating with zeolite Y generally results in materials with higher surface area. When ethylene is used as a carbon precursor, materials with high surface area up to 1300 $m^2\,g^{-1}$ are only obtained at lower CVD temperature of 550–750 °C. The ethylene-derived carbons retain some zeolite-type pore channel ordering but also exhibit significant levels of graphitisation even at low CVD temperature. It was found that the carbon materials retain the particle morphology of the zeolite templates, with solid-core particles obtained at 750–850 °C, while hollow shells are generated at higher CVD temperature ($\geq$900 °C).

Very recently, Yang *et al.* have successfully synthesised zeolite-like porous carbon materials that exhibit well-resolved powder XRD patterns

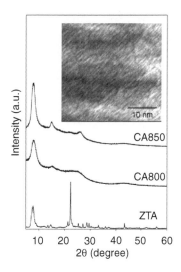

Figure 4.6 Powder XRD patterns of zeolite β template and corresponding carbon materials prepared at various CVD temperatures. The inset shows a representative TEM image of sample CA850.[71] Reprinted with permission from Z.X. Yang, Y.D. Xia and R. Mokaya, *J. Am. Chem. Soc.*, **129**, 1673. Copyright (2007) American Chemical Society

and very high surface areas.[71] The zeolite-like carbon materials are prepared *via* CVD at 800 or 850 °C using zeolite β as solid template and acetonitrile as carbon precursor. The zeolite-like structural ordering of the carbon materials is indicated by powder XRD patterns with at least two, well-resolved diffraction peaks and TEM images that reveal well-ordered micropore channels, as seen in Figure 4.6. The carbons possess surface area of up to $3200 \, \mathrm{m^2 \, g^{-1}}$ and pore volumes of up to $2.41 \, \mathrm{cm^3 \, g^{-1}}$. A significant proportion of the porosity in the carbons, up to 76 and 56% for surface area and pore volume, respectively, is from micropores. The porosity in the resulting material is dominated by pores of size 0.6–0.8 nm. Pacula and Mokaya have successfully used as-synthesized zeolites as templates for the preparation of well ordered zeolite-like carbon materials.[72] The obtained carbon materials have high surface areas and high pore volumes, and exhibit some zeolite-like structural ordering replicated from the zeolite template. Carbon materials prepared at 800 and 850 °C are nongraphitic and retain the particle morphology of the zeolite templates. Carbons prepared at 900 °C contains some graphitic domains and irregular particles that are dissimilar to the zeolite template particles. Hydrogen uptake of up to 5.3 wt% at −196 °C and 20 bar, and 2.3 wt% at 1 bar, for the carbon materials can be achieved. The use of as-synthesised rather than calcined zeolite β significantly the carbon yield improves and reduces the number of steps in the preparation of the templated carbons.

4.2.2 Clays as Hard Template

Clays have also been explored as templates to synthesise porous carbon materials. Carbonisation of organic polymers in the two-dimensional opening between the lamellae of a layered clay can produce thin graphite films. A typical nongraphitisable carbon precursor such as poly(furfuryl alcohol) can be graphitised very well by using layered clays as hard template. This is beyond the bounds of the conventional common knowledge of carbon science, where it was said that the final structures of a carbon material depended strongly on the nature of the original precursor rather than the conditions of the carbonisation process.

Sonobe *et al.* reported the preparation of carbon materials using the interlamellar spaces of montmorillonite as template and polyacrylonitrile, poly(furfuryl alcohol) and poly(vinyl acetate) as carbon precursor.[73–75] They found that carbon precursors were easily graphitised and the obtained carbon materials were two-dimensional film-like graphite, consisting of highly oriented layer planes, which were stacked thin and wide.[74, 75] Kyotani *et al.* produced carbons from poly(furfuryl alcohol) using three different layered clays including taeniolite, montmorillonite and saponite as templates.[76] It was found that carbons derived from different clays exclusively exhibited film-like shapes, a highly stacked structure of (002) planes and high graphitisability. Among the three carbons templated from different kinds of clays, the graphitisability of carbon from taeniolite was the highest and the one from saponite was the lowest. Bandosz *et al.* synthesised carbon materials *via* carbonisation of furfuryl alcohol within a sodium form of Wyoming smectite intercalated with hydroxyaluminium cations. They concluded that the inorganic matrix acts as a limiting pore size former in the carbons washed from smectite precursors.[77] Bandosz *et al.* also obtained carbons by carbonisation of poly(furfuryl alcohol) within the smectite and taeniolite matrices. They found that the resulting carbon materials presented sieving effects for molecular sizes between 3.6 Å and 6 Å.[78] The lithium form of taeniolite, intercalated with hydroxyaluminium and hydroxyaluminium-zirconium cations, also served as template to fabricate carbon materials.[58] The work of Bandosz *et al.* demonstrated that the structural properties of the carbon materials are also affected by the water content of the template inorganic matrix. Using a sodium taeniolite clay film as hard template and polyacrylonitrile as carbon precursor, a flexible graphite film with crystallised and oriented graphite characteristics can be prepared.[79] The commercial cationic clay bentonite was used as inorganic template and furfuryl alcohol, together with an additional treatment using propylene carbon vapour deposition, was used as a carbon source to generate templated carbon with little or no ordering evident at low 2θ angles and a relatively low surface area of $446\,m^2\,g^{-1}$.[59] The carbon produced with additional propylene showed more densely packed crystals than the carbon produced from the furfuryl alcohol impregnation only, which could be an indication of a stronger carbon network left after leaching the template. Mg-Al layered double hydroxides intercalated with 1,5-naphthalene disulfonate dianions have also been carbonised to produce carbon materials. The obtained carbon materials exhibit heterogeneous micropore structures with pore diameters no larger than the interlayer space of the clay.[80]

Pillared clays, layered silicates whose sheets have been permanently propped open by thermally stable molecular props, were used as templates to load various organic precursors.[81] The pore diameters of resulting carbon materials range from 8 to 22 Å, and the mass fractal dimension varies from 2.5 to 2.9, which are accessible to lithium ions when the intercalation process takes place in a lithium secondary battery. An approach to pyrolyse aromatic hydrocarbons such as pyrene within a pillared clay was also reported, in which the pillared clay serves two functions.[82] It acts as the inorganic template around which the carbon can be formed, and it also functions as an acid catalyst to promote condensation of the aromatics similar to the Scholl reaction. The resulting carbon materials have pore sizes from 15 to 50 Å.[82]

Recently, Pacula and Mokaya reported that porous carbon materials can be prepared using Mg-Al layered double hydroxides as template *via* the CVD method using acetonitrile as carbon precursor.[83] Depending on the CVD temperature, the carbons exhibit microporosity and/or mesoporosity and significant levels of graphitisation. The layer structural ordering of the layered double hydroxides is somewhat retained in the layered double hydroxides/carbon composite obtained after CVD, but is lost after removal of the template to generate carbon. The carbon materials exhibit significant porosity depending on the CVD temperatures. The morphology of the layered double hydroxides template is retained in the carbon materials, which in general consist of an assembly of small flaky particles.

4.2.3 Other Microporous Materials as Hard Template

Apart from zeolites and clays, other materials, such as metal-organic frameworks (MOFs), have also been explored as template to produce porous carbons. Recently, Liu *et al.* have synthesised porous carbon by heating the carbon precursor furfuryl alcohol within the pores of MOF-5. The resultant carbon exhibits high specific surface area up to 2872 m^2 g^{-1} and high pore volume of 2 cm^3 g^{-1} but possesses both micropores and mesopores. This porous carbon material shows good hydrogen uptake of 2.6 wt% at 760 Torr and -196 °C, as well as excellent electrochemical properties as an electrode material for an electrochemical double-layered capacitor.[84]

4.3 MESOPOROUS CARBON MATERIALS

Since the first report of the synthesis of structurally well-ordered mesoporous silica *via* the supramolecular self-assembly method using long chain organic ammonium bromide as surfactant,[85, 86] there has been rapid development in the synthesis of mesoporous silica with diverse structures and uniform pore sizes using various surfactants, including ionic surfactants, neutral amines and block polymers, as structure-directing agents.[85–89] The use of mesoporous silicas as solid template to nanocast porous carbon materials is a natural development of the fast expansion in the field of mesoporous materials. Due to the diverse structures of ordered mesoporous silica, nanocasting ordered mesoporous carbon from mesoporous silica is attractive and has developed rapidly in the past ten years. Different from their counterpart zeolites, mesoporous silicas usually have larger and controllable pore diameter, and tunable wall thickness with variable pore structure. Such excellent characteristics of ordered mesoporous silicas with three-dimensional pore channels clearly offer more flexibility to control the structures, pore diameters, morphologies and surface properties of templated mesoporous carbon materials.

4.3.1 Conventional Hard Template Synthesis Strategy

4.3.1.1 Ordered Mesoporous Silica as Hard Template

4.3.1.1.1 Mesoporous Silica MCM-48, SBA-15 and Analogues as Hard Template

Two Korean research groups, Ryoo et al.[25] and Lee et al.[5] first independently reported the synthesis of ordered mesoporous carbon CMK-1 and SNU-1, respectively, using cubically structured mesoporous silica MCM-48 ($Ia\bar{3}d$) as a hard template in 1999. Ryoo et al. impregnated the pores of MCM-48 with sucrose solution containing sulfuric acid, followed by polymerisation, carbonisation and removal of silica framework to obtain mesoporous carbon CMK-1.[25] As shown in Figure 4.7, the low angle XRD pattern of the carbon-silica composite was different from that of the template MCM-48 due to the lattice contraction and intensity loss. Lattice contraction usually occurs when mesoporous silica is heated to high temperatures, while intensity loss is simply due to the pores of MCM-48 filling with carbon. It is clear that the pore structure

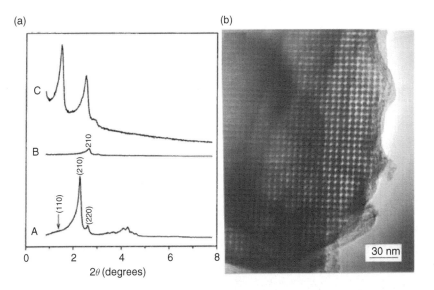

Figure 4.7 (a) Changes in powder XRD patterns during synthesis of CMK-1 using mesoporous silica MCM-48 as template: (A) MCM-48; (B) MCM-48 after completing carbonisation within pores; and (C) CMK-1 obtained by removing the silica wall after carbonisation. (b) TEM image of the ordered carbon molecular sieve CMK-1.[25] Reprinted with permission from R. Ryoo, S.H. Joo and S. Jun, *J. Phys. Chem. B*, **103**, 7743. Copyright (1999) American Chemical Society

can undergo a systematic transformation to a new ordered cubic $I4_1 32$ structure after removal of the silica template, as indicated in Figure 4.7. It was explained that the structural change was related to the disconnected nature of the two interwoven parts of the CMK-1 framework and might involve their mutual displacement to create some contacts between them.[90, 91] The resulting mesoporous carbon material (CMK-1) possesses high surface area and pore volume due to the presence of microporosity. It is believed that the micropores are formed due to the pyrolysis of carbon precursor prior to the removal of the silica pore wall, and mesopores are formed after the removal the template.[90] CMK-1 exhibits pore size of 3 nm, as evidenced by TEM in Figure 4.7. Lee *et al.* used aluminosilica MCM-48 as template and phenol resin as carbon precursor to produce the mesoporous carbon named SNU-1.[5] The surface implanted aluminium species acted as acidic sites to catalyse the polymerisation of phenol and formaldehyde in the pores of Al-MCM-48. The resulting mesoporous carbon SNU-1 was not a true negative replica of MCM-48 template and had regular three-dimensionally interconnected 2 nm pore arrays. This mesoporous carbon material exhibited excellent performance as an electrochemical double-layered capacitor.

Over the past decade, mesoporous silicas including MCM-48,[5] SBA-15,[20, 31, 34, 35] HMS[92] and MSU-H[93] have been widely explored as templates to prepare ordered mesoporous carbon materials. Unlike the mesoporous carbon templated from MCM-48, the mesoporous carbons templated from SBA-15, HMS and MSU-H are real inverse replicas of the silica template.[5, 25, 31, 92, 93] Interestingly, the successful preparation of mesoporous carbon SNU-2[92] from HMS and of CMK-3[31] from SBA-15 in turn shed some light on the understanding of the structure of the templates HMS and SBA-15. Originally, the HMS silica[87] was proposed to be MCM-41-like with a one-dimensional hexagonal channel structure, but it turned out to possess a wormhole-like, three-dimensional interconnected hexagonal pore structure.[92] SBA-15 was initially thought to have a two-dimensional hexagonal tubular pore structure which was disconnected,[89] similar to mesoporous silica MCM-41. However, later studies found that micropores or small mesopores exist between the primary cylindrical pore channels of SBA-15 due to the penetration of hydrophilic ethylene oxide groups into the silica framework.[94–96] The successful synthesis of mesoporous carbon CMK-3 also confirmed the existence of complementary micropores connecting the hexagonally packed mesopores in SBA-15 silica.[31]

Interestingly, depending on the degree of pore filling of the carbon precursor into the hexagonal (*P6mm* symmetry) pore system of mesoporous silica SBA-15, mesoporous carbon material with different structures can be obtained. If the pore system of the SBA-15 is completely filled with carbon precursor, an ordered mesoporous carbon CMK-3 with *P6mm* symmetry, in which parallel carbon fibres interconnected through thin carbon spacers, is generated.[31] The ordered structure of CMK-3 carbon is the exact inverse replica of the SBA-15 silica. However, if the pore system of the SBA-15 is partially coated by carbon precursor, a surface-templated mesoporous carbon, named CMK-5, with an array of hollow carbon tubes is obtained.[97] Moreover, due to the fact that the tubular structure exhibits both inner and outer surfaces, CMK-5 can reach very high surface areas and large pore volumes, thus making it a potentially useful material for adsorption and catalyst-support applications. It was found that partial coating of carbon precursor on the surface of SBA-15 silica was critical for the successful synthesis of CMK-5. A TEM image of CMK-5 is presented in Figure 4.8. The removal of the silica template then results in two different types of pores in the CMK-5 matrix. One type of pore is generated in the inner part of the channels that are not filled with carbon precursor. The other type of pore is obtained from the spaces where the silica walls of the SBA-15 template had previously been. Since there are two different mechanisms for pore generation, it should be possible to control the properties of the two pore systems independently. Lu *et al.* explored the possibility of controlling these two pore systems using Al-SBA-15 as template, and obtained ordered mesoporous carbon, NCC-1 with a bimodal pore system and high pore volume.[98] It was found that the crucial factors for the synthesis of such carbon are an ageing temperature of 140 °C for the template

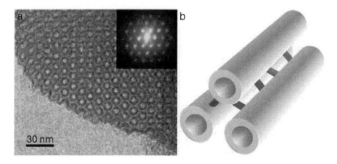

Figure 4.8 (a) TEM image of CMK-5 viewed along the direction of the ordered nanoporous carbon and the corresponding Fourier diffractogram. (b) Schematic model for the structure of CMK-5.[97] Reprinted by permission from Macmillan Publishers Ltd: S. H. Joo, S.J. Choi, I. Oh, J. Kwak, Z. Liu, O. Terasaki and R. Ryoo, *Nature*, **412**, 169. Copyright (2001)

SBA-15, a relatively low concentration of carbon precursor furfuryl alcohol (25 vol%) and a carbonisation temperature higher than 750 °C. Several groups have reported other synthesis strategies to CMK-5 mesoporous carbon, such as controlling the polymerisation temperature and time,[35, 99–102] introducing the carbon precursor by catalytic CVD[103] and varying the concentration of furfuryl alcohol.[104] To synthesise bimodal porous carbons, a new method was reported, *i.e.* a combination of the nanocasting and imprinting strategies.[105, 106] In principle, the pore sizes of the resulting carbons can be tuned by choosing different silica colloid particles and mesoporous silica.

Comparing the mesoporous carbon materials templated from MCM-48 with SBA-15, it is obvious that the pore diameter for CMK-1 templated from MCM-48 is difficult to adjust since the wall thickness for MCM-48 is not easy to tune. However, the pore size for CMK-3 can be controlled *via* changing the synthesis conditions of template SBA-15. The synthesis of larger pore mesoporous carbon materials with the cubic $Ia\overline{3}d$ structure is interesting due to their potential application as catalysts or catalyst supports. Using triblock copolymer P123 as the structure-directing agent, Ryoo and co-workers and Zhao and co-workers synthesised mesoporous silica materials with an $Ia\overline{3}d$ structure and large meospores from 3 to over 10 nm,[107–110] from which large pore carbons can be replicated. Ryoo and co-workers, using large pore cubic $Ia\overline{3}d$ mesoporous silica KIT-6 as template, synthesised both the rod-type and tube-type mesoporous carbon replica, [107, 108] by controlling the polymerisation of furfuryl alcohol inside the pores. Liu et al. reported the synthesis of large-pore three-dimensional bicontinuous cubic $Ia\overline{3}d$ mesoporous silica FDU-5 by a solvent evaporation method using P123 as surfactant and organosiloxane (3-mercaptopropyl)trimethoxy-silane and trimethylbenzene as a modifier.[109] Using cubically structured large-pore silica as hard template, both rod-like and tube-like mesoporous carbons with bicontinuous cubic $Ia\overline{3}d$ symmetry were synthesised.[110, 111] In contrast to CMK-1, carbon materials synthesised using KIT-6 and FDU-5 as templates show the same symmetry as their parent silica materials. According to Jun et al. the presence of porous bridges between the channel-like enantiomeric systems of the cubic KIT-6 are responsible for structure retention. As in the case of CMK-3, the structure retention was due to the interconnecting micropores/small mesopores among the hexagonal cylindrical pores of SBA-15 silica.[31] Liu et al. explained the structure retention as the result of a rigid carbon framework that prevents the symmetry change owing to the larger pore size of FDU-5 compared with MCM- 48.[109]

4.3.1.1.2   Other Ordered Mesoporous Silicas as Hard Template
Other mesoporous silicas including MCM-41, HMS, MSU-1, MSU-H, SBA-1, SBA-7, SBA-12 and SBA-16 have also been explored as hard templates to fabricate mesoporous carbons. Mesoporous silica MCM-41, which has hexagonally ordered cylindrical one-dimensional pores,[85, 86] was found to be unsuitable as a template for mesoporous carbon as its use yielded disordered high surface area microporous carbon.[47, 92] This is due to the absence of complementary micropores within the MCM-41 silica walls. However, Tian *et al.* have successfully produced self-supported, ordered, ultrathin carbon nanowire arrays by employing the mesoporous silica MCM-41 as template.[112] The carbon nanowire arrays exhibit high surface areas up to $1400 \, m^2 \, g^{-1}$, large pore volumes of $1.1 \, cm^3 \, g^{-1}$ and uniform mesopore size of *ca* 2.2 nm.

Mesoporous carbon templated by HMS (named SUN-2) displays a bimodal pore system centred at 2.0 and 0.6 nm, which indirectly confirms that HMS silica possesses a wormlike pore structure rather than the originally proposed MCM-41-like one-dimensional hexagonal channel structure.[92] Sevilla *et al.* found that the pore size and wall thickness of mesoporous HMS were tunable *via* changing the synthesis conditions,[113] and using variously prepared HMS silica as template, realised carbon with large pore volumes of up to $3.5 \, cm^3 \, g^{-1}$, high BET surface areas of up to $2300 \, m^2 \, g^{-1}$, and narrow pore size distribution in the range of 2–10 nm. They also adopted the same idea to tailor the pore size of mes-oporous carbon by using MSU-1 silica as template.[36] MSU-H, a mesoporous silica with structure similar to SBA-15, which is synthesised under near neutral conditions using sodium silicates as silica source,[114] is a cost-effective template for the synthesis of ordered mesoporous carbons.[93]

Ryoo *et al.* used mesoporous silica SBA-1 (which has $Pm\overline{3}n$ cubic structure with two different kinds of mesoporous cages)[47] as template for carbon formation. The resultant carbon, named CMK-2, exhibited a low degree of structural ordering, which manifested itself in the occurrence of only one peak in the XRD pattern. The poor structural ordering of the CMK-2 carbon was explained by the small apertures in the cage-like mesoporous silica SBA-1 template, which make it difficult for the carbon precursor to fill the cages and to form rigid carbon bridges between carbon nanoparticles prepared in the silica cages.[47] Li *et al.* also employed SBA-1 and SBA-7 as templates and sucrose as carbon source to synthesise mesoporous carbon.[115] They found that the pore size of cage-like mesoporous silica could be expanded under refluxing in acid solution without any addition of organic cosolvent. The obtained mesoporous silicas, SBA-1 and SBA-7, are good templates for the synthesis of mesoporous carbons with cubic $Pm\bar{3}n$ and three-dimensional hexagonal $P6_3/mmc$ mesostructures.

Xia and Mokaya explored the synthesis and characterisation of porous graphitic carbon materials via CVD using various mesoporous silicas including SBA-12, SBA-15, MCM-48, MCM-41 and HMS as solid template, and acetonitrile as carbon precursor.[116] They achieved structural replication and high surface area N-doped mesoporous carbon materials from SBA-12, MCM-48 and SBA-15 silica templates. The N-doped carbon materials exhibit both well ordered mesoporosity and high levels of graphitic character depending on the CVD temperature and the nature of the silica template. It was found that higher CVD temperatures (>900 °C) generated high levels of graphitic character but compromised the mesostructural ordering of the carbon materials. The mesostructural ordering of the carbon materials and replication of pore channel ordering from the silica template was found to be dependent on the nature of the mesoporous silica used as solid template.

Kim et $al.$ synthesised ordered mesoporous carbon using sucrose, furfuryl alcohol and acenaphthene as carbon source and mesoporous silica SBA-16 with cubic $Im\bar{3}m$ structure as template.[117] They found furfuryl alcohol and acenaphthene were more suitable carbon precursors than sucrose for the formation of rigidly interconnected carbon bridges through narrow apertures of the cage-like siliceous SBA-16 mesostructure. Therefore cubic $Im\bar{3}m$ structure was retained in the resulting mesoporous carbon product when furfuryl alcohol was used. Guo et $al.$ also synthesised mesostructured carbon materials with a cubic $Im\bar{3}m$ symmetry using mesoporous silica SBA-16 as template.[118] They found that the manner of surfactant removal in the SBA-16 silica prior to use as a template played an important role in the subsequent carbon mesostructure templating process; solvent-extracted SBA-16 produced well-ordered mesostructured carbon while calcined SBA-16 gave poorly ordered carbons.

4.3.1.2 Colloidal Silica Particles and Silica Gels as Hard Template

To synthesise mesoporous carbons with larger pore size, colloid silica particles and silica gels have been explored as hard templates. Hyeon's group pioneered the synthesis of mesoporous carbon using colloidal silica particles as hard templates. Initially, they synthesised mesoporous carbon using a silica sol solution with silica particle size of 12 nm as template and resorcinol/formaldehyde as carbon source.[119] It was found that the silica/resorcinol ratio had great effect on the pore size distribution of the resulting carbon material. In general, the pore size distribution of silica-particle-templated carbons was wide, suggesting that the carbon was not a true replica of the silica nanoparticles. To avoid agglomeration of silica particles, surfactant (cetryltrimethylammonium bromide) was used to stabilise them.[120] The resulting nanoporous carbon showed a very narrow pore size distribution centred at 12 nm, which matched well with the size of the silica particles. Later studies found that colloid silica with various particle sizes could be used as template, and the silica sol content and pH value had significant impact on the textural properties of the resulting mesoporous carbons.[119, 121, 122]

Li and Jaroniec reported the synthesis of mesoporous carbons by using spherical silica particles as template and mesophase pitch or acrylonitrile as carbon precursor.[123] The obtained porous carbon is dominated by mesoporosity with negligible microporosity. Li and Jaroniec developed a colloid imprinting technique to synthesise meso-porous carbons[17] by using commercially available colloidal silica as template to generate mesoporous carbons whose uniform pore size was determined by the size of the colloidal silica spheres. The key to this method is the incorporation of spherical silica colloids into mesophase pitch particles.[17, 18, 124] The colloid imprinted carbon can be further graphitised at high temperature with significant retention of textural properties but with a reduction in pore size from 24 to 16 nm. Recently, Jaroniec's group has synthesised mesoporous carbons with extremely large pore volume of up to 6 cm^3 g^{-1}[125] via formation of a thin carbon film on the pore walls of colloidal silica templates followed by template dissolution. The pore volume and pore size could be tailored by con-trolling the carbon film thickness and the size of silica colloids used. Carbons with bimodal distribution of uniform mesopores were also formed by co-imprinting of spherical silica colloids and hexagonally ordered mesoporous particles of SBA-15 into mesophase pitch particles.[105]

Making use of constrained polymerisation of divinylbenzene on sur-factant-modified colloid silica, Jang and Lim prepared carbon nanocap-sules and mesocellular foams.[126] Later, they reported that mesoporous carbons with highly uniform and tunable mesopores were fabricated by one-step vapour deposition polymerisation using colloidal silica nano-particles as template and polyacrylonitrile as carbon precursor.[127] Hampsey et al. recently reported the synthesis of spherical mesoporous carbons via an aerosol-based, one-step approach using colloidal silica particles and/or silicate clusters as template.[128]

Mesoporous carbons at three length scales of micrometric (2–8 mm), submicrometric (0.2–0.5 mm) and nanometric (10–20 nm) have been synthesised successfully.[20] The micrometric carbon shows a perfectly spherical morphology and a unimodal pore system made up of structural mesopores of 3 nm. The carbons synthesised at the submicrometric and nanometric length scales exhibit bimodal porosity made up of structural pores of 3 nm derived from the silica framework and textural pores corresponding to the interparticle voids. Using bimodal mesoporous silica composed of 30–40 nm sized nanoparticles with 3.5 nm-sized, three-dimensionally interconnected mesopores as template, bimodal mesoporous carbon having 4 nm-sized framework mesopores and approximately 30 nm-sized textural pores was synthesised.[129]

Mesocellular carbon foam with uniform mesopores has been produced by partially impregnating mesocellular aluminosilicate foam with phenol/formaldehyde.[130] As indicated in Figure 4.9, during the synthesis, the carbon precursor was impregnated into the complementary mesopores while the filling of the primary cellular space was avoided. Using mesocellular silica foam with a main cell diameter of 27 nm and window size of 11 nm as template, the resulting mesocellular carbon foam shows a primary cell diameter of 27 nm and window size of 14 nm. Mesocellular carbon foam composed of nanometre sized primary particles, and 40 nm pores, has also been prepared using hydrothermally synthesised MSU-F silica as a template and poly(furfuryl alcohol) as a carbon source.[131]

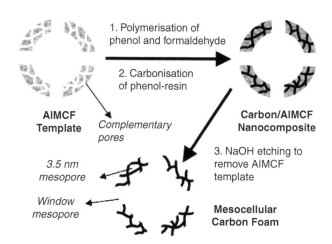

Figure 4.9 Schematic illustration for the synthesis of mesocellular carbon foam.[130] Reprinted with permission from J. Lee, K. Sohn and T. Hyeon, *J. Am. Chem. Soc.*, **123**, 5146. Copyright (2001) American Chemical Society

Tatsumi's group synthesised mesocellular carbon foam with a main cell size of 24 nm and window size of 18 nm *via* two successive impregnation of sucrose and subsequent carbonisation steps.[132] The obtained mesoporous carbon had closed hollow spherical pores, while the carbon obtained by single step impregnation of sucrose had open mesocellular pores.

4.3.2 Cost-Effective Strategies for the Synthesis of Mesoporous Carbons

The disadvantage of traditional hard template synthesis of porous carbon is that it usually involves several steps starting with the preparation of a mesoporous silica/surfactant mesophase followed by calcination to remove the surfactant (to generate the mesoporous silica hard template), introduction of carbon precursor into the mesoporous silica, carbonisation of the carbon precursor and finally silica etching (*i.e.* washing with HF or NaOH) to generate the mesoporous carbon.[25, 31] Therefore, recently there have been some attempts to prepare mesoporous carbon *via* more direct methods involving fewer steps.

Yoon *et al.*[133] reported the synthesis of mesoporous carbon using as-synthesised MCM-48 silica/surfactant mesophase as the template, followed by introduction of carbon precursor (divinylbenzene), carbonisation and removal of silica. Hyeon and co-workers[134, 135] reported the synthesis of mesoporous carbon by the carbonisation of composites containing silica, P123 triblock copolymer and phenol resin, followed by removal of silica. The synthesis was achieved by treating the as-synthesised silica/triblock copolymer nanocomposite with sulfuric acid to cross-link the triblock copolymers followed by carbonisation.[134] Kim *et al.* reported the synthesis of carbon nanotubes using P123 surfactant inside mesoporous silica, although ordered arrays of carbon nanotubes were not observed.[136] Kawashima *et al.* synthesised mesoporous carbon *via* copolymerisation of tetraethoxysilane (TEOS) and furfuryl alcohol.[137] It was observed that the furfuryl alcohol/TEOS ratio significantly affected the mesopore size of resulting carbon. Moriguchi *et al.* reported the direct synthesis of mesoporous carbon material by *in situ* polymerisation of divinylbenzene in the hydrophobic phase of a hexagonally arrayed micelle/silicate nanocomposite and its subsequent carbonisation and HF treatment.[138] The resulting carbon material had poor structural ordering with only wormhole-like mesopores of 2 nm. The synthesis of mesoporous carbon *via* one-step vapour deposition polymerisation using colloidal silica particles as template and polyacrylonitrile as carbon precursor has recently been reported by Jang *et al.*[139] Han *et al.* presented a one-step, *in situ* polymerisation preparation route to nanoporous carbon *via* carbonisation of a cyclodextrin-templated silica mesophase, followed by removal of silica.[140] Hampsey *et al.* have used a direct, one-step aerosol process to prepare mesoporous carbon from sucrose solutions containing colloidal silica particles.[141] Mesoporous carbons have also been fabricated *via* direct carbonisation of organic–organic nanocomposites comprising of a thermosetting polymer and thermally decomposable surfactant.[142]

Pang *et al.*[143] have reported the direct synthesis of mesoporous carbon by carbonisation of phenyl-bridged mesoporous organosilica/surfactant mesophases followed by silica removal. The resulting carbon has an average pore size of 2.5 nm. Yang *et al.* also reported the direct preparation of nanostructured carbon materials from mesoporous, ethyl-bridged organosilica/surfactant mesophases.[144] By varying the pyrolysis temperature, the textural properties of the resulting carbon are tunable. The mesoporous carbon materials retained the particle morphology of the organosilica mesophases. Furthermore, mesoporous silica/carbon composites, mesoporous silica and silicon carbide materials can be also prepared from mesoporous ethyl-bridged organosilica/surfactant mesophases depending on the synthesis conditions.[144]

4.3.3 Soft-Template Synthesis Strategy for Ordered Mesoporous Carbons

It is noteworthy that, more recently, significant progress has been achieved on the direct synthesis of ordered mesoporous carbon materials by self-assembly of copolymer surfactant and carbon precursor, which opens a new way for the preparation of ordered mesoporous carbon materials with reduced synthesis steps. The self-assembly of organic–organic species *via* soft-template approach represents a breakthrough allowing the efficient synthesis of mesoporous polymers and mesoporous carbons with controlled pore structures. Liang *et al.* first reported the preparation of highly ordered and well-oriented mesoporous carbon thin films through the carbonisation of a nanostructured resorcinol-formaldehyde resin and self-assembled block copolymer polystyrene-*block*-poly(4-vinylpyridine) nanocomposite.[145] The resulting mesoporous carbon thin film possesses oriented cylindrical pores perpendicular to the substrate with dimensions of 33.7 ± 2.5 nm. They also prepared ordered mesoporous carbon structures in the form of monoliths, fibres, sheets, and films *via* self-assembly of triblock copolymer (F127) with a mixture of phloroglucinol and formaldehyde.[146] It is believed that hydrogen bonding between soft templates and carbon precursors is the driving force for the successful organic–organic self-assembly. Tanaka *et al.* reported the synthesis of ordered mesoporous carbon film with ordered hexagonal structure *via* direct carbonisation of organic–organic nanocomposite obtained by using resorcinol/formaldehyde and triethyl orthoacetate as the carbon co-precursor and triblock copolymer Pluronic F127 as surfactant.[142] Although there was no evidence to support triethyl orthocetate contributing to the carbon content of the final carbon material, it was shown to increase the periodicity of porous carbon film.

Zhao and co-workers have made major advances in the soft-template synthesis of mesoporous carbons. They reported the self-assembly of poly(ethylene oxide)-poly(propylene oxide)-poly(ethylene oxide) (PEO-PPO-PEO) and resol mixtures and successful removal of the templates including F127, F108 and P123 at a series of temperatures to produce mesoporous polymer and carbon materials.[147–150] Figure 4.10 illustrates the five-step synthesis procedure reported by Meng *et al.*[149] Low-molecular-weight phenolic resol is first mixed with PEO-PPO-PEO triblock copolymer in an ethanolic solvent, followed by the evaporation of the solvent which induces the self-assembly of the block copolymer into an ordered structure. Driven by the hydrogen-bonding interaction between the PEO block and phenolic resol, the ordered mesostructure of the phenolic resol/block copolymer composite is formed. Curing of the resol at 100 °C solidifies the polymeric framework. Because of the difference in chemical and thermal stability between the resin and triblock copolymer, the template can be removed either by calcination at 350–450 °C or by extraction with 48 wt % sulfuric acid solution, leaving the Bakelite framework with ordered aligned voids. Heating at a high temperature above 600 °C transforms the polymeric framework to homologous carbon mesostructures. The increase in both the resol/triblock copolymer ratio and the PEO content in triblock copolymers results in the mesophase transformation from lamellar, bicontinuous $Ia\bar{3}d$, columnar $P6mm$ to globular $Im\bar{3}m$ mesophases. The carbon mesostructures are highly stable and can be retained at a temperature as high as 1400 °C under a nitrogen atmosphere.[148]

Deng *et al.* further explored other PEO-containing block copolymers as template to produce self-assembled mesostructures.[151] They found that large-molecular-weight poly(ethylene oxide)-polystyrene (PEO-PS)

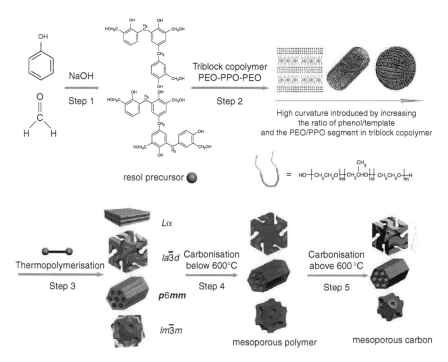

Figure 4.10 Scheme for the preparation of ordered mesoporous polymer resins and carbon frameworks.[149] Reprinted with permission from Y. Meng, D. Gu, F. Zhang, Y. Shi, L. Cheng, D. Feng, Z. Wu, Z. Chen, Y. Wan, A. Stein and D.Y. Zhao, *Chem. Mater.*, **18**, 4447. Copyright (2006) American Chemical Society

block copolymer (PEO$_{125}$-PS$_{230}$) and low-molecular-weight phenolic resin can assemble to an ordered face-centred cubic $Fm\overline{3}m$ mesophase. The pore size depends on the length of the hydrophobic PS blocks. A selective swelling of the PEO phase with phenolic resols can also be achieved in the reversed triblock copolymer PPO-PEO-PPO self-assembled structure.[150] Two outer PPO blocks in a chain participate in two different micelles or aggregates, forming interconnected micelles. At a certain molar ratio, the mixture of reversed triblock copolymer PPO-PEO-PPO with large PEO fraction of 45 wt% and phenolic resol self-assembles to face-centred cubic packed spheres of PPO blocks with $Fd\overline{3}m$ symmetry in the PEO/phenolic resin matrix. A slight decrease in the phenolic resol/triblock copolymer ratio leads to the formation of the two-dimensional hexagonal mesostructure. Interestingly, highly ordered mesoporous polymer and carbon frameworks with $Fd\overline{3}m$ symmetry exhibit a bimodal pore-size distribution centred at 3.2–4.0 and 5.4–6.9 nm, respectively.

4.3.4   Ordered Mesoporous Carbons with Graphitic Pore Wall

Despite the excellent structural ordering and narrow pore size distribution, the framework of most reported mesoporous carbons is generally not graphitic, *i.e.* the carbon framework is amorphous. It is, therefore, highly desirable to prepare mesoporous carbons with graphitic pore wall. Mesoporous carbon material (CMK-3G) with graphitic framework structures was prepared by Kim *et al. via in situ* conversion of aromatic compounds, including acenaphthene, acenaphthylene, indene, indane and naphthalene, to mesophase pitch inside mesoporous silica templates.[152] The XRD pattern of CMK-3G is characterised by peaks at 2θ of 26, 45, 53 and 78°, which correspond to the (002), (101), (004) and (110) diffractions of graphitic carbon. The TEM results show that the carbon framework consists of discoid graphene sheets which are self-aligned perpendicularly to the template walls. The CMK-3G material exhibits remarkably improved thermal and mechanical stabilities. C. H. Kim *et al.* have synthesised graphitic mesoporous carbon with high electrical conductivity using MSU-H silica as template and various aromatic hydrocarbons as carbon precursors in the presence of a catalyst.[153] They found that carbon materials obtained from pyrene and naphthalene precursors were more graphitic, whereas those from benzene exhibited the highest degree of conductivity. The methods reported by T. Kim *et al.*[152] and C. H. Kim *et al.*[153] required catalysts to convert the carbon precursor to mesoporous carbon. Fuertes and Alvarez synthesised graphitic mesoporous carbons by liquid impregnation of poly(vinyl chloride) followed by carbonisation.[154] Further heat treatment of the graphitisable carbon at a high temperature up to 2300 °C allows it to convert into a graphitised porous carbon with a relatively high BET surface area and a porosity made up of mesopores in the 2–15 nm range. Later, they also produced ordered mesoporous graphitic carbon from $FeCl_3$ impregnated pyrrole at a lower temperature of 900 °C.[155] $FeCl_3$ acted not only as an oxidant for the polymerisation of pyrrole, but also a catalyst to promote the formation of graphitic structure during the carbonisation step. Using mesophase pitch as carbon precursor, Yang *et al.* group prepared ordered graphitic mesoporous carbon materials with two-dimensional hexagonal *P6mm* symmetry or three-dimensioal bicontinuous cubic *Ia3̄d* structure *via* the melting–impregnation method.[156]

The methods mentioned above for the preparation of ordered graphitic mesoporous carbon involve the use of liquid impregnation strategies (repeated infiltration and polymerisation) followed by the carbonisation procedure and are time consuming. CVD infiltration offers obvious advantages over liquid impregnation, such as time saving, high degree of pore filling, enabling the formation of a dense pore wall and avoiding formation of undesired microporosity. Xia and Mokaya reported the synthesis of graphitic mesoporous carbon materials *via* a simple noncatalytic CVD method in which mesoporous SBA-15 is used as the solid template and styrene or acetonitrile as the carbon precursor.[43] As indicated in Figure 4.11, all the carbon materials show low-angle XRD peaks replicated from the template and high-angle XRD peaks contributed by the graphitic pore wall. The degree of graphitisation depends on the CVD temperature with higher temperature resulting in higher levels of graphitisation. Compared with styrene, the use of acetonitrile as carbon precursor generates more highly graphitised carbon.[43] The CVD method can be generalised for the preparation of various mesoporous, nitrogen-doped graphitic carbon materials using mesoporous silica templates including SBA-12,

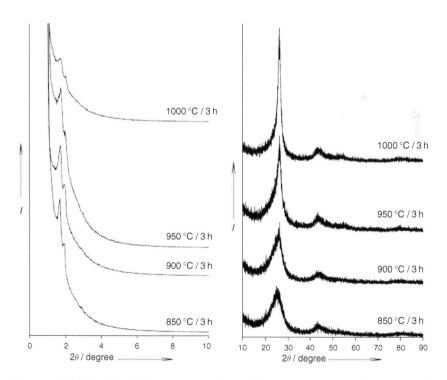

Figure 4.11 Powder XRD patterns of graphitic mesoporous carbon materials prepared *via* the CVD method at various temperatures.[43] Reprinted with permission from Y. Xia, R. Mokaya, *Adv. Mater.*, **16**, 1553. Copyright (2004) John Wiley and Sons

SBA-15, MCM-48, HMS and MCM-41.[116] Xia *et al.* also used CVD with acetonitrile as precursor to prepare graphitic mesoporous carbon materials with diverse morphologies, such as sphere, hollow sphere, rod and nanotube.[32, 33, 157, 158] Su *et al.* used a similar CVD method and benzene as carbon source to fabricate graphitic mesoporous carbon materials.[159]

4.3.5 Mesopore Size Control

The pore size of mesoporous carbon is of importance with respect to practical applications. When mesoporous carbon is synthesised *via* soft-template methods, the self-assembly of organic–organic species and pore size can be influenced by synthesis conditions, including surfactant type and concentration, and synthesis temperature. For example, Meng *et al.* observed that the pore size of mesoporous carbon derived from soft-templated mesoporous polymer composites decreased from 7.4 to 5.9 nm when the pyrolysis temperature increased from 400 to 800 °C.[148] However, it is more difficult to control the pore size of hard-templated mesoporous carbons. This is because control of pore size depends on changes in the wall thickness of the mesoporous silica templates.

Lee *et al.* were the first to report the control of the pore size of ordered mesoporous carbons by controlling the pore wall thickness of the silica template.[160] By varying the ratio of two surfactants, cetyltrimethylammonium bromide (CTAB) and polyoxyethylene hexadecylether $C_{16}H_{33}$ $(C_2H_5O)_8OH$ ($C_{16}EO_8$), in the synthesis of mesoporous silica, the wall thickness in the mesoporous silica changed between 1.4 nm and 2.2 nm, which resulted in mesoporous carbons with pore sizes in the range 2.2–3.3 nm. Xia *et al.* have reported control of the pore size in the structurally well-ordered mesoporous carbons by the CVD nanocasting route using rod SBA-15 as hard templates.[158] The pore size was tunable from 2.0 to 4.3 nm by varying the synthesis temperature of the SBA-15 templates. Fuertes and co-workers reported the synthesis of mesoporous carbons with tailorable pore size and porosity using SBA-15, MSU-1 and HMS silica as template, respectively.[35, 36, 113] The pore size of the mesoporous carbon could be tuned continuously within a range of 2–10 nm, by varying the synthesis temperature of MSU-1 or HMS. Moreover, by controlling the way the mesopores were filled by the carbon precursor, mesoporous carbon with unimodal or bimodal pore size distribution is generated. As mentioned earlier, based on the two different mechanisms for the pore generation of CMK-5, Lu *et al.* demonstrated that the intra-nanotube pore system can be tuned independently of the inter-nanotube pore system.[98] More recently, Lee *et al.* reported the synthesis of ordered mesoporous carbons with controllable mesopore sizes in the range of 3–10 nm, using boric acid as a pore expanding agent.[161] They used a procedure similar to that reported by Ryoo *et al.*[25] except that boric acid was added along with sucrose as carbon precursor. A phase separation of boron species during the carbonisation process was proposed as the genesis of the pore expansion.

4.3.6 Morphology Control

The morphology of ordered mesostructured carbons is another important factor with respect to their practical applications. Various macroscopic morphologies are required, for example, films (in sensor, separation and optical applications), uniformly sized spheres (in chromatography) or transparent monoliths. Using suitable synthesis strategies, it is possible to control the external shape of the templated mesoporous carbon materials to generate powders, films and membranes, spheres, hollow spheres, rods, fibres, nanowires, nanotubes and monoliths.

4.3.6.1 Morphology Control in Soft-Template Synthesis

In the case of soft-template self-assembly synthesis, mesostructure assembly and morphology growth can be controlled concurrently. Due to the versatility of the solvent-based soft-template self-assembly process, highly ordered mesoporous carbons can be produced relatively easily with different morphologies such as thin film and membrane,[142, 145, 148, 151] monolith,[146] fibre,[146] sphere,[162] rod,[163] single-crystal,[164] and discus-like crystal.[164] Initially, mesoporous carbons synthesised with soft templates were exclusively in the form of films.[142, 145, 146, 148, 149, 151] Usually mesoporous carbon films obtained from the soft-template self-assembly method need evaporation-induced self-assembly, solvent annealing or the spin coating technique. Polydispersed mesoporous carbon spheres with diameters at the micrometric scale can be prepared *via* the aerosol-assisted self-assembly route.[162] To get fibres with well-aligned mesopores, the shear-aligned block copolymer/polymeric matrix is used.[146] The phloroglucinol/Pluronic F127 complex reacts with formaldehyde to form a phenolic resin/F127 composite. Macroscopic alignment by shearing force such as spin coating and fibre extrusion produces mesostructured films and fibres. Flexible carbon sheets can be woven by the fibres.[146] The aqueous cooperative assembly route from phenol/formaldehyde and triblock copolymer F127 and P123 under weakly basic conditions can produce pellet-like mesoporous carbons in the size range of 1–5 mm, rod-like particles ranging from 5 to 200 μm, perfect rhombdo-decahedral single crystals and discus-like crystals.[164]

4.3.6.2 Morphology Control in Hard-Template Synthesis

In hard-template synthesis, since mesoporous carbons are obtained as inverse replicas of the silica templates, the carbons generally retain the particle morphology of the templates.[5, 25, 31, 92, 93, 130, 165, 166] Accordingly, control of the particle morphology is generally achieved by choosing the mesoporous silica template with the desired morphology.[32]

4.3.6.2.1 Rods and Spheres

Zhao and co-workers synthesised highly ordered mesoporous SBA-15 with high-yield rod-like morphology by using inorganic salts, and they obtained ordered mesoporous carbon rods using the SBA-15 rods as templates.[167] Sphere-shaped mesoporous carbon materials with controlled particle diameters ranging from 10 nm to 10 μm were prepared by Fuertes using the corresponding mesoporous silica spheres synthesised at various conditions.[20] Xia et al. also synthesised spherical mesoporous carbon using spherical mesoporous silica SBA-15 as template via a CVD route.[32, 33] Later, they synthesised mesoporous carbon nanorods via a CVD method using SBA-15 rods as templates (see Figure 4.12). When rod-like mesoporous silica templates synthesised at temperatures lower than 70 °C were used, hollow mesoporous nanotubules were obtained rather than solid core rods (Figure 4.12).[158]

4.3.6.2.2 Hollow Spheres

Porous carbon materials in the form of hollow spheres may find application in catalysis, controlled delivery, sensing and storage owing to their large volume and low density.[168] Making use of hollow spherical mesoporous aluminosilicate as template, Li et al. found that a bicontinuous mesostructure was faithfully and directly replicated to the resulting hollow spherical mesoporous carbon via a simple incipient-wetness impregnation technique.[169] Depending on the synthesis conditions, hollow spherical carbon can be also obtained from mesoporous silica spheres. For example, carbon capsules with hollow core/mesoporous

(a) (b)

Figure 4.12 SEM images of mesoporous carbon materials nanocast *via* CVD at 950 °C using mesoporous silica SBA-15 rods synthesised at different temperatures as templates. SBA-15 template synthesised at (a) 40 °C and (b) 100 °C.[158] Reprinted with permission from Y. Xia, Z. Yang and R. Mokaya, *Chem. Mater.*, Simultaneous Control of Morphology and Porosity in Nanoporous Carbon: Graphitic Mesoporous Carbon Nanorods and Nanotubules with Tunable Pore Size, 18, 140. Copyright (2006) American Chemical Society

shell have been obtained using solid core mesoporous silica spheres as templates, and the diameter of the hollow core and the mesoporous shell thickness can be controlled by using appropriate solid core/mesoporous shell silica sphere templates.[170] The selective deposition of phenol into the mesopores of solid core mesoporous silica is important to preserve the spherical morphology of the template silica. Xia and Mokaya have also successfully prepared well-ordered mesoporous carbon hollow spheres *via* a simple CVD method using conventionally synthesised SBA-15 as template.[32] As shown in Figure 4.13, the pyrolysis/carbonisation temperature is very important and should be at least 950 °C for

(a) (b)

Figure 4.13 SEM images of mesoporous carbon CMK-3 prepared at various CVD temperatures: (a) 900 °C; and (b) 950 °C.[32] Reprinted with permission from Z. Yang, Y. Xia and R. Mokaya, Zeolite ZSM-5 with unique supermicropores synthesized using mesoporous carbon as a template. *Adv. Mater.*, 16, 727–732 (2004). Copyright WILEY-VCH Verlag GmbH & Co. KGaA

the successful formation of hollow spherical carbons. CVD temperature lower than 950 °C results in the formation of mesoporous carbon with solid spheres. Later they reported a synthesis route that utilises spherical solid core mesoporous silica SBA-15 as a template and optimises the morphology of the resulting mesoporous carbon toward hollow spheres.[33] By changing the synthesis temperature of the template SBA-15, hollow spherical mesoporous carbon with pore size from 2 to 5 nm can be produced.[157] Yu *et al.* also synthesised hollow core/mesoporous shell carbon using silicate-1 zeolite core/mesoporous silica shell structures as template.[171]

4.3.6.2.3 Films

Continuous mesoporous carbon thin films have been synthesised by Pang *et al.* through direct carbonisation of sucrose/silica nanocomposite films *via* a spin-coating technique and subsequent removal of the silica to create a mesoporous carbon network.[172] The mesoporous carbon contains an average pore wall thickness of around 2.0 nm and pore diameter of around 2.4 nm. The obtained continuous mesoporous carbon thin films display uniform-sized and interconnected pore channels, high surface area up to $2600 \, m^2 \, g^{-1}$ and high pore volume of $1.39 \, cm^3 \, g^{-1}$. Recently, Chen *et al.* have synthesised ordered mesoporous thin film carbon material with short channels vertical to the film by the replication of mesoporous silica SBA-15 film template, which has perpendicular channels obtained by using a ternary surfactant system $C_{16}TMAB/SDS/P123$ as template.[173] The resulting mesoporous carbon film was deposited with a PtRu nanocatalyst as an anodic material in the direct methanol fuel cell.[174] The much enhanced methanol electrochemical oxidation activity in PtRu/mesoporus carbon film was ascribed to increased utilisation efficiency of nanocatalysts with the short nanochannels of thin mesoporous carbon film.

4.3.6.2.4 Monoliths

Monolithic carbons are easier to handle than powdered materials. Direct shaping of monolithic mesoporous carbons during their preparation is highly desirable. Mesoporous carbon monoliths may be fabricated by using mesoporous silica monoliths as template. Carbon monoliths with well-developed and accessible porosity have been produced using silica monoliths with a hierarchical structure containing macropores and mesopores as templates and furfuryl alcohol or sucrose as a carbon precursor.[175–177] The obtained carbon monolith is a positive replica of the silica monolith on the micrometre scale, and a negative replica on the nanometre scale. Interestingly, the pore system of the carbon monoliths can be varied to three- or four-modal porosity by varying the loading of the carbon precursor.[175–177] Shi *et al.* prepared carbon monoliths with bi- or trimodal porosity from monolithic silica templates,[178, 179] while monolith carbon has been fabricated from monolithic colloidal silica.[180] By integrating gel casting with CVD, Wang *et al.* reported the synthesis of hierarchically porous carbon monoliths with either hexagonal or cubic mesostructures starting from well-ordered mesoporous silica powders SBA-15 and KIT-6.[181]Powdery silica particles are first fused together to form silica monoliths by the gel-casting method. Furfuryl alcohol at various concentrations in trimethylbenzene is used as carbon precursor. The use of a low concentration of furfuryl alcohol, together with a secondary loading of carbon *via* CVD enabled greater control over the hierarchical porosity of the carbon monoliths. Monolithic mesoporous carbon with a bicontinuous cubic structure ($Ia\bar{3}d$ symmetry) prepared by using mesoporous silica monoliths as the hard template[111] exhibits a uniform pore size of 4.6 nm and a surface area of $1530 \, m^2 \, g^{-1}$, and is a promising electrode for electrochemical double-layered capacitors. Xia and Mokaya have prepared ordered mesoporous carbon monoliths using mesoporous silica monoliths as template *via* CVD.[182] The size and shape of the silica monolith are well retained in the carbon monolith, making it mechanically robust. A relatively high level of mesostructural ordering with a small proportion of graphitic character is observed in the carbon monolith.

4.3.6.2.5   Other Morphologies

Mesoporous carbon materials in forms with diverse morphology including monodisperse nanocubes and uniform nanospheres or tetrapods have been prepared by Wang *et al.* using block copolymer surfactant as the mesopore-directing agent and colloidal crystals to mould the external shape of the particles.[183] Mesoporous carbon nanowires, fibres and nanotubes with high aspect ratios and low defect density may be fabricated using porous alumina membranes as confinement matrices.[184–187] Such confined growth of mesostructures inside the voids of porous alumina membranes is an effective approach to make fibres or nanowire since the macroscopic morphology of porous alumina membranes is maintained in the resulting mesoporous nanowires or fibres. Indeed the confinement synthesis of mesostructured porous materials has opened new opportunities for mesoporous materials.[188–190]

4.4 MACROPOROUS CARBON MATERIALS

Spherical submicrometre-sized particles such as polystyrene and silica, can self-organise to form colloidal crystals, known as opals, which are excellent templates to prepare three-dimensional macroporous carbon materials with hollow or core/shell structures. The pore size of the resulting macroporous carbon materials is tunable *via* simply changing the size of spherical silica or polymer particles. To obtain macroporous carbon materials, colloid crystals are first formed *via* packing uniform spherical silicas into two- or three-dimensional arrays (as shown in Figure 4.14).[191] Filling of carbon precursors into the interstitial space of the colloid crystals, carbonisation and finally removal of spherical templates generate a carbon skeleton in the location of the former interstitial space and interconnected voids where the spheres were originally located. It is worth mentioning that before the infiltration of the carbon precursors, sintering is usually performed to create necks between the silica spheres, which provide interconnections between the spherical pores in the macroporous carbons.

4.4.1   Silica Colloidal Crystals as Hard Template

The synthesis of macroporous carbon materials was first realised by Zakhidov *et al.* in 1998.[37] Macroporous carbon materials with inverse opal structures, as shown in Figure 4. 14, were obtained using silica opals as hard templates and phenol resin and/or propylene gas as carbon precursor. The macroporous carbons had different structures depending

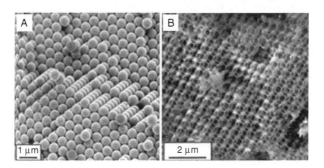

Figure 4.14   SEM image of (A) porous silica opal and (B) macroporous carbon prepared from silica spheres[19]

on the synthesis conditions; diamond and glassy carbon inverse opals were formed *via* volume filling, while graphite inverse opals, comprising 4-nm-thick layers of graphite sheets tiled on spherical surfaces were produced by surface templating.

In principle, the morphology of macroporous carbon materials is largely dependent on the degree of void infiltration of the opal template. In order to maximise the filling of the interstitial voids of the colloid crystal with carbon precursors, liquid phase carbon precursors such as phenolic resin and sucrose solution are usually used to achieve better replication.[37, 38, 192–194] A variety of carbon precursors, including propylene gas, benzene and divinylbenzene can also be successfully utilised to make three-dimensional macroporous carbon materials using colloid crystals as hard templates.[37, 195, 196] The resulting macroporous carbons can exhibit large pore volumes and high surface areas, possibly due to the presence of micropores in the framework. For example, macroporous carbon synthesised using phenol resin as the carbon precursor showed close packed uniform spherical pores with a diameter of 62 nm, a total pore volume of $1.68 \, cm^3 \, g^{-1}$, and a BET surface area of $750 \, m^2 \, g^{-1}$.[3] This macroporous carbon exhibits good gas adsorption properties that have been attributed to adsorption into mesopores of size below 50 nm. However, during the carbonisation and pyrolysis processing, the carbon phase can shrink remarkably. To achieve highly ordered carbon structures, it is therefore necessary to perform repeated impregnation cycles. Perpall *et al.* reported the synthesis of inverse carbon opal using a low-shrinkage precursor bis-*ortho*-diynyl arene (BODA) monomer.[197] Infiltration of the sphercial silica template with the melted monomer, *in situ* thermal polymerisation and pyrolysis, followed by removal of the silica with HF acid afford a carbon inverse opal structure that conserves the original dimensions of the template.

The surface properties of the colloid spheres and the chemical nature of the carbon precursors are important parameters in carbon growth. Consequently, the morphology of macroporous carbon materials can be controlled by altering the surface properties of colloid spheres and choosing suitable carbon precursors. Usually, good adhesion of the carbon precursor to the colloid spherical surface results in a surface-templating mechanism. For this reason, the position of carbon growth can be controlled *via* chemical modification of the colloid surface. Yu *et al.* reported the synthesis of three-dimensionaly ordered macroporous carbon materials with various morphologies.[38] They used Al-impregnated silica spheres as template, which allowed control of the initiation sites of the

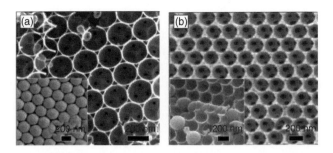

Figure 4.15   SEM images of macroporous carbon materials fabricated (a) by surface templating and (b) by volume templating using 250 nm silica spheres as template. The insets show carbon-colloidal silica composites.[38] Reprinted with permission from J.S. Yu, S. Kang, S.B. Yoon and G. Chai, *J. Am. Chem. Soc.*, **124**, 9382. Copyright (2002) American Chemical Society

acid-catalysed carbonisation reaction of phenol and formaldehyde. An Al-grafted silica array results in surface coating, namely surface templating, as shown in Figure 4.15(a). In this case, polymerisation is initiated at the acidic sites on the Al-grafted silica surface. However, polymerisation is thought to occur everywhere to fill up the entire space between the particles when an acid catalyst is mixed with a carbon solution. Figure 4.15(b) clearly shows the formation of an ordered macroporous carbon framework by the volume-templating mechanism with complete filling of all the voids in the silica spheres. Each of the spherical pores is interconnected though small channels. Chai *et al.* also reported the synthesis of an ordered uniform porous carbon framework with pore size in the range of 10–1000 nm using crystalline silica spheres as templates *via* carbonisation of phenol with formaldehyde.[192] The resulting porous carbons show a variety of porous structures, which can be easily controlled by tuning the size of colloid silica spheres.

4.4.2   Polymer Microspheres as Template

The use of polymer microspheres as template has the advantage of eliminating the chemical dissolution step since the polymer template can simply wash away or burn off during the process. The use of polystyrene microspheres as template for the synthesis of ordered macroporous carbons was realised by Baumann and Satcher.[39] Following infiltration of hydrogel into the interstitial spaces between polystyrene colloid spheres, the polystyrene sphere template was removed by washing with toluene, after which heat treatment generated an ordered macroporous carbon with 100 nm cavities and 6 nm interconnections. Macroporous carbon doped with various metal nanoparticles could be prepared when metal-ion-doped hydrogel was used as carbon source since metal ions were reduced to metal nanoparticles during the carbonisation step. Recently, Adelhelm *et al.* reported a soft-templating-based method for the synthesis of carbon with meso- and macroporosity in a one-step process, taking advantage of the phase separation (spinodal decomposition) of mesophase pitch, which acted as the carbon precursor, and a commercially available organic polymer polystyrene(PS), which acted as a soft template.[198] Due to the use of soluble mesophase pitch and PS, it is possible to control the morphology of the resulting carbon materials to monoliths or films.[198]

4.4.3 Dual Template Method

Chai *et al.* developed a dual template method for the fabrication of macrostructurally patterned, highly ordered, three-dimensionally interconnected, porous carbon with uniform mesopore walls.[196] They used both monodisperse PS spheres and silica particles as templates and divinylbenzene as carbon precursor. Both macroporosity and mesoporosity could be manipulated in the resulting carbon by controlling the sphere size of PS and silica particles. The introduction of secondary mesopores into a three-dimensionaly ordered macroporous carbon skeleton by templating with secondary silica nanoparticles led to improved electrochemical activity when the biporous carbon product was employed as a catalyst support in a direct methanol fuel cell.[196] Woo *et al.* also reported the synthesis of hierarchical three-dimensional ordered macroporous carbons with walls composed of hollow mesosized spheres *via* a dual templating strategy.[199] Macropores were created by using poly(styrene-*co*-2-hydroxyethyl methacrylate) polymeric colloids as hard template and hollow mesosized spheres were templated by smaller silica colloids. Thus, two types of colloids of significantly different sizes were used to create a macrosized polymeric colloidal crystal with voids filled with mesosized smaller silica colloids. The large macropores were interconnected and three-dimensionally ordered, and the walls of the large macropores were composed of small, hollow, carbon spheres, as shown in Figure 4.16. The macropore size and the hollow carbon sphere size can be controlled easily by choosing the size of polymer and silica particles, respectively.

(a) (b)

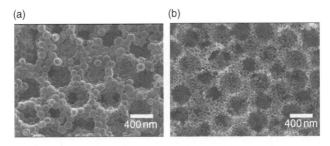

Figure 4.16 SEM images of porous carbons prepared using hard template consisting of large polymer colloids (450 nm diameter) and small silica colloids. The particle size of silica colloids used as template for the small spherical pores is (a) 70–100 nm and (b) 40–50 nm.[199] Reprinted with permission from S.W. Woo, K. Dokko, K. Sasajima, T. Takei and K. Kanamura, Three-dimensionally ordered macroporous carbons having walls composed of hollow mesosized spheres. *Chem. Commun.*, 4099–4101. Copyright (2006) Royal Society of Chemistry

The use of dual templating strategies to synthesise three-dimensional macroporous carbon materials has been explored by the groups of Zhao and of Stein.[200–202] In this case, the two templates play different roles: one as hard template to control the macroscopic structures, and the other as soft template for self-assembly of structurally ordered mesopores. Deng *et al.* prepared hierarchically ordered macro-/mesoporous carbons using monodispersed silica colloidal crystals as hard template, amphiphilic triblock copolymer PEO-PPO-PEO as a soft template, and soluble resols as a carbon source.[200] As shown in Figure 4.17, the porous

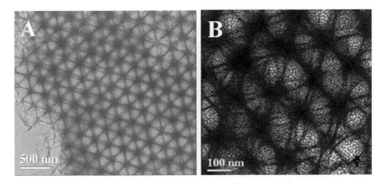

Figure 4.17 (a) TEM image of ordered macro-/mesoporous carbons with macropore arrays viewed along (111) planes and (b) HRTEM image showing the carbon mesostructure.[200] Reprinted with permission from S.W. Woo, K. Dokko, K. Sasajima, T. Takei and K. Kanamura, *Chem. Commun.*, 4099. Copyright (2006) Royal Society of Chemistry

carbons obtained have a highly ordered, face-centred cubic macrostructure with tunable pore sizes of 230–430 nm and interconnected windows with a size of 30–65 nm. The rigid silica hard templates prevent shrinkage of the mesostructure during the thermosetting and carbonisation process, resulting in large cell parameters of 18 nm and pore sizes of 11 nm, as seen in Figure 4.17(b). The bimodal porous carbon materials have large BET surface area up to $760 \, m^2 \, g^{-1}$, high pore volume of up to $1.25 \, cm^3 \, g^{-1}$, and partially graphitised frameworks. With the increase in the silica sphere diameter, the BET surface areas and the window sizes increase. Stein Wang and group, on the other hand, reported the synthesis of three-dimensionally ordered macro-/mesoporous carbon monoliths using poly(methyl methacrylate) (PMMA) colloidal crystals as template and a concentrated triconstituent precursor solution (a soluble phenolformaldehyde prepolymer, tetraethoxysilane, and the nonionic triblock copolymer F127) as carbon precursor.[202] The colloidal PMMA crystal template permits control over the external morphology of the carbon products. It is possible to produce either monoliths with hierarchical porosity (ordered macropores from PMMA spheres and large mesopores from F127) or cubic and spherical mesoporous nanoparticles. The specific morphology depends on the concentration of F127 and on the presence of 1,3,5-trimethyl benzene as an additive. Wang et al. also found that hierarchically ordered macroporous polymer and carbon monoliths with walls containing face-centred cubic or two-dimensional hexagonal mesopores can be synthesised via a dual-templating technique using PMMA colloidal crystals and amphiphilic triblock copolymer surfactants as templates.[201] The mesostructures could be conveniently controlled by tuning the concentration of the copolymer surfactant. A two-step thermal curing strategy was adopted to achieve a highly cross-linked mesostructure composed of phenolic resin and block copolymer. This ensured the formation of robust mesopore walls containing phenolic resin that survived during the decomposition of the block copolymer template. The ordered mesoporous phenolic resin was subsequently transformed to mesoporous carbon after heat treatment at high temperatures under an inert atmosphere. The growth of mesopores was significantly influenced by the confinement effect of the colloidal crystal template. Both spherical and cylindrical mesopores were aligned parallel to the surface of PMMA spheres, and therefore the obtained mesostructures exhibited apparent curvatures near the surface of macropore walls. The marocroporous carbons produced via this method were mechanically more stable than hierarchically porous carbon monoliths synthesised by nanocasting.

Three-dimensional macroporous carbon materials usually possess lower or no crystallinity due to the relatively low carbonisation temperature which is typically lower than 1000 °C. To increase the crystallinity of macroporous carbon to form graphitic carbon, high temperature annealing is required. Recently, Yoon *et al.* reported the preparation of highly ordered graphitised meso-/macroporous carbons using commercial mesophase pitch as a carbon precursor and silica colloidal crystals as templates.[203] The synthesis of the graphitised ordered nanoporous carbon was carried out by the incorporation of mesophase pitch dissolved in quinoline in the interstitial space of the silica templates under a static vacuum. After carbonisation and the removal of the silica template, the resulting carbon was further heated at a high temperature of up to 2500 °C in argon atmosphere to generate highly graphitised carbon. The XRD patterns of the carbon material after the graphitisation at 2500 °C showed a sharp (002) reflection, which gives an interlayer spacing of about 0.33 nm, and two further peaks, corresponding to graphitic reflections of (101) and (004). After graphitisation, the carbon material possesses both mesopores and macropores with size in the range of 40–100 nm that are interconnected, and exhibits relatively large graphite crystallites in the carbon pore walls.

REFERENCES

[1] Y. Mastai, S. Polarz and M. Antonietti, *Adv. Funct. Mater.*, **12**, 197 (2002).
[2] M.E. Davis, *Nature*, **417**, 813 (2002).
[3] M. Kang, S.H. Yi, H.I. Lee, J.E. Yie and J.M. Kim, *Chem. Commun.*, 1944 (2002).
[4] Z. Yang, Y. Xia and R. Mokaya, *Adv. Mater.*, **16**, 727 (2004).
[5] J. Lee, S. Yoon, T. Hyeon, S.M. Oh and K.B. Kim, *Chem. Commun.*, 2177 (1999).
[6] B. Sakintuna and Y. Yueruem, *Ind. Eng. Chem. Res.*, **44**, 2893 (2005).
[7] T. Kyotani, *Carbon*, **38**, 269 (2000).
[8] H.C. Foley, *Microporous Mater.*, **4**, 407 (1995).
[9] C.R. Bansal, J.B. Donnet and F. Stoeckli, *Active Carbon*, Marcel Dekker, New York, 1988.
[10] K. Miura, J. Hayashi and K. Hashimoto, *Carbon*, **29**, 653 (1991).
[11] H. Nakagawa, K. Watanabe, Y. Harada and K. Miura, *Carbon*, **37**, 1455 (1999).
[12] F. Cheng, J. Liang, J. Zhao, Z. Tao and J. Chen, *Chem. Mater.*, **20**, 1889 (2008).
[13] A. Oya, S. Yoshida, J. Alcaniz-Monge and A. Linares-Solano, *Carbon*, **33**, 1085 (1995).
[14] N.R. Khalili, M. Campbell, G. Sandi and J. Golas, *Carbon*, **38**, 1905 (2000).
[15] J. Ozaki, N. Endo, W. Ohizumi, K. Igarashi, M. Nakahara, A. Oya, S. Yoshida and T. Iizuka, *Carbon*, **35**, 1031 (1997).
[16] H. Tamon, H. Ishizaka, T. Araki and M. Okazaki, *Carbon*, **36**, 1257 (1998).

[17] Z. Li and M. Jaroniec, *J. Am. Chem. Soc.*, **123**, 9208 (2001).

[18] Z. Li and M. Jaroniec, *J. Phys. Chem. B*, **108**, 824 (2004).

[19] Z. Li, M. Jaroniec, Y. Lee and L.R. Radovic, *Chem. Commun.*, 1346 (2002).

[20] A.B. Fuertes, *J. Mater. Chem.*, **13**, 3085 (2003).

[21] T. Kyotani, *Bull. Chem. Soc. Jpn.*, **79**, 1322 (2006).

[22] Y. Wan, Y.F. Shi and D.Y. Zhao, *Chem. Mater.*, **20**, 932 (2008).

[23] F. Schueth, *Angew. Chem. Int. Ed.*, **42**, 3604 (2003).

[24] J.H. Knox, B. Kaur and G.R. Millward, *J. Chromatogr.*, **352**, 3 (1986).

[25] R. Ryoo, S.H. Joo and S. Jun, *J. Phys. Chem. B*, **103**, 7743 (1999).

[26] C. Vix-Guterl, S. Saadallah, L. Vidal, M. Reda, J. Parmentier and J. Patarin, *J. Mater. Chem.*, **13**, 2535 (2003).

[27] T. Kyotani, T. Nagai, S. Inoue and A. Tomita, *Chem. Mater.*, **9**, 609 (1997).

[28] T. Kyotani, Z.X. Ma and A. Tomita, *Carbon*, **41**, 1451 (2003).

[29] J. Rodriguez-Mirasol, T. Cordero, L.R. Radovic and J.J. Rodriguez, *Chem. Mater.*, **10**, 550 (1998).

[30] Z.X. Ma, T. Kyotani and A. Tomita, *Chem. Commun.*, 2365 (2000).

[31] S. Jun, S.H. Joo, R. Ryoo, M. Kruk, M. Jaroniec, Z. Liu, T. Ohsuna and O. Terasaki, *J. Am. Chem. Soc.*, **122**, 10712 (2000).

[32] Y. Xia and R. Mokaya, *Adv. Mater.*, **16**, 886 (2004).

[33] Y. Xia, Z. Yang and R. Mokaya, *J. Phys. Chem. B*, **108**, 19293 (2004).

[34] A. Lu, A. Kiefer, W. Schmidt and F. Schueth, *Chem. Mater.*, **16**, 100 (2004).

[35] A.B. Fuertes, *Microporous Mesoporous Mater.*, **67**, 273 (2004).

[36] S. Alvarez and A.B. Fuertes, *Carbon*, **42**, 433 (2004).

[37] A.A. Zakhidov, R.H. Baughman, Z. Iqbal, C.X. Cui, I. Khayrullin, S.O. Dantas, I. Marti and V.G. Ralchenko, *Science*, **282**, 897 (1998).

[38] J.S. Yu, S. Kang, S.B. Yoon and G. Chai, *J. Am. Chem. Soc.*, **124**, 9382 (2002).

[39] T.F. Baumann and J.H. Satcher, *Chem. Mater.*, **15**, 3745 (2003).

[40] M. Alvaro, P. Atienzar, J.L. Bourdelande and H. Garcia, *Chem. Commun.*, 3004 (2002).

[41] C.J. Meyers, S.D. Shah, S.C. Patel, R.M. Sneeringer, C.A. Bessel, N.R. Dollahon, R.A. Leising and E.S. Takeuchi, *J. Phys. Chem. B*, **105**, 2143 (2001).

[42] F.B. Su, X.S. Zhao, L. Lu and Z.C. Zhou, *Carbon*, **42**, 2821 (2004).

[43] Y. Xia and R. Mokaya, *Adv. Mater.*, **16**, 1553 (2004).

[44] Z.X. Yang, Y.D. Xia and R. Mokaya, *Microporous Mesoporous Mater.*, **86**, 69 (2005).

[45] P.X. Hou, H. Orikasa, T. Yamazaki, K. Matsuoka, A. Tomita, N. Setoyama, Y. Fukushima and T. Kyotani, *Chem. Mater.*, **17**, 5187 (2005).

[46] C. Liang, Z. Li and S. Dai, *Angew. Chem. Int. Ed.*, **47**, 3696 (2008).

[47] R. Ryoo, S.H. Joo, M. Kruk and M. Jaroniec, *Adv. Mater.*, **13**, 677 (2001).

[48] J. Lee, J. Kim and T. Hyeon, *Adv. Mater.*, **18**, 2073 (2006).

[49] A.H. Lu and F. Schuth, *Adv. Mater.*, **18**, 1793 (2006).

[50] A. Stein, Z.Y. Wang and M.A. Fierke, *Adv. Mater.*, **21**, 265 (2009).

[51] F.B. Su, Z.C. Zhou, W.P. Guo, J.J. Liu, X.N. Tian and X.S. Zhao, *Chem. Phys. Carbon*, **30**, 63 (2008).

[52] H. Yang and D. Zhao, *J. Mater. Chem.*, **15**, 1217 (2005).

[53] J. Lee, S. Han and T. Hyeon, *J. Mater. Chem.*, **14**, 478 (2004).

[54] B. Sakintuna and Y. Yurum, *Ind. Eng. Chem. Res.*, **44**, 2893 (2005).

[55] G. Ferey and A.K. Cheetham, *Science*, **283**, 1125 (1999).

[56] A. Corma, *Chem. Rev.*, **97**, 2373 (1997).

[57] P. Enzel and T. Bein, *Chem. Mater.*, **4**, 819 (1992).
[58] T.J. Bandosz, J. Jagiello, K. Putyera and J.A. Schwarz, *Chem. Mater.*, **8**, 2023 (1996).
[59] P.M. Barata-Rodrigues, T.J. Mays and G.D. Moggridge, *Carbon*, **41**, 2231 (2003).
[60] A. Garsuch and O. Klepel, *Carbon*, **43**, 2330 (2005).
[61] A. Garsuch, O. Klepel, R.R. Sattler, C. Berger, R. Glaser and J. Weitkamp, *Carbon*, **44**, 593 (2006).
[62] S.A. Johnson, E.S. Brigham, P.J. Ollivier and T.E. Mallouk, *Chem. Mater.*, **9**, 2448 (1997).
[63] Z.X. Ma, T. Kyotani, Z. Liu, O. Terasaki and A. Tomita, *Chem. Mater.*, **13**, 4413 (2001).
[64] P.X. Hou, T. Yamazaki, H. Orikasa and T. Kyotani, *Carbon*, **43**, 2624 (2005).
[65] C.O. Ania, V. Khomenko, E. Raymundo-Pinero, J.B. Parra and F. Beguin, *Adv. Funct. Mater.*, **17**, 1828 (2007).
[66] F.O.M. Gaslain, J. Parmentier, V.P. Valtchev and J. Patarin, *Chem. Commun.*, **29**, 991 (2006).
[67] S. Lei, J.I. Miyamoto, T. Ohba, H. Kanoh and K. Kaneko, *J. Phys. Chem. C*, **111**, 2459 (2007).
[68] P. Srinivasu, A. Vinu, N. Gokulakrishnan, S. Anandan, A. Asthana, T. Mori and K. Ariga, *J. Nanosci. Nanotechnol.*, **7**, 2913 (2007).
[69] A.P. Wang, F.Y. Kang, Z.H. Huang and Z.C. Guo, *New Carbon Mater.*, **22**, 141 (2007).
[70] Z.X. Yang, Y.D. Xia, X.Z. Sun and R. Mokaya, *J. Phys. Chem. B*, **110**, 18424 (2006).
[71] Z.X. Yang, Y.D. Xia and R. Mokaya, *J. Am. Chem. Soc.*, **129**, 1673 (2007).
[72] A. Pacula and R. Mokaya, *J. Phys. Chem. C*, **112**, 2764 (2008).
[73] N. Sonobe, T. Kyotani and A. Tomita, *Carbon*, **26**, 573 (1988).
[74] N. Sonobe, T. Kyotani and A. Tomita, *Carbon*, **28**, 483 (1990).
[75] N. Sonobe, T. Kyotani and A. Tomita, *Carbon*, **29**, 61 (1991).
[76] T. Kyotani, H. Yamada, N. Sonobe and A. Tomita, *Carbon*, **32**, 627 (1994).
[77] T.J. Bandosz, K. Putyera, J. Jagiello and J.A. Schwarz, *Carbon*, **32**, 659 (1994).
[78] T.J. Bandosz, J. Jagiello, K. Putyera and J.A. Schwarz, *Langmuir*, **11**, 3964 (1995).
[79] T. Kyotani, T. Mori and A. Tomita, *Chem. Mater.*, **6**, 2138 (1994).
[80] K. Putyera, T.J. Bandosz, J. Jagieo and J.A. Schwarz, *Carbon*, **34**, 1559 (1996).
[81] G. Sandi, P. Thiyagarajan, K.A. Carrado and R.E. Winans, *Chem. Materials,Materials Research Society Conference Proceedings* **11**, 235 (1999).
[82] G. Sandi, K.A. Carrado, R.E. Winans, J.R. Brenner and G.W. Zajac, *Microporous Macroporous Materials, Materials Research Society Conference Proceedings*, **431**, 39 (1996).
[83] A. Pacula and R. Mokaya, *Microporous Mesoporous Mater.*, **106**, 147 (2007).
[84] B. Liu, H. Shioyama, T. Akita and Q. Xu, *J. Am. Chem. Soc.*, **130**, 5390 (2008).
[85] J.S. Beck, J.C. Vartuli, W.J. Roth, M.E. Leonowicz, C.T. Kresge, K.D. Schmitt, C.T.W. Chu, D.H. Olson, E.W. Sheppard, S.B. McCullen, J.B. Higgins and J.L. Schlenker, *J. Am. Chem. Soc.*, **114**, 10834 (1992).
[86] C.T. Kresge, M.E. Leonowicz, W.J. Roth, J.C. Vartuli and J.S. Beck, *Nature*, **359**, 710 (1992).
[87] P.T. Tanev and T.J. Pinnavaia, *Science*, **267**, 865 (1995).
[88] P.T. Tanev and T.J. Pinnavaia, *Science*, **271**, 1267 (1996).
[89] D. Zhao, J. Feng, Q. Huo, N. Melosh, G.H. Frederickson, B.F. Chmelka and G.D. Stucky, *Science*, **279**, 548 (1998).

[90] M. Kruk, M. Jaroniec, R. Ryoo and S.H. Joo, *J. Phys. Chem. B*, **104**, 7960 (2000).

[91] M. Kaneda, T. Tsubakiyama, A. Carlsson, Y. Sakamoto, T. Ohsuna, O. Terasaki, S.H. Joo and R. Ryoo, *J. Phys. Chem. B*, **106**, 1256 (2002).

[92] J. Lee, S. Yoon, S.M. Oh, C. Shin and T. Hyeon, *Adv. Mater.*, **12**, 359 (2000).

[93] S.S. Kim and T.J. Pinnavaia, *Chem. Commun.*, 2418 (2001).

[94] R. Ryoo, C.H. Ko, M. Kruk, V. Antochshuk and M. Jaroniec, *J. Phys. Chem. B*, **104**, 11465 (2000).

[95] M. Imperor-Clerc, P. Davidson and A. Davidson, *J. Am. Chem. Soc.*, **122**, 11925 (2000).

[96] M. Kruk, M. Jaroniec, C.H. Ko and R. Ryoo, *Chem. Mater.*, **12**, 1961 (2000).

[97] S.H. Joo, S.J. Choi, I. Oh, J. Kwak, Z. Liu, O. Terasaki and R. Ryoo, *Nature*, **412**, 169 (2001).

[98] A. Lu, W. Schmidt, B. Spliethoff and F. Schueth, *Adv. Mater.*, **15**, 1602 (2003).

[99] L.A. Solovyov, T. Kim, F. Kleitz, O. Terasaki and R. Ryoo, *Chem. Mater.*, **16**, 2274 (2004).

[100] S.N. Che, K. Lund, T. Tatsumi, S. Iijima, S.H. Joo, R. Ryoo and O. Terasaki, *Angew. Chem. Int. Ed.*, **42**, 2182 (2003).

[101] M. Kruk, M. Jaroniec, T. Kim and R. Ryoo, *Chem. Mater.*, **15**, 2815 (2003).

[102] H. Darmstadt, C. Roy, S. Kaliaguine, T.W. Kim and R. Ryoo, *Chem. Mater.*, **15**, 3300 (2003).

[103] W. Zhang, C. Liang, H. Sun, Z. Shen, Y. Guan, P. Ying and C. Li, *Adv. Mater.*, **14**, 1776 (2002).

[104] A. Lu, W. Li, W. Schmidt, W. Kiefer and F. Schueth, *Carbon*, **42**, 2939 (2004).

[105] K.P. Gierszal and M. Jaroniec, *Chem. Commun.*, 2576 (2004).

[106] H.I. Lee, C. Pak, C.H. Shin, H. Chang, D. Seung, J.E. Yie and J.M. Kim, *Chem. Commun.*, **28**, 6035 (2005).

[107] T.W. Kim, F. Kleitz, B. Paul and R. Ryoo, *J. Am. Chem. Soc.*, **127**, 7601 (2005).

[108] F. Kleitz, S.H. Choi and R. Ryoo, *Chem. Commun.*, 2136 (2003).

[109] X.Y. Liu, B.Z. Tian, C.Z. Yu, F. Gao, S.H. Xie, B. Tu, R.C. Che, L.M. Peng and D.Y. Zhao, *Angew. Chem. Int. Ed.*, **41**, 3876 (2002).

[110] S. Che, A.E. Garcia-Bennett, X. Liu, R.P. Hodgkins, P.A. Wright, D. Zhao, O. Terasaki and T. Tatsumi, *Angew. Chem. Int. Ed.*, **42**, 3930 (2003).

[111] H. Yang, Q. Shi, X. Liu, S. Xie, D. Jiang, F. Zhang, C. Yu, B. Tu and D. Zhao, *Chem. Commun.*, 2842 (2002).

[112] B.Z. Tian, S.N. Che, Z. Liu, X.Y. Liu, W.B. Fan, T. Tatsumi, O. Terasaki and D.Y. Zhao, *Chem. Commun.*, 2726 (2003).

[113] M. Sevilla, S. Alvarez and A.B. Fuertes, *Microporous Mesoporous Mater.*, **74**, 49 (2004).

[114] S.S. Kim, T.R. Pauly and T.J. Pinnavaia, *Chem. Commun.*, 1661 (2000).

[115] H.C. Li, Y. Sakamoto, Y.S. Li, O. Terasaki, M. Thommes and S.A. Che, *Microporous Mesoporous Mater.*, **95**, 193 (2006).

[116] Y. Xia and R. Mokaya, *Chem. Mater.*, **17**, 1553 (2005).

[117] T. Kim, R. Ryoo, K.P. Gierszal, M. Jaroniec, L.A. Solovyov, Y. Sakamoto and O. Terasaki, *J. Mater. Chem.*, **15**, 1560 (2005).

[118] W. Guo, F. Su and X.S. Zhao, *Carbon*, **43**, 2423 (2005).

[119] S.J. Han, K. Sohn and T. Hyeon, *Chem. Mater.*, **12**, 3337 (2000).

[120] S. Han and T. Hyeon, *Chem. Commun.*, 1955 (1999).

[121] S. Han, K.T. Lee, S.M. Oh and T. Hyeon, *Carbon*, **41**, 1049 (2003).
[122] S.J. Han, S. Kim, H. Lim, W.Y. Choi, H. Park, J. Yoon and T. Hyeon, *Microporous Mesoporous Mater.*, **58**, 131 (2003).
[123] Z. Li and M. Jaroniec, *Carbon*, **39**, 2080 (2001).
[124] Z.J. Li and M. Jaroniec, *Chem. Mater.*, **15**, 1327 (2003).
[125] K.P. Gierszal and M. Jaroniec, *J. Am. Chem. Soc.*, **128**, 10026 (2006).
[126] J. Jang and B. Lim, *Adv. Mater.*, **14**, 1390 (2002).
[127] J. Jang, B. Lim and M. Choi, *Chem. Commun.*, 4214 (2005).
[128] J.E. Hampsey, Q.Y. Hu, L. Rice, J.B. Pang, Z.W. Wu and Y.F. Lu, *Chem. Commun.*, **28**, 3606 (2005).
[129] J. Lee, J. Kim and T. Hyeon, *Chem. Commun.*, 1138 (2003).
[130] J. Lee, K. Sohn and T. Hyeon, *J. Am. Chem. Soc.*, **123**, 5146 (2001).
[131] J. Lee, K. Sohn and T. Hyeon, *Chem. Commun.*, 2674 (2002).
[132] Y. Oda, K. Fukuyama, K. Nishikawa, S. Namba, H. Yoshitake and T. Tatsumi, *Chem. Mater.*, **16**, 3860 (2004).
[133] S.B. Yoon, J.Y. Kim and J. Yu, *Chem. Commun.*, 1536 (2002).
[134] J. Kim, J. Lee and T. Hyeon, *Carbon*, **42**, 2711 (2004).
[135] J. Lee, J. Kim, Y. Lee, S. Yoon, S.M. Oh and T. Hyeon, *Chem. Mater.*, **16**, 3323 (2004).
[136] S.S. Kim, D.K. Lee, J. Shah and T.J. Pinnavaia, *Chem. Commun.*, 1436– (2003).
[137] D. Kawashima, T. Aihara, Y. Kobayashi, T. Kyotani and A. Tomita, *Chem. Mater.*, **12**, 3397 (2000).
[138] I. Moriguchi, Y. Koga, R. Matsukura, Y. Teraoka and M. Kodama, *Chem. Commun.*, 1844 (2002).
[139] J. Jang, B. Lim and M. Choi, *Chem. Commun.*, **28**, 4214 (2005).
[140] B. Han, W. Zhou and A. Sayari, *J. Am. Chem. Soc.*, **125**, 3444 (2003).
[141] J. Eric Hampsey, Q. Hu, L. Rice, J. Pang, Z. Wu and Y. Lu, *Chem. Commun.*, **28**, 3606 (2005).
[142] S. Tanaka, N. Nishiyama, Y. Egashira and K. Ueyama, *Chem. Commun.*, **28**, 2125 (2005).
[143] J. Pang, V.T. John, D.A. Loy, Z. Yang and Y. Lu, *Adv. Mater.*, **17**, 704 (2005).
[144] Z.X. Yang, Y.D. Xia and R. Mokaya, *J. Mater. Chem.*, **16**, 3417 (2006).
[145] C. Liang, K. Hong, G.A. Guiochon, J.W. Mays and S. Dai, *Angew. Chem. Int. Ed.*, **43**, 5785 (2004).
[146] C. Liang and S. Dai, *J. Am. Chem. Soc.*, **128**, 5316 (2006).
[147] F. Zhang, Y. Meng, D. Gu, Y. Yan, C. Yu, B. Tu and D. Zhao, *J. Am. Chem. Soc.*, **127**, 13508 (2005).
[148] Y. Meng, D. Gu, F. Zhang, Y. Shi, H. Yang, Z. Li, C. Yu, B. Tu and D. Zhao, *Angew. Chem. Int. Ed.*, **44**, 7053 (2005).
[149] Y. Meng, D. Gu, F. Zhang, Y. Shi, L. Cheng, D. Feng, Z. Wu, Z. Chen, Y. Wan, A. Stein and D.Y. Zhao, *Chem. Mater.*, **18**, 4447 (2006).
[150] Y. Huang, H.Q. Cai, T. Yu, F.Q. Zhang, F. Zhang, Y. Meng, D. Gu, Y. Wan, X.L. Sun, B. Tu and D.Y. Zhao, *Angew. Chem. Int. Ed.*, **46**, 1089 (2007).
[151] Y.H. Deng, T. Yu, Y. Wan, Y.F. Shi, Y. Meng, D. Gu, L.J. Zhang, Y. Huang, C. Liu, X.J. Wu and D.Y. Zhao, *J. Am. Chem. Soc.*, **129**, 1690 (2007).
[152] T. Kim, I. Park and R. Ryoo, *Angew. Chem. Int. Ed.*, **42**, 4375 (2003).
[153] C.H. Kim, D.K. Lee and T.J. Pinnavaia, *Langmuir*, **20**, 5157 (2004).
[154] A.B. Fuertes and S. Alvarez, *Carbon*, **42**, 3049 (2004).

[155] A.B. Fuertes and T.A. Centeno, *J. Mater. Chem.*, **15**, 1079 (2005).

[156] H. Yang, Y. Yan, Y. Liu, F. Zhang, R. Zhang, Y. Meng, M. Li, S. Xie, B. Tu and D. Zhao, *J. Phy. Chem. B*, **108**, 17320 (2004).

[157] Y.D. Xia, Z.X. Yang and R. Mokaya, *Stud. Surf. Sci. Catal.*, **156**, 565 (2005).

[158] Y. Xia, Z. Yang and R. Mokaya, *Chem. Mater.*, **18**, 140 (2006).

[159] F. Su, J. Zeng, X. Bao, Y. Yu, J.Y. Lee and X.S. Zhao, *Chem. Mater.*, **17**, 3960 (2005).

[160] J.S. Lee, S.H. Joo and R. Ryoo, *J. Am. Chem. Soc.*, **124**, 1156 (2002).

[161] H.I. Lee, J.H. Kim, D.J. You, J.E. Lee, J.M. Kim, W.S. Ahn, C. Pak, S.H. Joo, H. Chang and D. Seung, *Adv. Mater.*, **20**, 757 (2008).

[162] Y. Yan, F.Q. Zhang, Y. Meng, B. Tu and D.Y. Zhao, *Chem. Commun.*, **30**, 2867 (2007).

[163] F. Zhang, Y. Meng, D. Gu, Y. Yan, Z. Chen, B. Tu and D. Zhao, *Chem. Mater.*, **18**, 5279 (2006).

[164] F.Q. Zhang, D. Gu, T. Yu, F. Zhang, S.H. Xie, L.J. Zhang, Y.H. Deng, Y. Wan, B. Tu and D.Y. Zhao, *J. Am. Chem. Soc.*, **129**, 7746 (2007).

[165] C. Vix-Guterl, S. Boulard, J. Parmentier, J. Werckmann and J. Patarin, *Chem. Lett.*, **31**, 1062 (2002).

[166] A.B. Fuertes and D.M. Nevskaia, *J. Mater. Chem.*, **13**, 1843 (2003).

[167] C.Z. Yu, J. Fan, B.Z. Tian, D.Y. Zhao and G.D. Stucky, *Adv. Mater.*, **14**, 1742 (2002).

[168] A. Kasuya, H. Takahashi, Y. Saito, T. Mitsugashira, T. Shibayama, Y. Shiokawa, I. Satoh, M. Fukushima and Y. Nishina, *Mater. Sci. Eng. A*, **217/218**, 50 (1996).

[169] Y.S. Li, Y.Q. Yang, J.L. Shi and M.L. Ruan, *Microporous Mesoporous Mater.*, **112**, 597 (2008).

[170] S.B. Yoon, K. Sohn, J.Y. Kim, C.H. Shin, J.S. Yu and T. Hyeon, *Adv. Mater.*, **14**, 19 (2002).

[171] J.S. Yu, S.B. Yoon, Y.J. Lee and K.B. Yoon, *J. Phys. Chem. B*, **109**, 7040 (2005).

[172] J. Pang, X. Li, D. Wang, Z. Wu, V.T. John, Z. Yang and Y. Lu, *Adv. Mater.*, **16**, 884 (2004).

[173] B.C. Chen, H.P. Lin, M.C. Chao, C.Y. Mou and C.Y. Tang, *Adv. Mater.*, **16**, 1657 (2004).

[174] M.L. Lin, C.C. Huang, M.Y. Lo and C.Y. Mou, *J. Phys. Chem. C*, **112**, 867 (2008).

[175] A. Taguchi, J.H. Smatt and M. Linden, *Adv. Mater.*, **15**, 1209 (2003).

[176] A.H. Lu, J.H. Smatt, S. Backlund and M. Linden, *Microporous Mesoporous Mater.*, **72**, 59 (2004).

[177] A.H. Lu, J.H. Smatt and M. Linden, *Adv. Funct. Mater.*, **15**, 865 (2005).

[178] Z. Shi, Y. Feng, L. Xu, S. Da and Y. Liu, *Mater. Chem. Phys.*, **97**, 472 (2006).

[179] Z. Shi, Y. Feng, L. Xu, S. Da and M. Zhang, *Carbon*, **41**, 2677 (2003).

[180] S. Alvarez, J. Esquena, C. Solans and A.B. Fuertes, *Adv. Eng. Mater.*, **6**, 897 (2004).

[181] X.Q. Wang, K.N. Bozhilov and P.Y. Feng, *Chem. Mater.*, **18**, 6373 (2006).

[182] Y. Xia and R. Mokaya, *J. Phys. Chem. C*, **111**, 10035 (2007).

[183] Z.Y. Wang, F. Li and A. Stein, *Nano Lett.*, **7**, 3223 (2007).

[184] G.W. Zhao, J.P. He, C.X. Zhang, J.H. Zhou, X. Chen and T. Wang, *J. Phys. Chem. C*, **112**, 1028 (2008).

[185] K. Wang, W. Zhang, R. Phelan, M.A. Morris and J.D. Holmes, *J. Am. Chem. Soc.*, **129**, 13388 (2007).

[186] D.J. Cott, N. Petkov, M.A. Morris, B. Platschek, T. Bein and J.D. Holmes, *J. Am. Chem. Soc.*, **128**, 3920 (2006).

[187] J.T. Chen, K. Shin, J.M. Leiston-Belanger, M.F. Zhang and T.P. Russell, *Adv. Funct. Mater.*, **16**, 1476 (2006).

[188] J. Fan, S.W. Boettcher, C.K. Tsung, Q. Shi, M. Schierhorn and G.D. Stucky, *Chem. Mater.*, **20**, 909 (2008).

[189] Y.Y. Wu, T. Livneh, Y.X. Zhang, G.S. Cheng, J.F. Wang, J. Tang, M. Moskovits and G.D. Stucky, *Nano Lett.*, **4**, 2337 (2004).

[190] Y.Y. Wu, G.S. Cheng, K. Katsov, S.W. Sides, J.F. Wang, J. Tang, G.H. Fredrickson, M. Moskovits and G.D. Stucky, *Nat. Mater.*, **3**, 816 (2004).

[191] A. Stein, *Microporous Mesoporous Mater.*, **44–45**, 227 (2001).

[192] G.S. Chai, S.B. Yoon, J.S. Yu, J.H. Choi and Y.E. Sung, *J. Phys. Chem. B*, **108**, 7074 (2004).

[193] J.S. Yu, S.B. Yoon and G.S. Chai, *Carbon*, **39**, 1442 (2001).

[194] Z.B. Lei, Y.G. Zhang, H. Wang, Y.X. Ke, J.M. Li, F.Q. Li and J.Y. Xing, *J. Mater. Chem.*, **11**, 1975 (2001).

[195] F.B. Su, X.S. Zhao, Y. Wang, J.H. Zeng, Z.C. Zhou and J.Y. Lee, *J. Phys. Chem. B*, **109**, 20200 (2005).

[196] G.S. Chai, I.S. Shin and J.S. Yu, *Adv. Mater.*, **16**, 2057 (2004).

[197] M.W. Perpall, K.P.U. Perera, J. DiMaio, J. Ballato, S.H. Foulger and D.W. Smith, *Langmuir*, **19**, 7153 (2003).

[198] P. Adelhelm, Y.S. Hu, L. Chuenchom, M. Antonietti, B.M. Smarsly and J. Maier, *Adv. Mater.*, **19**, 4012 (2007).

[199] S.W. Woo, K. Dokko, K. Sasajima, T. Takei and K. Kanamura, *Chem. Commun.*, **29**, 4099 (2006).

[200] Y.H. Deng, C. Liu, T. Yu, F. Liu, F.Q. Zhang, Y. Wan, L.J. Zhang, C.C. Wang, B. Tu, P.A. Webley, H.T. Wang and D.Y. Zhao, *Chem. Mater.*, **19**, 3271 (2007).

[201] Z.Y. Wang, E.R. Kiesel and A. Stein, *J. Mater. Chem.*, **18**, 2194 (2008).

[202] Z.Y. Wang and A. Stein, *Chem. Mater.*, **20**, 1029 (2008).

[203] S.B. Yoon, G.S. Chai, S.K. Kang, J.S. Yu, K.P. Gierszal and M. Jaroniec, *J. Am. Chem. Soc.*, **127**, 4188 (2005).

5

Synthetic Silicate Zeolites: Diverse Materials Accessible Through Geoinspiration

Miguel A. Camblor[1] and Suk Bong Hong[2]

[1]Instituto de Ciencia de Materiales de Madrid (CSIC), Madrid, Spain
[2]Department of Chemical Engineering and School of Environmental Science and Engineering, POSTECH, Pohang, Korea

5.1 INTRODUCTION

Zeolites are outstanding materials for a number of reasons, including their usefulness in a wide range of industrial and commercial applications, the beauty of their open crystalline structures, and their ample diversity of properties and chemical compositions. In fact, zeolites are so diverse that two particular zeolites may exhibit totally different, and even completely opposite, properties: one may be extremely hydrophilic while another may be strictly hydrophobic; one may be strongly acidic (or rather basic) while another may have no acid–base properties; one may have a high capacity for cation exchange, while another may present no ion exchange ability whatsoever; and one may adsorb large quantities of relatively large molecules, while another may not adsorb even the smallest molecules. But still, within the current most common view of what a zeolite is, they may both be zeolites in their own right. Zeolites are diverse materials.

Porous Materials   Edited by Duncan W. Bruce, Dermot O'Hare and Richard I. Walton
© 2011 John Wiley & Sons, Ltd.

All this diversity does not primarily come from nature. Although zeolites exist as minerals and have been known for more than 150 years, natural zeolites cover only a small fraction of the structures, properties, chemical compositions and applications that zeolite scientists have been able to confer on their synthetic zeolites during chiefly around half a century of efforts. Undoubtedly, these efforts have been inspired by the knowledge of natural zeolites and promoted by the early success of synthetic zeolites in the petrochemical industry, but later on also by the challenge and fun of discovering new zeolite structures, compositions, properties and applications.

Zeolite literature has been growing during the last 50 years at a quite impressive pace (Figure 5.1). Currently, journals indexed in ISI publish roughly 2500 papers each year searchable under the Topic zeolit* in ISI

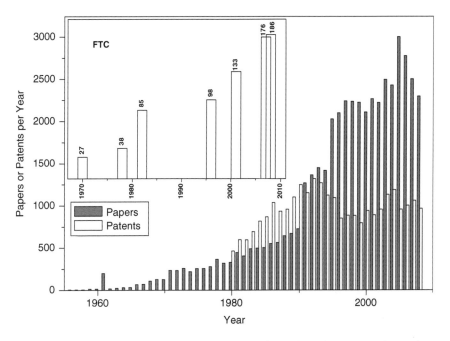

Figure 5.1 Evolution of the scientific and technological productivity in the zeolite field, measured as the number of papers (grey) or patents (white) indexed in ISI Web of Knowledge® per year under the topic zeolit.* Copyright Thomson Reuters® 2009. All rights reserved
Source: Web of Science® (papers) and Derwent Innovation Index® (patents), October 2008, used with permission. The inset shows the number of Framework Type Codes (FTCs) collected in the different editions of the *Atlas of Zeolite Framework Types*,[4] plus the number of FTCs already approved by the Structure Commission of the International Zeolite Association in November 2008 (http://www.iza-structure.org/)

Web of Science® and around 1000 zeolite related patents per year can be found under the same Topic search in the Derwent Innovation Index®. The number of new Zeolite Framework Type Codes (FTCs, see below) also grows fast, with an average of over 7 FTCs per year for the last 12 years. This chapter tries to cover a vast portion of zeolite science at an intermediate level, with a focus on and more in-depth coverage of zeolite synthesis. The chapter is organised in, perhaps, a less usual way, since the general sections on structural and compositional chemistry and applications go before the more specific sections on synthesis. We think the early sections will show that the vast richness of current zeolite science and applications heavily relies on an enormous synthetic effort that we will try to summarise and rationalise in the later sections.

5.2 ZEOLITES: SOME DEFINITIONS

The early, traditional, and strict definition of zeolites is based on the natural minerals and states that zeolites are crystalline microporous aluminosilicates belonging to the tectosilicate group and comprising within their structures channels and cavities of molecular dimensions. Dissecting this definition we find three important aspects:

- zeolites are crystalline tectosilicates. This means they are constructed from silicon (aluminium) oxide tetrahedra by sharing every oxygen atom once and only once between two adjacent tetrahedra, yielding an ordered three-dimensional framework. This may be represented by a (4;2)-3D net, *i.e.* a three-dimensional net in which 4-connected vertices (Si or Al) are bridged by 2-connected vertices (O)[1] (Figure 5.2).
- zeolites are microporous materials, with channels and cavities of molecular dimensions. Here we have some arbitrariness because, while it means zeolites have pores with diameter smaller than 20 Å (IUPAC arbitrary definition of 'micropores'),[2] these are of course 'of molecular dimensions' if we only consider relatively small molecules.
- zeolites are aluminosilicates. One problem we may encounter here is the limit between an aluminosilicate of very low Al content and an aluminium-free silicate. It does not appear to make sense that only one of two materials with essentially the same structure and a small difference in composition (say Si/Al = 500 compared with an

essentially pure SiO_2 material) may fall within the 'zeolite' definition. The same would be true, for instance, for the materials in which Al or Si is substituted by a 'heteroatom' (see below), such as the Ti-substituted zeolites.

We believe zeolite scientists currently have a broader and less restrictive concept of what a zeolite is, entirely based on its structure: a zeolite is a crystalline porous material made up of tetrahedra that share all the vertices once and only once with the four adjacent tetrahedra, building a

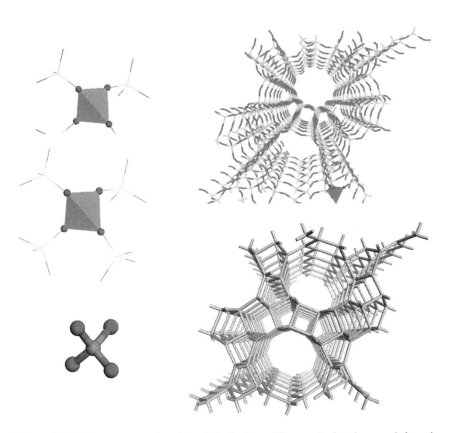

Figure 5.2 The structure of zeolites is built from AB_4 tetrahedra (bottom left and middle) sharing all vertices once and only once with neighbouring tetrahedra (top left) to yield a (4;2)-3D net in which connection between 4-connected nodes (A, light grey) occurs through 2-connected nodes (B, dark grey). Thus, the composition of the framework is AB_2. In the structure of SiO_2-ZSM-12 (top right), A=Si, B=O and the connectivity of tetrahedra is defined by the MTW Framework Type Code, a (4;2)-3D net shown at the bottom right

three-dimensional (4;2) framework. This definition is valid for the natural zeolites (including non-aluminosilicates such as, for instance, beryllosilicates) and for the synthetic ones irrespective of their composition [aluminosilicates, pure silica microporous materials, substituted (alumino)silicates, aluminophosphates and substituted aluminophosphates]. The condition of being porous is somewhat loose, although dense and porous frameworks can be distinguished by their framework densities (FDs, the number of tetrahedra T per nm^3), which for zeolites is typically below around $21\,T\,nm^{-3}$, while for dense materials is above that value.

In this chapter we will entirely focus on 'silicate zeolites', meaning any material primarily based on silica with a zeolitic structure as defined above. Recent reviews on other zeolitic materials such as microporous aluminophosphates are available in the literature.[3]

5.3   ZEOLITE STRUCTURES

As stated above, zeolites are diverse materials, showing a huge variety of structures, chemical compositions, properties and applications. From a structural point of view, zeolites may be classified by the so-called Zeolite Framework Types[4] (formerly known as Zeolite Structure Types). A Zeolite Framework Type represents the topology of a three-dimensional host structure composed of fully linked, corner-sharing tetrahedra, *i.e.* it represents a (4;2)-3D net. The Structure Commission of the International Zeolite Association (SC-IZA) has the IUPAC authority to assign a unique three-letter code (the Zeolite Framework Type Code, FTC) to each confirmed Zeolite Framework Type. Thus, an FTC stands for a mathematical abstraction, a (4;2)-3D net,[5] whose topology (the way the tetrahedra are connected to each other yielding a three-dimensional framework) has been confirmed in a real material. Since the code represents in fact a mathematical abstraction it is independent of the chemical composition and structural details. For instance, the code **SOD** stands for a Zeolite Framework Type that is shared by a legion of materials with different chemical composition, space group and unit cell dimensions (see below).

The number of confirmed zeolitic frameworks keeps growing at a significant rate (Figure 5.1, inset). As of November 2008, the SC-IZA recognises 186 codes, seven of which correspond to interrupted frameworks. Of the 179 truly zeolitic codes, 52 are known in mineral form, 164 as synthetic materials and 37 as both natural and synthetic materials, 8 of which were first discovered in the laboratory, then recognised as rare minerals. Thus, although 15 mineral zeolites still defy synthesis, the field is clearly dominated by the synthetic zeolites.

Zeolites have pore systems with several different ring sizes, but they have traditionally been classified as small-, medium-, large-, or extra-large-pore structures according to the smallest ring size of the largest pore that restricts access to the pore system. Ring size is the number of 4-coordinated vertices or tetrahedral atoms (T-atoms) that circumscribes the pore. Small-pore zeolites are limited by 8-rings, medium-pore by 10-rings, large-pore by 12-rings and extra-large-pore by 14-rings or larger, regardless of their variations in pore shape, size, dimensionality, and void volume due to the presence of channels, cages, or both. This pore size classification is based on the general consensus that interactions of zeolites with guest species are primarily constrained by the size and shape of the zeolite pore opening as opposed to the pore volume. Thus, even if a certain zeolite has large cages that are circumscribed by 14-rings or larger but are accessible only through 8-rings, it is classified as a small-pore zeolite.

It is remarkable that there are few examples of pores with an odd number of tetrahedra. Zeolites with 7-, 9- and 11-ring pores are scarce, and there are no examples of 13- or 15-ring zeolites. Apparently, there is no thermodynamic reason for this.[6] Considering not just pores but any kind of ring in zeolites, 3-rings are also scarce and, as it will be discussed below, with the quite peculiar exception of zeolite ZSM-18 (**MEI**), all 3-ring silicate zeolites crystallise in the presence of Be^{2+}, Zn^{2+} or Li^+. For a silicate, a 3-ring window is highly stressed. By contrast, 5-rings abound in high-silica zeolites.

Other important features of zeolites are cavities or cages, which can be named by the size and number of rings limiting the inner space (Figure 5.3). For instance, a small cage traditionally called double 4-ring (D4R), in which T atoms are close to the vertices of a cube, can be referred to as a $[4^6]$ cage, while the so-called β-cage, with T-atoms close to the vertices of a truncated octahedron, can be termed a $[4^6 6^8]$ cage. Silicate zeolites that have pores smaller than an 8-ring window but contain cavities are frequently called clathrasils.

5.4   CHEMICAL COMPOSITION OF SILICATE ZEOLITES

The vast majority of the 179 zeolite FTCs have silica-based representatives (131 codes, excluding silicoaluminophosphates). This number is to

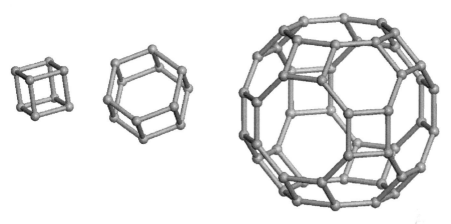

Figure 5.3 Examples of three types of cages that can be recognised in some zeolites. Only the T-atom connectivity is shown, and the cages are denoted by the number m of rings with n vertices limiting the cavity as $[m^n m'^{n'} m''^{n''} m'''^{n'''}\ldots]$: $[4^6]$, $[4^6 6^2]$ and $[4^{12} 6^8 8^6]$ cages (from left to right)

be compared with 66 FTCs for which phosphate-based materials (including silicoaluminophosphates) are known. Obviously, a number of them have both kinds of chemical compositions. Few topologies are represented only by germanates with no Si. There are also some exotic materials with zeolitic structures and totally different chemical compositions, including a hydrated copper sulfate with the **LTA** topology typical of zeolite A,[7] a family of Ga, In, Ge, Sn sulfides and selenides,[8] and an FTC possessed only by some nitrides (oxonitridophosphate,[9] oxonitridosilicate[10] and tantalum nitrides[10]).

The variability in chemical composition of silicate zeolites is just amazing. Considering first just the framework, the number of heteroatoms (*i.e.* atoms other than Si that can occupy tetrahedral positions in the framework) that have been (or have been claimed to be) incorporated into zeolites is formidable: Al^{3+}, Ga^{3+}, B^{3+}, Fe^{2+}, Fe^{3+}, Co^{2+}, Be^{2+}, Zn^{2+}, Li^+, Ti^{4+}, Ge^{4+}, V^{4+}, Sn^{4+}, Ni^{2+}, *etc.* (for a review on the synthesis of zeolites containing transition metals, see Perego *et al.*).[11] The extent of this heteroatom substitution is also surprising, although it strongly depends on the FTC considered.

Chemical variability in zeolites may also be related to the extraframework open space. The definition given above for zeolites implies a stoichiometry AB_2 for the framework, which is constructed by corner sharing of AB_4 tetrahedra. However, depending on the oxidation state of A and B, the framework may be electrically charged. For silicate zeolites with substitution of Si by heteroatoms, net negative charges arise when the oxidation state (OS) of the heteroatom is lower than +4, so the composition of the framework is $[Si_{(1-x)}T^{OS}_xO_2]^{-x(4-OS)}$. In this case, cations are needed for charge balance and are located at the open extraframework spaces, which may also contain neutral species (water, organic molecules, gases). There may also be additional negative charges, arising from connectivity defects between tetrahedra (generating Si-O⁻ groups), or occluded anions (like F⁻ in pure silica zeolites, see below, or a large number of anions in **SOD** and **CAN** materials).

5.4.1 Naming Zeolites

The chemical composition of zeolites may be quite complicated, and they may possess a large variety of different structures. This has turned into a myriad of names for zeolites, where each material, natural or synthetic, claimed or supposed to be different in terms of structure, composition and/or properties has received a specific and arbitrary name. In addition to mineral names (such as faujasite, mordenite, natrolite, *etc.*) there is a vast variety of names for synthetic materials. Very frequently, synthetic zeolites are named with a trivial code related to the claimed inventor or owner of patent protected rights followed by a sequential number. For instance, ZSM-5, SSZ-24 and RUB-13 are zeolites discovered by researchers at Mobil, Chevron and Ruhr University Bochum, respectively. There are also zeolites with other names, such as zeolite Beta (despite its discovery by Mobil researchers), silicalite (a pure SiO_2 version of ZSM-5) and Theta-1 (discovered by BP researchers).

The IUPAC recommendations on the nomenclature of microporous and mesoporous inorganic solids[12] define a rather complex naming for porous materials based on its crystal chemical formula, which for zeolites with a confirmed FTC can be safely reduced to

$$|\text{guest composition}|[\text{host composition}] - \textbf{FTC}$$

Given that the FTC already gives a wealth of information on the structure of the host and pore structures, frequently no need for additional information is required. The guest and host compositions should be provided on a per cell basis. For instance, the mineral sodalite can be ideally formulated as $|Na_8Cl_2|[Al_6Si_6O_{24}]$ − **SOD**, while the material with a pure silica SOD framework prepared with ethylene glycol has a formula $\text{leg}_2\text{I}[\text{Si}_{12}\text{O}_{24}]$ - **SOD**, where eg stands for ethylene glycol.

5.4.2 Loewenstein's Rule

Loewenstein's rule states that 'whenever two tetrahedra are linked by one oxygen bridge, the centre of only one of them can be occupied by aluminium; the other centre must be occupied by silicon, or another small ion of electrovalence four or more, e.g. phosphorus'.[13] This rule, also known as the Al avoidance rule, was established based on Pauling's rules and subsequently justified by the energy cost of the repulsive inter-action between two adjacent $[\text{AlO}_{4/2}]^-$ charged tetrahedra. The rule would imply a lower limit for the Si/Al ratio equal to 1 in zeolites and other tectosilicates (maximum extension of Si by Al substitution of 50 %) and also the need for an ordered distribution of Al and Si when this limit is attained, since in this case it implies the strict alternation of Si and Al tetrahedra. For zeolites with Si/Al = 1 (and also for zeolitic aluminopho-sphates, where P/Al = 1), the rule has also a topological implication: a strict alternation of Al and Si (P) requires that no windows with an odd number of tetrahedra exist.

Loewenstein's rule appears to apply frequently, and it has been extended as to the avoidance of any pair of connected tetrahedra with the same sign of charge. However, there are many exceptions to the rule, which in our opinion fall within two different categories: (1) materials that break the rule extensively; and (2) materials that break the rule at a low percentage. Among the first category, bicchulite[14] is a natural mineral with the **SOD** framework and Si/Al = 0.5 ($|\text{Ca}_8(\text{OH})_8|[\text{Al}_8\text{Si}_4\text{O}_{24}]$ – **SOD**), which has also synthetic analogues (including $|\text{Ca}_8(\text{OH})_8|[\text{Ga}_8\text{Si}_4\text{O}_{24}]$ – **SOD**).[15] Even for these bicchulite-type materials it has been argued that the number of Al-O-Al pairs is minimised.[16] Also in the first category, the so-called aluminate sodalites have no Si at all and a general composi-tion $|\text{M}_8\text{X}_8|[\text{Al}_{12}\text{O}_{24}]$ – **SOD**, where M is a divalent cation (Ca^{2+}, Sr^{2+} ...) and X is a divalent anion ($[\text{SO}_4]^{2-}$, $[\text{CrO}_4]^{2-}$, $[\text{MoO}_4]^{2-}$, $[\text{WO}_4]^{2-}$ but also S^{2-}, Te^{2-}). All the materials in the first category crystallise at temperatures well above 200 °C, often in nonhydrothermal conditions, and it may be argued that such high temperatures are necessary to over-come the energy cost associated with Al-O-Al pairs. In fact, the energy penalty for such a violation does not appear to be exceedingly large. For instance, quantum mechanical calculations on clusters with 4-ring tetramers point to an energy penalty in the range of 60–80 kJ mol^{-1} of Al-O-Al (evaluated as the energy difference between the alternating and paired isomers).[17]

The second category comprises materials where a small percentage of violations to Loewenstein's rule has been detected, for instance by ^{17}O triple quantum MAS NMR.[18] In this case, where entropic effects may be largely important, a violation can be understood as a defect arising from a kinetic control and does not necessarily require a high-temperature crystallisation, as found, for instance, in synthetic gallosilicate **NAT** materials with disordered structures.[19] Another exception occurs in a molecular compound, an aluminosilsesquioxane comprising two [4^6] cages, where the cages are linked through two Al atoms.[20] Thus, what is important to realise is that Loewenstein's rule may be useful but is not unavoidable

5.5   ZEOLITE PROPERTIES

Typically, zeolites are characterised by (1) a high cation exchange ability, derived from the presence of mobile cations in their pores, (2) a high internal surface area allowing selective adsorption of molecules according to their sizes relative to the regular size of zeolitic pores, which awarded them also the term *molecular sieves*, and (3) catalytic properties provided by countercations or framework constituents and modulated by the size of pores and internal cavities. However, given the current large variety of zeolite structures and compositions, the properties of zeolites are extremely diverse (Table 5.1). The more traditional zeolites (*i.e.*, zeolites with low Si/Al ratios) are highly polar and hydrophilic and, typically, present a high cation-exchange capacity, although this largely depends on the pore size. Depending on the nature of the countercation, they may be strong acids (when the cation is H$^+$) or moderately basic (when the cation is, for instance, Cs$^+$) which may give them interesting catalytic properties. By contrast, defect-free pure silica zeolites, *i.e.* zeolites with SiO$_2$ composition (see below), are strictly hydrophobic and have no cation-exchange capacity and no catalytic properties. For both types of materials, isomorphous substitutions may impart the zeolite with new catalytic properties.

With regard to the structure, zeolites may have pores too small to allow the passage of even the smallest molecule, or large enough to adsorb relatively large molecules. They may also contain cavities and

Table 5.1 Zeolite properties as a function of chemical composition[a]

Chemical composition	Properties
Al-rich aluminosilicate	Hydrophilic Polar High-cation exchange capacity Relatively low chemical, thermal and hydrothermal stability Catalytic activity Relatively low acid strength in H^+ form Moderately high basic strength in cationic form (Cs^+)
Si-rich aluminosilicate	Increased hydrophobicity Decreased polarity Low cation-exchange capacity Relatively high chemical, thermal and hydrothermal stability Catalytic activity High acid strength in H^+ form Low basic strength in cationic form
Defect-free pure silicate	Strictly hydrophobic Highly apolar No cation exchange ability No catalytic activity No acidity, no basicity Very high chemical, thermal and hydrothermal stability
Heteroatom-substituted (alumino)silicate	Specific catalytic properties, depending on the type of heteroatoms

[a] Some properties such as cation exchange ability and catalytic activity depend also on the structure (pore size) or specific composition (size of countercations).

pores in which complex chemical species such as metal clusters, organometallic complexes, or molecules with photochemical properties are contained.

5.6 ZEOLITE APPLICATIONS

Zeolites find widespread commercial applications and new opportunities emerge as new structures and compositions are attained. In terms of bulk quantities, synthetic zeolites are most commonly sold as cation exchangers (*builders*) in the detergent industry: zeolites are able to soften water by capturing Ca^{2+} and Mg^{2+} through exchange with zeolitic Na^+, thus enhancing (*building*) the effect of surfactants. Zeolites also ease the formation of particles in compact detergents. It is claimed that zeolites are environmentally more benign than other builders, like sodium tripolyphosphate, which may cause water eutrophication, and also that their use has allowed a decrease in detergent dosage by 65 % in the last 20 years.[21] Other important applications, such as radioactive waste management (by sequestering and immobilising Cs^+, and Sr^{2+} radionuclides)[22] and wastewater management (by sequestering heavy metal ions), also exploit the cation-exchange properties of zeolites, although natural zeolites may be preferred.[23]

However, in terms of economic value, zeolites are far more important as catalysts and adsorbents in the chemical and petrochemical industries. In fact, the large success of zeolite catalysts and adsorbents in the petrochemical industry, following commercialisation by Union Carbide in the 1950s and early 1960s,[24] has been likely the more important driving force for zeolite research during half a century.

Zeolites are key components of catalysts used worldwide by the petrochemical industry in, among others processes, catalytic cracking, hydrocracking, dewaxing and hydroisomerisation, as well as in the synthesis of ethylbenzene, cumene, *p*-xylene, styrene and aromatics and in the production of light olefins from methanol.[25,26] The petrochemical industry benefits not only from the strong acidity that H^+-exchanged zeolites present, in addition to being easily handled and recyclable solids, and from the good adsorption properties they possess, but also and more fundamentally from the so-called *shape selectivity* they present. The concept of shape selectivity, which is already almost 50 years old,[27] refers to the marked selectivity imparted to a reaction by imposing a steric constraint to reactants to diffuse in, transition states to be formed within or products to diffuse out of the pores in which the active sites are located (Figure 5.4).[28]

In addition to their uses as catalysts, zeolites are also used as adsorbents in chemical and petrochemical plants, petroleum refineries and natural gas plants in a large number of purification processes, including the removal of H_2S, mercaptans, organic chlorides, CO and mercury from different industrial streams, where the high affinity of aluminosilicate zeolites for polar compounds allows removal to very low levels.[25] Other important industrial uses of zeolites as adsorbents include the noncryogenic separation of air into its components, the purification of air prior to cryogenic separation, the water removal from air conditioning circuits and from the air brake systems of heavy trucks and locomotives, as well as from double-glazed insulating windows. In these cases, removal of

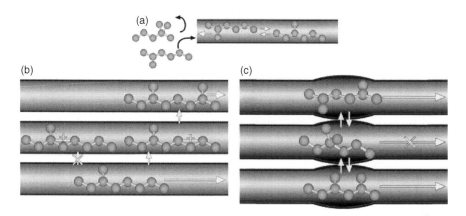

Figure 5.4 A schematic illustration of the *shape selectivity* concept, classically referring to a significant modification in the distribution of products in a catalytic reaction due to the localisation of the active sites within a confined space of the catalyst. *Reactant* shape selectivity (a) occurs when some but not all the components in a reaction mixture can reach the inner active sites and react. When the size and shape of the inner void modify the distribution of products by limiting the number of possible transition states, the effect is called *transition state* shape selectivity (b). And, finally, when a molecule can be formed but cannot be desorbed due to its size and shape, the product distribution is altered by *product* shape selectivity (c). In addition to these broad concepts, thermodynamic effects are also of importance. Reproduced by permission from Macmillan Publishers Ltd: *Nature*, B. Smit and T.L.M. Maesen, **451**, 671 (20). Copyright (2008) Macmillan Publishers Ltd.

water up to a very low dew-point is necessary to avoid clogging of the circuits or fogging between the glass panels as the temperature drops.[29]

Since there are recent comprehensive reviews on current industrial uses of synthetic zeolites,[25] as well as natural zeolites,[30] we shall now elaborate on more recent potential or actual applications. Calzaferri's group, for instance, has proposed zeolite L (**LTL**) as a convenient host for organic-inorganic composite materials that may serve as optical antennas for light harvesting and transport:[31,32] cation exchange affords cationic dyes to be introduced into the zeolite channels, where molecular aggregation is prevented, allowing high quantum yields. Isomerisation and bimolecular reactions may also be prevented, enhancing the dye stability. When the channels are filled with a donor dye (Pyronine$^+$) and the channels ends are modified with a luminescent acceptor dye (Oxonine$^+$) acting as a trap, the system behaves as a bi-directional antenna: light is collected by the donor and energy migrates up to the acceptor, where no return is possible because its excitation energy is lower than that of the donor (Figure 5.5).

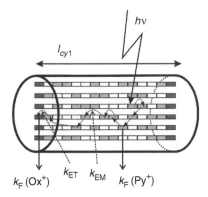

Figure 5.5   A zeolite-based bidirectional antenna system for light harvesting. The unidimensional large pores of zeolite L (one crystal represented here as a cylinder of length l_{cyl}) were loaded by cation exchange with pyronine (Py^+, light grey rectangles) as donor molecules, and modified at both ends with oxonine (Ox^+, dark grey rectangles) to act as traps (luminescent acceptors). Light excites a Py^+ molecule, which can relax by fluorescence or transfer its excitation energy to another Py^+ until the energy eventually reaches an Ox^+ molecule, from which there is no way back because of its lower excitation energy (k_F, k_{EM} and k_{ET} are rate constants for fluorescence of each dye, energy migration between pyronine molecules, and energy transfer from pyronine to oxonine, respectively). Reproduced with permission from S. Megelski and G. Calzaferri, *Adv. Funct. Mater.*, *Tuning the Size and Shape of Zeolite L-Based Inorganic–Organic Host–Guest Composites for Optical Antenna Systems*, **11**, 277 (2001). Copyright Wiley-VCH Verlag GmbH & Co. KGaA. Reproduced with permission.

Dye and co-workers have shown that pure silica zeolites devoid of connectivity defects are promising hosts for the preparation of fully inorganic electrides by adsorption of alkali metal vapours.[33,34] The host–guest compounds obtained have physically separated alkali ions and electrons (which can be considered as anions, hence, the name electrides). While these materials are currently unstable to temperature and air and may show potential as powerful reducing agents, it is believed that stable zeolite electrides will eventually be obtained, and a range of other potential applications based on electron emission induced by light, heat or electric fields can be envisaged.[35]

Recently, Comes *et al.* developed a versatile strategy to covalently anchor dyes inside siliceous zeolites and demonstrated its potential use as chromogenic sensors for the visual discrimination of closely related molecules by size and polarity.[36]

Pure-silica materials have been shown to have also a large potential in separation reactions. For instance, their hydrophobicity could afford the recovery of alcohols from dilute aqueous solutions.[37] A combination of silica zeolites with varying pore sizes has been proposed also for the separation and on-line monitoring of dioxins and related highly toxic and persistent pollutants.[38] Pure silica zeolites with small pores have been shown to be promising adsorbents for the energetically efficient separation of propene and propane at room or moderate temperatures.[39,40] Their hydrophobic character avoids competition of water in the adsorption process, while the lack of catalytically active sites prevents pore blockage by propene polymerisation.

There are also examples of emerging applications in biology and medicine, as shown by Wheatley *et al.*, who demonstrated the possibility of using zeolites for controlled NO release under physiologically relevant conditions and showed that NO-loaded Co-exchanged zeolite A inhibits platelet aggregation *in vitro*.[41] Moreover, it is possible to simultaneously deliver and produce NO from a feed of NO_2^- in larger amounts and over longer periods of time using NO-preloaded Cu-exchanged zeolite X.[42]

Finally, the authors find very interesting, appealing and refreshing one recent application of zeolites: beer kegs with 5–70 L capacity are able to self-cool down to a drinkable temperature on account of the high water affinity of low-silica zeolites.[43] Presumably, two hermetic reservoirs, one containing water and the other containing a dehydrated zeolite, are separated from the main beer container and from each other by an adequate choice of thermally insulating and thermally conducting walls. When the environment of both reservoirs is left in contact water evaporates, adsorbing heat (enthalpy of vaporisation) from the beer reservoir, and adsorbs into the dehydrated zeolite, where heat is released (enthalpy of adsorption), so a net heat flow from the beer vessel to the exterior is produced. Compared with beer kegs that need to be cooled in the freezer, we find this a significant advance.

5.7 ZEOLITE SYNTHESIS

5.7.1 The Synthetic Zeolites as Geoinspired Materials

The synthesis of zeolites can be traced back to the nineteenth century, when Jean-Baptiste Guimet was able to produce an artificial version of the highly valued and expensive natural pigment ultramarine, an aluminosilicate sodalite containing polysulfides in its cavities. For this discovery he was awarded a French national prize in 1828.[44] In 1862 another French chemist, St Claire Deville, synthesised an analogue of levyne

(**LEV**).[45] The scientific study of the synthesis of zeolites, mainly driven by the interesting adsorption properties of natural zeolites, and its extraordinary subsequent scientific and commercial success, however, is a genuine twentieth century venture and was independently pioneered by Barrer, first at Aberdeen University and later at Imperial College London, and by Milton (and later on also by Breck and Flanigen) at Union Carbide.

According to Rees, the natural formation of zeolites 'guided the laboratory techniques developed by Barrer'.[46] The first zeolite syntheses reported are reminiscent of the natural crystallisation of zeolites through alteration of volcanic glasses or aluminosilicate minerals by saline water at moderately high temperature and alkalinity, frequently in the presence of alkaline earth cations. The vast majority of natural zeolites have Ca^{2+} as a main countercation, which limits the maximum attainable pH in the mother liquors [due to the low solubility of $Ca(OH)_2$]. This, together with the low solubility of volcanic glasses and silicate minerals, leads to a very long crystallisation time, affordable by nature but not so easily by scientists, who were forced to increase temperature.[47] However, a higher crystallisation temperature favours more thermodynamically stable dense phases, thus limiting the range of structures attainable.

The first major, and probably most important ever, breakthrough in zeolite synthesis came by an apparently slight change in conditions, bringing about drastic consequences: Milton started to use alkaline hydroxides, commercial silica and alumina sources of high reactivity, working at higher alkalinities and using a lower crystallisation temperature, in an attempt to shorten the crystallisation time while favouring high hydration-water content materials.[48] This approach really opened the way to the field of synthetic zeolites by affording the crystallisation of less stable (metastable) materials, including in a first stage (1949–1953) six analogues of natural zeolites plus fourteen new structures.[49] Milton's synthesis of zeolites A (**LTA**) and X (**FAU**), followed by their rapid and withstanding commercial success as adsorbents and catalysts, is probably the most important milestone in zeolite science. Further modifications have followed, ever departing from natural conditions, by substituting key elements by others that may play a similar role but which, in their turn, impart their own peculiarities to the phase selectivity of the crystallisation (Table 5.2).

Considering the history of zeolite synthesis and the evolution of synthetic strategies as a whole we propose that it was, and still is, *geomimetic*, in the very same sense that Vincent proposed for biomimetic materials: synthetic zeolites realise 'the abstraction of good design from

Table 5.2 The geoinspired strategy: components in the hydrothermal crystallisation of natural zeolites are substituted by others, frequently displaying structure-directing effects (in italics)

Natural crystallisation	Role	Synthetic crystallisation
H_2O	Solvent	H_2O $HOCH_2CH_2OH$ and other alcohols $H_2NCH_2CH_2NH_2$ and other amines Ionic liquids None
Si, Al (Be, Mg, P, Fe, Zn)	T-atoms (from volcanic glasses or minerals)	Si, Al, P, *Ge, Ga*, Ti, *Be*, V, Fe, *Zn, Li,*... (from reactive sources)
Ca, Na, K (Sr, Ba, Mg)	Countercations	*Na, K (Rb, Cs, Li)* Organic (*organoammonium, organophosphonium, organometallic*) cations
[OH⁻] (at moderate alkalinity)	Mineraliser	[OH⁻] (at moderate to high alkalinity) F^- (at close to neutral pH)
H_2O, small gases	Uncharged pore fillers	H_2O, amines, organo-oxygenated compounds, small gases

nature',[50] although not from the realm of biology but from that of geology. Furthermore, we think that the synthetic strategies to make new zeolites are truly *geoinspired*,[51] as the natural crystallisation of zeolites was first emulated (Barrer and others), and then modified (Milton, Breck, Barrer, Flanigen, and then all of us) to discover new materials of the same class and with the same overall *good design* (microporous framework) but with new structures, compositions, properties and applications. Considering the key elements that are likely of importance in a natural crystallisation, zeolite scientists have replaced each of them by others (Table 5.2), and in each case, with a less or more marked character, some degree of peculiar *structure-direction* was found, as will be shown below.

5.7.2 Thermochemistry of Zeolite Synthesis

The main issue in zeolite synthesis is the control of the specific phase that crystallises, *i.e.* the *phase selectivity* of the crystallisation. Nobody can tell for sure whether a new given set of conditions (chemical composition, temperature, time, preparation, stirring and other synthesis parameters) will yield a new zeolite, a known zeolite or a dense phase, although most currently active zeolite scientists, will be willing to bet (and lose!). As Davis and Zones put it, this uncertainty 'becomes an enjoyable challenge' for zeolite synthesisers.[52] On the other hand, it may be quite disappointing for zeolite scientists, as it still shows a profound lack of knowledge for a scientific discipline that, after at least 60 years, should be already mature. Furthermore, the *a priori* design of new zeolite structures for targeted applications, a matter that seemed within reach more than a decade ago,[53] still remains elusive. Arguably, for a good portion of the last 60 years zeolite scientists have been more interested in discovering new structures, compositions and applications than in a true scientific understanding of the crystallisation of zeolites. Hopefully, things have been slowly changing.

There are many ways in which tetrahedra can be fully connected to yield a (4;2)-3D net. Of course not all of them may be stable and, if we consider pure silica materials, for instance, all of them would be metastable compared with quartz (the most stable silica phase). However, an under-construction database of hypothetical but energetically feasible structures already lists several hundred thousand zeolite structures. Although this database contains duplicates, at least 333 energetically feasible unique topologies have been so far identified.[54] Importantly, known pure-silica zeolite materials are quite stable, with enthalpies of transition from quartz below 15 kJ mol^{-1}.[55] An important finding is that, as expected, there seems to be a correlation between the stability of SiO_2 zeolites and their framework density: more dense phases appear to be more stable than more open zeolites.[55] This refers, of course, to the pure silica framework without considering host–guest stabilisation and entropic effects (Figure 5.6).

From those studies, the synthesis of a zeolite does not appear to be strongly hindered by thermodynamics and it may be afforded either by kinetic control, by entropically favoured conditions, or by a strong host–guest interaction between the (organic) cations and the framework. Interestingly, entropy measurements on a limited number of zeolites also show little thermodynamic control on the phase selectivity of the crystallisation, with entropies roughly 4 kJ mol^{-1} above that of quartz, and with overall Gibbs free energy of transformation from quartz that is within twice the available thermal energy at room temperature.[56] Finally, host–guest interactions in an even more limited number of pure silica zeolites revealed relatively low stabilisation, suggesting the phase selectivity of zeolite synthesis is, generally,[57] kinetically controlled. Of course, this does not imply any universal conclusion, as it is still possible that a

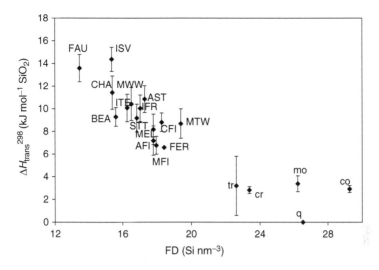

Figure 5.6 Enthalpy of transition from quartz *vs* framework density (FD) for several dense and microporous SiO_2 polymorphs. Zeolite phases are denoted by their FTC, while minor case codes correspond to the dense phases quartz (q), tridymite (tr), cristobalite (cr), moganite (mo) and coesite (co). The figure suggests the stabilities of silica phases decrease when their densities decrease (except for the high-pressure phase coesite, which is compressed beyond the density of quartz; the enthalpy for moganite deviates from the trend, but it was not directly measured and is not as reliable as the rest). The figure also shows pure-silica zeolites are not highly destabilised. Reproduced with permission from P.M. Piccione, C. Laberty, S. Yang, M.A. Camblor, A. Navrotsky and M.E. Davis, *J. Phys. Chem. B*, **104**, 10001. Copyright (2000) American Chemical Society.

particular host–guest interaction, for a zeolite system (cation framework) not considered in previous studies, could override the energetic costs of a low density SiO_2 framework (this would be best demonstrated by the transformation of quartz into a zeolite).

The metastability of zeolites allows for phase transformations in the crystallising mixture, a phenomenon in which, of course, the well known Ostwald rule is obeyed: the sequence of phases follows an order of increasing stability. A very typical example is the transformation of zeolite A into hydroxysodalite (**SOD**) or that of synthetic faujasite zeolite X or Y (**FAU**) into gismondine-like zeolite P (**GIS**) or ZSM-4 (**MAZ**).[58] A more peculiar case is that of the gallosilicate natrolites with Si/Ga ratios slightly above 1.5, in which there is a phase transformation that involves no change in chemical composition and no change in topology, but just a change in T-ordering:[59] the initial crystallisation of disordered natrolites (containing violations of Loewestein's rule, most likely allowed for entropic reasons) is followed by a dissolution/recrystallisation process in which more ordered natrolites, with the same composition and topology but with a lower concentration of Ga-O-Ga pairs occur.[19]

5.7.3   Organic Structure-Directing Agents

A second major breakthrough in the synthesis of zeolites was the introduction of organic cations in the synthesis, partly or fully replacing alkali cations. The use of alkylammonium cations in zeolite synthesis was pioneered by Barrer's group,[60] quickly followed by Kerr and co-workers at Mobil.[61,62]

The first obvious consequence of the use of organic cations was an increase in the Si/Al ratio of the zeolite,[62] which was especially striking in the case of zeolite A, for which a consistently constant Si/Al ratio was previously found.[63] The organic cations, together with alkali cations, were found occluded in the synthesised zeolites, a fact that explained the observed increase in Si/Al ratios: alkylammonium cations are more voluminous than alkali cations (Figure 5.7), so their occlusion in the pores of the zeolite reduced the overall positive charge within the channels, thus requiring a lower negative charge density in the framework and a higher Si/Al ratio.[60,62] Such a change in chemical composition was found to be of extraordinary practical importance, bringing about an increased chemical, thermal and hydrothermal stability of the zeolite, an increased acid strength of its proton form and a larger available void space within the less crowded channels. The occluded organics could be burned out by calcination.

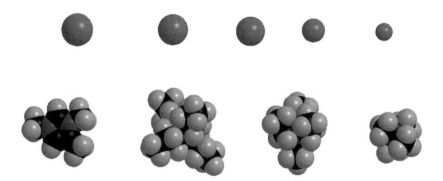

Figure 5.7   A van der Waals size comparison between the alkaline series of cations (top, from right to left: Li^+, Na^+, K^+, Rb^+ and Cs^+) and some organoammonium cations used as OSDAs in the synthesis of zeolites (bottom, from right to left: TMA^+, TEA^+, TPA^+ and 1,3,4-trimethylimidazolium). The size shown for TEA^+ and TPA^+ corresponds to one among many possible conformations. Note that in addition to large differences in their size and rigidity, alkali cations are considered featureless spheres

Soon afterwards, a second implication revolutionised the field as new zeolite structures were obtained. The synthesis by the Mobil group of zeolite Beta[64] (*BEA,[65] with interconnected large pores)[66] and later on of ZSM-5[67] (MFI, interconnected medium pores),[68] both with a Si/Al ratio much higher than those previously known for any zeolite were a consequence of using the larger tetraethylammonium (TEA$^+$) and tetra-propylammonium (TPA$^+$) cations, respectively. These two zeolites have shown extraordinary properties as catalysts and found important industrial applications, and they both remain as the most widely studied zeolites. Also relevant was the early synthesis of zeolite ZK-5 (KFI), with a lower Si/Al ratio and a small pore size but with a very low framework density.[69,70]

The synthesis of new, previously unknown, zeolite structures by using organic cations brought about the term templating: purportedly the size and shape of the organic cation somehow shaped the size and shape of the internal void space of the crystallised zeolite which had to accommodate it. For several decades organic cations used in zeolite synthesis were generally termed *templates*, and their use proved extraordinarily helpful in producing new zeolite structures, mainly by using increasingly sophisticated templates.

However, it also became obvious that, very frequently there was a general lack of specificity with respect to the ability of a given template to yield a given zeolite, with respect to the astringent necessity of a zeolite to be crystallised in the presence of a specific template and, finally, with respect to the geometric correspondence between the size and shape of the organic cation and the size and shape of the void space in which it resides. Some templates were quite unspecific, in the sense that they could produce different zeolites depending on the synthesis parameters. Some zeolite structures were also unspecific, in the sense that they could be synthesised with a large number of different organic cations showing large differences in size and shape (these zeolite structures have been termed 'default structures': they are intrinsically stable and crystallise when a lack of specificity of the template precludes the formation of a different phase).[71] Frequently, there was no close geometric correspondence between cation and void space. For all these frequent situations the terms template and templating seemed to be quite inadequate, and the function of the organic cation was, apparently, a mere effect of pore filling without a strong stabilisation.

By contrast, there were some examples in which the cation deserved the term template, because it was quite specific for a given zeolite, and because it showed a close correspondence between the size and shape of the cation and the size and shape of the zeolitic void. A good example of templating was the crystallisation of zeolite ZSM-18 containing strained 3-rings by a triquat (a triply charged organic cation).[72,73]

Probably the most important issue in zeolite synthesis is that of structure-direction, *i.e.* the ability to control the crystallisation to a specific structure, which eventually should allow the *a priori* design of zeolite structures with specific features for targeted applications (*designer zeolites*). Among the many different factors that contribute to realising a particular zeolite, organic additives have proved to be particularly fruitful. In general, organic entities that are used in zeolite synthesis and that finally reside in its voids are nowadays called structure-directing agents (SDAs) or organic structure-directing agents (OSDAs) and their role during the crystallisation can go from a loose pore filling effect to a strong and specific true templating effect bearing a close geometric correspondence and a presumably large energetic stabilisation of the zeolitic host by the guest.

Several groups have tried to understand structure-direction by OSDAs, especially with regard to the key aspects that determine a highly specific action. Gies *et al.* studied the synthesis of pure-silica materials using mainly amines and also organoammonium molecules, and tried to identify those key aspects.[74,75] For instance, the shape and size of the organic molecule are of much importance: spherical (globular) OSDAs tend to yield materials containing cavities, rather than pores, while chain-like additives generally lead to one-dimensional porous materials and molecules with branched chains afford intersecting channels. Of course, larger OSDAs need to be accommodated in larger cavities. The conformational flexibility of the OSDA is highly important, since a multitude of stable conformers would result in a multitude of differently shaped entities and hence a less specific action. Gies *et al.* then discussed basicity and amphiphilicity of OSDAs in zeolite synthesis: basicity helps to mobilise silica, while large amphiphilic additives tend to self-aggregate thus disfavouring the action of the individual molecule. They also discussed other aspects interplaying with the properties of OSDAs in determining the phase selectivity of the zeolite crystallisation. With regard to temperature, for instance, Gies *et al.* showed that amines fitting tightly in a clathrasil obtained at a given temperature tend to produce a clathrasil with larger cavities at higher temperatures due to the increased effective size of the amine. Finally, Gies *et al.* concluded from a set of syntheses in the presence of amines, that the larger the concentration of the OSDA amine the larger the concentration of cavities in the obtained structure, an effect that we will discuss later on for the case of organoammonium cations in synthesis of pure silica phases by the fluoride route (see below).

Focusing more on organoammonium cations for the synthesis of high silica alumino- and borosilicates, Davis and co-workers have also studied structure-direction during zeolite synthesis.[76,77] They also recognised that a high conformational freedom of the OSDA generally determines a decreased specificity in its structure-direction action, meaning that the OSDA may help crystallise several different phases and/or that the conditions for a given crystallisation are confined within a very narrow window. They have also argued that the shape of the OSDA is highly significant[78] and have shown how to go from a unidirectional channel system to a multidirectional one by laterally expanding the size of the OSDA (at the same time the size and rigidity increased).[79] One may go further and argue that a less defined shape may be more easily accommodated within different possible pores, without a need for a specific correspondence between host and guest and also that a less defined shape may be more likely shared by a range of molecules.

And, importantly, Davis and co-workers also considered the size of the OSDA, and more specifically, its hydrophobic character in determining the behaviour of the OSDA: considering that high-silica zeolites are believed to possess a considerable hydrophobic nature, a good host–guest interaction would require a certain degree of hydrophobicity. In fact, it has been proposed that, in the synthesis of pure-silica ZSM-5 using TPA$^+$, a preorganisation of inorganic-organic entities that would subsequently assemble into the zeolite is formed by overlapping of the hydrophobic hydration spheres of the inorganic (silicate oligomers) and organic (TPA$^+$) species in solution.[80] By contrast a too hydrophobic molecule would imply a low solubility in water, or a tendency to aggregation, hence detracting from its action as an OSDA. The partition between water and chloroform was taken as an experimental method to determine the relative hydrophobicity of different organic cations and when the percentage of transfer to the organic phase was plotted against the ratio of carbon atoms to charged nitrogen atoms (C/N$^+$), an S-shaped curve was obtained, irrespective of the shape of the molecules. According to these authors, OSDAs that 'work well' in the synthesis of high-silica zeolites have intermediate percentage of transfer and generally have C/N$^+$ ratios between 10 and 16.

As a conclusion, a moderate hydrophobicity of the OSDA was believed to be necessary for the synthesis of pure silica zeolites. Using this idea, this group has attempted the synthesis of pure silica **MFI** zeolite by using OSDAs derived from TPA$^+$, the most typical OSDA for **MFI**, by substituting one or two -CH$_3$ ending groups by -OH groups, hence decreasing the OSDA hydrophobicity without altering its size and shape much.[81] With one substitution, the OSDA was still able to crystallise **MFI**, although at a slower rate. With two substitutions, the OSDA was not able to crystallise **MFI** unless seeds or a small amount of TPA$^+$ was added. In a prior work the tetrasubstituted OSDA, *i.e.* tetraethanolammonium, was shown to be unable to crystallise any pure silica zeolite even in the presence of enough TPA$^+$ to nucleate **MFI**,[82] which may be attributed to the lack of a hydrophobic hydration sphere due to the strongly hydrophilic character of the alcohol groups at the end of the alkyl chains.

As a rule of thumb, bulky rigid cations with intermediate hydrophobicity and limited conformational freedom lead to more specific structure-direction. These concepts have allowed many of the discoveries of new zeolite structures, specially by Zones and co-workers, who combined the design and synthesis of new OSDAs deemed to display a high specificity in its structure-direction action with a careful screening of synthesis conditions. Among these conditions, T-atom substitution was found to be of large importance and Zones and co-workers have concluded that, generally, adding Al or B to a synthesis mixture that otherwise would yield a unidirectional channel zeolite results in the production of a multi-dimensional pore system.[76]

Finally, while most frequently an OSDA acts as a single molecule it is also possible to use a supramolecular assembly. Most interestingly, a π–π supramolecular assembly of two identical cations afforded the synthesis in fluoride medium at high concentration of the pure-silica analogue of zeolite **LTA**, which had defied synthesis for half a century.[83]

The works commented on above generally consider that the main host-guest interactions relevant to structure-direction by OSDAs are van der Waals contacts, essentially implying a cavity filling effect. Thus, little attention has been paid so far to other interactions, including the coulombic interaction between the OSDA cation and the anionic framework (which may be of different types: $[TO_{4/2}]^{-4+n}$ for T-atoms with oxidation state of $n = 3$, 2 or 1, occluded fluoride, Si-O$^-$ connectivity defects). However, there are indications that suggest the concept of structure-direction should be broadened to include other important interactions. Behrens *et al.* have shown a Si-O⋯H-C hydrogen bond of considerable strength [$d(\text{H}\cdots\text{O}) = 2.31$ Å, with a C-H⋯O angle of 167°] in nonasil (**NON**) synthesised using cobaltocinium $[Co^{III}(\eta^5\text{-}C_5H_5)_2]^+$ and fluoride anions and pointed out that this interaction may be of importance during the nucleation stage.[84] Also, Zicovich-Wilson *et al.* found by periodic quantum-mechanical calculations that there are complex electronic host-guest interactions in as-made ITQ-12 (**ITW**) zeolites that should be also considered to explain structure-direction.[85]

While most of the works commented on so far refer to the synthesis of high and pure silica zeolites, for which rigid OSDAs give the most typically better results, it has been shown that flexible organic cations can be specific and may allow the discovery of new zeolites when combined with the appropriate inorganic cations and T-atom substitution, especially at high levels of substitution, or through a multi-OSDA approach at proper inorganic conditions, as will be shown below.

5.7.4 Structure-Direction by Flexible, Hydrophilic OSDAs

In essence, hydrothermal zeolite crystallisation is a complex self-assembly process that can be dictated by many weak interactions such as van der Waals forces, and others, between the organic and inorganic fractions of the resulting composite material. Therefore, it is not surprising that flexible, hydrophilic OSDAs, as well as rigid, hydrophobic ones, can produce a range of new zeolite structures. This is well illustrated by the synthesis of the medium-pore zeolite ZSM-57 (**MFS**) that requires the use of a flexible, linear diquaternary N,N,N,N',N',N'-hexaethylpentane-diammonium (Et_6-diquat-5) ion as an OSDA.[86] Since certain levels of Na^+ ions and lattice negative charges in the synthesis mixture are also necessary to successfully crystallise ZSM-57 in the presence of Et_6-diquat-5,[87] however, the structure-directing ability of such a flexible OSDA cannot be strong enough to dominate over the impact of inorganic gel chemistry.

It was in the mid 1980s when Bibby and Dale synthesised pure-silica sodalite using ethylene glycol both as a solvent and as an OSDA, which is the first step towards the use of organic solvents in the synthesis of zeolites and related microporous materials,[88] with the exception of a BASF patent on the ether/water-mediated synthesis of ZSM-5.[89] Since each β-cage in pure-silica sodalite contains one ethylene glycol molecule, the intermolecular hydrogen bonds between the encapsulated ethylene glycol molecules are not possible, making a particular type of conformation dominant. This has led Hong *et al.* to focus their interests on the host–guest interactions within sodalite materials with a wide range of SiO_2/Al_2O_3 ratios ($10–\infty$) containing ethylene glycol as a guest molecule. On the basis of a combination of multinuclear MAS NMR and vibrational spectroscopic results, they were able to conclude that the molecular conformation of ethylene glycol in a pure-silica sodalite (0 Al atoms per β-cage) is tGg' stabilised only by one intramolecular hydrogen bond while

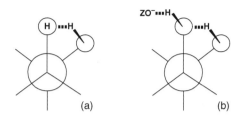

Figure 5.8   (a) tGg' and (b) gGg' conformers in the Newman projection for ethylene glycol within the pure-silica and aluminosilicate ($SiO_2/Al_2O_3 = 10$) β-cages, respectively. Reproduced with permission from S.B. Hong, M.A. Camblor and M.E. Davis, *J. Am. Chem. Soc.*, **119**, 761. Copyright (1997) American Chemical Society.

that of the guest molecule in a material with $SiO_2/Al_2O_3 = 10$ (1 Al atom per β-cage) is gGg' stabilised by one hydrogen bond to the framework in addition to one intramolecular hydrogen bond,[90] as seen in Figure 5.8. More interestingly, the intermolecular hydrogen bonds between the framework and the OH groups of the guest molecule in the β-cage with $SiO_2/Al_2O_3 = 10$ was found to remain intact even at 200 °C,[91] which is higher than the crystallisation temperature (175 °C) of the zeolite host. Since the calculated free energy of the gGg' conformer for ethylene glycol in the gas phase is higher by 1.3 kJ mol^{-1} than that of the most stable tGg' conformer,[92] the higher energy conformer gGg' could also play a role during nucleation at the crystallisation temperature (175 °C) in the aluminosilicate synthesis mixture. Thus, if OSDAs much longer and hence more flexible than ethylene glycol are used in zeolite synthesis, the phase selectivity of the crystallisation could then be sensitive to the type of the conformation of organic molecules dominant at zeolite crystallisation conditions, which should be greatly affected by the nature and extent of interactions between the organic and inorganic components in synthesis mixtures. Therefore, Hong's group has taken interest in the possibility that the conformation control of highly flexible OSDAs by varying the concentrations of inorganic components in the aluminosilicate system could be an alternative strategy for the discovery of new zeolites.

A piece of experimental evidence to support the above idea can be drawn from zeolite syntheses in the presence of N,N,N,N',N',N'-hexamethylpentanediammonium (Me_6-diquat-5) ion, a slightly shorter OSDA than Et_6-diquat-5, the only known OSDA leading to the crystallisation of ZSM-57. This flexible diquaternary cation was reported to give at least four different zeolites [*i.e.* EU-1 (**EUO**), ZSM-48 (***MRE***), ZSM-12 (**MTW**), MCM-22(P)] depending on the Al content and type and concentration of alkali cations in the synthesis mixtures.[93] Since neither of these zeolites has the same framework topology as that of any of the materials [*i.e.* P1 (**GIS**), SSZ-16 (**AFX**), SUZ-4 (**SZR**) and ZSM-57] that can be synthesised using Et_6-diquat-5,[71,94] the phase selectivity of the crystallisation in the presence of flexible diquaternary ions may differ significantly according to the length of the groups on the ammonium ion, as well as to that of the spacing alkyl chain. However, the structure-directing abilities of these organic species are not strong enough to dominate the impact of inorganic gel chemistry. In fact, a large number of linear diamines and diquaternary ammonium cations with different spacing alkyl chain lengths and aliphatic or cyclic alkylammonium moieties have been used as OSDAs for more than two decades,[74,95–103] leading to the discovery of many new materials including EU-1, NU-87 (**NES**), MCM-47, GUS-1 (**GON**) and TNU-10 (**STI**).

Among such OSDAs used in zeolite syntheses thus far, of particular interest are the $(C_5H_{11})N^+(CH_2)_nN^+(C_5H_{11})$ ions with $n = 4$–6 which are formed of two 1-methylpyrrolidinium groups connected by tetra-, penta- and hexamethylene bridging units. The 1,4-bis(*N*-methylpyrrolidinium)butane (1,4-MPB) ion directs the synthesis of TNU-9 (**TUN**), a new three-dimensional 10-ring channel system whose pore connectivity is much more complex than that of ZSM-5.[103–105] The longer pentane (1,5-MPP) and hexane (1,6-MPH) analogues, on the other hand, give IM-5 (**IMF**) and SSZ-74 (-**SVR**) that contain a two-dimensional 10-ring channel system with a very complicated connectivity and a three-dimensional 10-ring channel system with ordered T-atom vacancies, respectively.[106–109] Since the OSDA varies only with the length of spacing alkyl chains, these three new medium-pore zeolites can be regarded as a synthesis series. The framework structures and channel systems of TNU-9, IM-5 and SSZ-74, all of which were solved by combining synchrotron powder diffraction and transmission electron microscopy, are shown in Figure 5.9, together with the OSDA used in the synthesis of each zeolite. While both TNU-9 and IM-5 have 24 crystallographically distinct tetrahedral sites (T-sites) in the asymmetric unit, SSZ-74 possesses 23 T-sites with an unexpected T vacancy. This makes them the three most complex zeolite structures known to date, although SSZ-74 does not have a fully connected (4;2)-3D net but, more properly, an interrupted framework. Clearly, new multi-dimensional medium-pore zeolites can expand the successful shape selectivity observed for the same group of already known materials and hence are of great importance from both academic and industrial points of view. However, there has been little success in the synthesis of medium-pore zeolites for nearly two decades. Since TNU-9, IM-5 and SSZ-74 show different catalytic properties from ZSM-5, one of the most widely

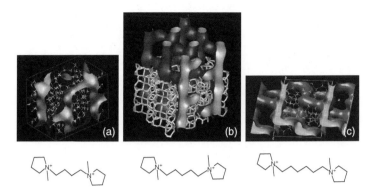

Figure 5.9 Framework structures and channel systems of (a) TNU-9, (b) IM-5 and (c) SSZ-74 with the OSDA used in the synthesis of each zeolite shown below. Reprinted with permission from Ch. Baerlocher, D. Xie, L.B. McCusker, S.-J. Hwang, K. Wong, A.W. Burton and S.I. Zones, *Nat. Mater.*, 7, 631. Copyright (2008) Macmillan Publishers Ltd

studied and industrially important zeolites, together with excellent hydrothermal stability, we look forward to new chemical technologies based on their own shape selectivity effects.

The synthesis of TNU-9 in the presence of 1,4-MPB was found to be possible only from synthesis mixtures with a very narrow range of SiO_2/Al_2O_3 and $NaOH/SiO_2$ ratios, which is also the case of the 1,5-MPP-mediated synthesis of IM-5,[110] mainly due to the flexible nature of the OSDA employed. As seen in Table 5.3,[103] furthermore, the 1,4-MPB ion can direct the synthesis of seven other different zeolite structures with significant organic contents ($\geq$10 wt %, according to TG/DTA analysis), *i.e.* TNU-10, mordenite (**MOR**), MCM-22(P), MCM-47, ZSM-12, offretite (**OFF**) and NU-87, depending on the oxide composition of the synthesis mixtures. Thus, there is a specific level of alkali cations in the gel, together with a reasonable amount of lattice charge, in the 1,4-MPB-mediated synthesis of this series of zeolites. A remarkable diversity in the phase selectivity was also observed for the N,N,N,N',N',N'-hexamethyl-hexanediammonium (Me$_6$-diquat-6) or hexamethonium ion. Until recently, Me$_6$-diquat-6 had been reported only to give EU-1 and ZSM-48 in alkaline media. In combination with the structure-directing effects of F- and/or Ge (see below), however, five more phases [ITQ-13 or IM-7 (**ITH**), ITQ-22 (**IWW**), ITQ-24 (**IWR**), IM-10 (**UOZ**) and ITQ-33] have been synthesised using this diquarternary cation.[111–115] It thus appears that the conformation of flexible OSDAs dominant at zeolite crystallisation conditions may be greatly affected by the nature and extent of

Table 5.3   Representative products obtained using 1,4-bis(N-methylpyrrolidinium)-butane (1,4-MPB) as an OSDA.[a] Reprinted with permission from S.B. Hong, *Catal. Surv. Asia*, **12**, 131. Copyright (2008) Springer

Run no.	Gel composition		Time (days)	Product[b]
	SiO_2/Al_2O_3	$NaOH/SiO_2$		
1	60	1.00	14	TNU-10
2[c]	60	1.00	14	Mordenite
3	15	1.00	14	Analcime
4	30	1.00	7	TNU-10
5	120	1.00	14	TNU-10
6	240	1.00	14	Analcime + (TNU-10)
7	∞	1.00	7	—[d]
8	60	1.13	14	Analcime
9	60	0.87	14	TNU-9 + IM-5
10	60	0.73	14	TNU-9
11	60	0.73	7	MCM-22(P)
12	30	0.73	7	Analcime + mordenite
13	40	0.73	14	TNU-9
14	∞	0.73	7	MCM-47
15[e]	∞	0.73	14	MCM-47
16	60	0.60	7	ZSM-12
17	60	0.47	7	ZSM-12
18	∞	0.47	7	MCM-47
19	60	0.33	7	ZSM-12 + amorphous
20	60	1.00[f]	7	D[g]
21	60	1.00[h]	7	Offretite
22	60	0.73[f]	14	Amorphous
23	60	0.73[h]	14	NU-87

[a] The oxide composition of the synthesis mixture is 4.5(1,4-MPB)·xNa$_2$O·yAl$_2$O$_3$·30SiO$_2$·1200H$_2$O, where x and y are varied (5.0 ≤ x ≤ 17.0 and 0.0 ≤ y ≤ 2.0, respectively). All the syntheses were carried out under rotation (100 rpm) at 160 °C, unless otherwise stated.
[b] The product appearing first is the major phase, and the product obtained in a trace amount is given in parentheses.
[c] Run performed under static conditions.
[d] No solids produced.
[e] Run performed after adding a small amount (2 wt % of the silica in the gel) of the previously prepared TNU-9 sample as a seed to the synthesis mixture.
[f] KOH/SiO$_2$ ratio.
[g] Unknown, probably dense material.
[h] (NaOH + KOH)/SiO$_2$ ratio with Na/K = 1.0.

interactions between the organic and inorganic components in the synthesis mixtures. If such were the case, combining a careful screening of inorganic synthesis parameters with the structure-directing ability of such organic additives could then be a viable alternative route for the discovery of novel zeolite structures.

Another unusual feature during the synthesis of TNU-9 is that it follows the initial formation of MCM-22(P), a layered precursor of MCM-22 (**MWW**), in the crystallisation medium. Given the high framework density (17.6 *vs* 15.9) of TNU-9 compared with MCM-22,[116] its crystallisation can be rationalised by considering Ostwald ripening through a dissolution/recrystallisation to a more stable phase. In contrast, the MCM-22(P) material synthesised in the presence of hexamethyleneimine (HMI), a well-known OSDA for this layered phase, is reported to transform into ZSM-35 (**FER**), when heated for a prolonged time in the crystallisation medium.[117] Since 1,4-MPB yields MCM-22(P) and TNU-9 in certain high-silica synthesis conditions and MCM-47, a layered precursor of ferrierite (**FER**) in the absence of Al (Table 5.3), the reason 1,4-MPB-MCM-22(P) transforms to TNU-9 rather than to ZSM-35 may likely be related to the energetic difference at synthesis temperatures between 1,4-MPB-TNU-9 and 1,4-MPB-MCM-47, depending on their Al content. It is interesting to note here that like TNU-9 and ZSM-35, IM-5 also crystallises at the expense of MCM-22(P) in the presence of 1,5-MPP and Na^+ ions as SDAs.[118] The *in situ* transformation of MCM-22(P) layered materials prepared using HMI, 1,4-MPB and 1,5-MPP into these three different zeolites with prolonged heating, respectively, are illustrated in Figure 5.10. Thus, we believe that when the metastable MCM-22(P) phase is obtained using OSDAs other than 1,4-MPB or HMI and its thermodynamic

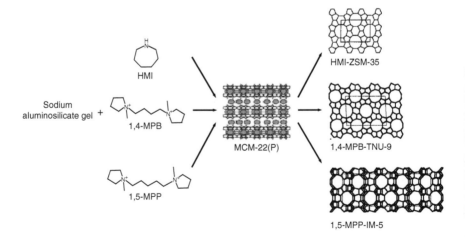

Figure 5.10 Transformation of MCM-22(P) layered precursors prepared using different flexible OSDAs into different zeolites with prolonged heating in the crystallisation medium

stability is significantly different than 1,4-MPB-MCM-22(P) or HMI-MCM-22(P), this composite material is capable of transforming into a new zeolite structure in the crystallisation medium. Unlike the case of TNU-9 and IM-5, on the other hand, the synthesis of SSZ-74 requires the use of F^- ions together with 1,6-MPH in the highly concentrated gel system where the H_2O/SiO_2 ratio is between 3.5 and 7.0.[108] Also, it is interesting to note that 1,5-MPP can yield a new large-pore zeolite ITQ-22 with intersecting 12-, 10- and 8-ring pores, when Ge is added as a partial Si substitute to silicate synthesis mixtures with low water contents $[H_2O/(Si+Ge)O_2 \leq 15]$.[112] The effects of water content in synthesis mixtures and of the presence of Ge on the phase selectivity of the crystallisation will be further discussed below.

5.7.5 Double OSDA Strategies

There are several examples in which two organic entities are used in zeolite synthesis, following two distinct types of strategy. In the first strategy the idea is simply that each OSDA could fit a different type of void, as exemplified by the synthesis of **MWW** zeolites by Camblor et al.[119] and Zones et al.[120] Zeolite **MWW** has two kinds of void spaces, one comprises large cavities connected to each other through 10-ring windows, and the other is a sinusoidal 10-ring system. While both systems are two-dimensional 10-ring pores, they are independent, i.e. there is no connection allowing the passage of molecules from one system to the other. Importantly, the available space in each system is radically different, in such a way that while the large cavities could host the bulky and rigid cations used before in the synthesis of **MWW** zeolites (typically, N,N,N-trimethyl-1-adamantammonium, 1-TMAda$^+$, for pure silica ITQ-1 and N,N,N-trimethyl-2-adamantammonium, 2-TMAda$^+$, for high silica SSZ-25), the sinusoidal 10-ring channel could not. Since high-alumina **MWW** materials (such as MCM-22) were typically synthesised with the sole aid of smaller amines like hexamethyleneimine, the use of a combination of OSDAs appeared logical. Camblor et al. showed that addition of hexamethyleneimine (or other small and flexible amines) to a 1-TMAda$^+$-containing synthesis mixture yielded pure-silica ITQ-1 materials with enhanced crystallinity at a faster crystallisation rate, with a better reproducibility and even in the presence of Na$^+$ ions, which were shown before to promote SSZ-31 rather than **MWW**.[121] These results were rationalised as a 'cooperative structure-directing effect', with the large and rigid adamantylammonium stabilising the large cavities and the small and flexible amine stabilising the smaller 10-ring pores, while in the absence of amines stabilisation of the smaller pores would occur by products of the degradation of 1-TMAda$^+$, a conclusion that was supported by ^{13}C MAS NMR experiments. Zones *et al.* showed a similar improvement in the synthesis of high-silica SSZ-25 when using 2-TMAda$^+$ in combination with several amines, particularly isobutylamine, which afforded a cheaper synthesis by much reducing the amount of the adamantylammonium cation, which is preferentially taken up by the growing zeolite, and even allowed the substitution of this cation by other polar adamantyl derivatives or different bulky polycyclic cations. Zones and Hwang further extended the strategy to other amines and organocations and were able to synthesise in a convenient and cheaper way several other zeolites, including the new SSZ-47.[122]

The second strategy was recently developed by researchers at UOP and is called the 'Experimental Charge Density Matching' (ECDM) approach.[123] In this case, two OSDAs are also used but the conditions are adjusted seeking to force a dual structure-directing action. This is done by first preparing a crystallisation mixture that cannot crystallise because of a large mismatch between the charge density of the alumino-silicate gel (*i.e.* a low Si/Al ratio at a high pH) and that of the first OSDA, with a low charge density. A first run of experiments should determine the correct range of conditions (pH, Si/Al ratio, temperature and time) that prevent crystallisation of zeolites or dense phases with this OSDA. These conditions ensure the success of the *mismatched step* of the synthesis. In a second step (the *matching step*), a second OSDA (the 'crystallisation-inducing template', with a higher charge density, more appropriate to the 'inorganic composition' prepared) is added in stoichiometric or substoichiometric amounts with respect to the heteroatom introducing a charge in the framework (Al). This clever approach has produced very interesting results: applied to two of the most extensively investigated OSDAs [TEA$^+$, acting at the mismatched step, and tetramethylammonium (TMA$^+$) as 'the crystallising template'] the strategy has allowed the discovery of a new zeolite, UZM-5 [**UFI**, a two-dimensional 8-ring pore zeolite containing large cavities],[124] and the synthesis of two known zeolites with unusual chemical compositions, *i.e.* UZM-4 (a silica-rich **BPH** zeolite) and UZM-9 (a silica-rich **LTA** zeolite). UZM-12 (a silica rich **ERI** zeolite)[125] and UZM-22 (an alumina-rich **MFI** zeolite)[126] are another two successful examples. Adding interest to the strategy, it tends, inherently, to yield zeolites with a significant concentration of framework charges, an aspect that bears interest for catalytic applications.

5.7.6 Structure-Direction by T-Atoms

As described above, the use of a wide array of alkylamines and alkylammonium ions with variations in size, shape, charge density, *etc.*, as OSDAs in the synthesis of zeolites and related microporous materials has greatly expanded their structural and compositional regimes. However, recent advances in this research area have shown that the substitution of Al and/or Si by other atoms in zeolite frameworks during the crystallisation process is another rational synthetic strategy for the discovery of previously unobserved topologies, which may be related to differences in the relative solubility and ratio of nonbonded radius to T-O bond length for a given heteroatom, possibly favouring its preferential siting into particular crystallographically distinct T-sites.[127] In fact, a number of novel zeolite structures have been synthesised by heteroatom lattice substitution in alkaline or fluoride media, both with and without OSDA present.[116]

5.7.6.1 Boron

Taramasso *et al.* were probably the first to introduce B instead of Al as a trivalent, tetrahedral lattice-substitute into zeolite frameworks during the crystallisation process: in 1980 they reported the synthesis of a borosilicate NU-1 (**RUT**) denoted BOR-A using TMA$^+$ as an OSDA.[128] Since then, a number of different borosilicate zeolites [*e.g.* B-ZSM-5 (**MFI**), ERB-1 (**MWW**), B-Beta (**BEA*), *etc.*] have been synthesised.[129] Unlike the case of aluminosilicate zeolites, however, the synthesis of borosilicate materials frequently requires the addition of a large excess of B to the synthesis mixture compared with that actually incorporated in the crystallised product, mainly due to the much higher solubility of this trivalent atom than Al in alkaline media. While a high degree of B substitution has not been found yet, in particular, its use as a phase selectivity factor in conjunction with rather sophisticated OSDAs has led the Chevron group to discover several new high-silica zeolites that are not obtained from aluminosilicate and silicate gel systems.[130] An excellent example is SSZ-58 (**SFG**) with two intersecting 10-ring channels synthesised in the presence of 1-butyl-1-cyclooctylpyrrolidinium and Na$^+$ ions.[131] The introduction of B into T-sites of silicate frameworks was postulated to promote the formation of structures with more acute T-O-T angles compared with Al substitution and hence to give rise frequently to unprecedented building units.[130] However, it is still unclear why some new high-silica structures can only be prepared from borosilicate systems.

5.7.6.2 Gallium

Due to the close similarity in their chemical properties, substitution of Al by Ga in zeolite synthesis mixtures has long been considered to give microporous products whose topologies are usually the same as those for the aluminosilicate crystallisation[132] and as a matter of fact, one of the earliest reports on zeolite synthesis is on a gallosilicate analogue of the natural zeolite thomsonite (**THO**).[133] However, the discovery of two novel gallosilicate zeolites with **CGS** and **ETR** topologies,[134–137] neither of which has a counterpart among those of aluminosilicate zeolites, has demonstrated that the structure-directing effect of Ga can be substantially different from that of Al in certain low-silicate synthesis conditions. However, the presence of a small amount of Al in the synthesis mixture, together with appropriate levels of Ga, Na^+, K^+ and TEA^+ concentrations, appears to be essential for crystallising the latter extra-large-pore material (*i.e* ECR-34). Another example supporting this is TNU-7 (**EON**) prepared without the aid of any OSDA, a novel large-pore gallosilicate zeolite consisting of strictly alternating layers of the well-known **MOR** and **MAZ** structures linked *via* chains of 5-rings.[138,139] This topology was previously proposed as the model ECR-1A for aluminosilicate zeolite ECR-1, first synthesised in the presence of various OSDAs such as bis(2-hydroxyethyl)dimethylammonium and subsequently under wholly inorganic conditions.[140,141] Very recently, the crystal structure of direnzoite, a mineral form of TNU-7 and ECR-1, has been reported.[142]

It is also remarkable that the crystallisation of TNU-7 depends critically on the presence of Ga in the synthesis mixture, as well as on its content. As seen in Table 5.4,[138] variation of the Ga concentration results in the sequential formation of gallosilicate analcime (**ANA**), mazzite (**MAZ**), TNU-7 and mordenite as the Ga concentration decreases. However, substitution of Ga by Al results in the formation of mordenite at intermediate and low Al contents. Thus, TNU-7 can be regarded as a new example of the boundary phase strategy proposed by Vaughan for the discovery of novel designer zeolite structures,[137,140] and also a good example of the structure-direction exerted by Ga. More interestingly, extensive Rietveld analyses of the synchrotron diffraction data for various cation-exchanged forms of TNU-7 have revealed the existence of

Table 5.4 Representative products obtained using gallosilicate synthesis mixtures with different Ga contents.[a] Reprinted with permission from S.J. Warrender, P.A. Wright, W. Zhou, P. Lightfoot, M.A. Camblor, C.-H. Shin, D.J. Kim and S.B. Hong, *Chem. Mater.*, **17**, 1272. Copyright (2005) American Chemical Society

Si/Me ratio in the gel	Product[b,c]			
	M = Na		M = K	
	Me = Ga	Me = Al	Me = Ga	Me = Al
2.5	Ga-ANA (2.52)	Amorphous	Ga-MER[e]	Amorphous
3.0	Ga-ANA + Ga-MAZ	Amorphous		
3.0[d]	Ga-MAZ (3.03)			
4.0	Ga-MAZ + TNU-7			
5.0	TNU-7 (3.85)	Al-MOR (4.56)	Feldspar	Al-MER
6.0	Ga-MOR			
7.5	Ga-MOR			
10	Ga-MOR (5.89)	Al-MOR (9.61)	Feldspar	Amorphous
20	Ga-MOR + (kenyaite)	Al-MOR	Feldspar	Al-FER
40	Ga-MOR + kenyaite		Feldspar	Al-FER + (L)
∞	Kenyaite		Quartz	

[a] The oxide composition of the synthesis mixture is $2.0M_2O\ xMe_2O_3\ 10SiO_2\ 150H_2O$, where M is Na or K, Me is Ga or Al, and x is varied ($0 \leq x \leq 2.0$). Crystallisation was performed under rotation (60 rpm) at 150 °C for 7 days, unless otherwise stated.
[b] The phase appearing first is the major phase, and the product obtained in a trace amount is given in parentheses: ANA, analcime; MAZ, mazzite; MOR, mordenite; MER, merlinoite; FER, ferrierite; L, unknown but probably layered phase.
[c] The values given in parentheses are Si/Ga or Si/Al ratios of the product, as determined by elemental analysis.
[d] A small amount (2 wt% of silica in the synthesis mixture) of the calcined gallosilicate with Si/Ga = 2.60 or aluminosilicate mazzite with Si/Al = 3.17 prepared in the presence of tetramethylammonium ion was added as seeds prior to heating at 150 °C for 3 days.
[e] The material obtained after 4 days of heating at 150 °C.

structural chemical zoning in this gallosilicate zeolite, *i.e.* an alternation of Ga-rich **MAZ** and Ga-poor **MOR** layers in its framework, probably due to the local fluctuation of composition of the mother liquor caused by the crystallisation of the previous layer.[139]

Another important but somewhat unexpected finding achieved in the gallosilicate chemistry is that the spatial distribution of heteroatoms over the available T-sites in a given silicate framework can be controlled by systematically varying the synthesis conditions. Cho *et al.* have previously synthesised two different versions of a gallosilicate zeolite with the same topology (**NAT**) and chemical composition (Si/Ga ∼1.5) but with different T-atom distributions: one is a quite (but not completely) disordered tetragonal ($I\overline{4}2d$) member and the other a quite (but not completely) ordered orthorhombic ($Fdd2$) form.[59] Since these two materials were obtained at different crystallisation temperatures (100 *vs* 150 °C) but from the same gel composition (Si/Ga = 5.0) and crystallisation time (10 days), however, their difference in the ordering of Si and Ga atoms, related to the splitting of a crystallographically distinct T-site, could not be governed only by the extended Loewenstein's rule (*i.e.* avoidance of Ga-O-Ga bonds). Our further studies on this issue have demonstrated that at a given temperature the initial disordered crystalline material slowly transforms *in situ* into the ordered version, with no noticeable signs of phase segregation and changes in chemical composition, and hence materials with any intermediate degree of order can also be synthesised.[59] *Ab initio* calculations strongly suggest that two important driving forces for this transformation may be a preferential siting of Ga to occupy half of the high multiplicity sites in the **NAT** topology to avoid strain, and a thermodynamic tendency to decrease the concentration of violations to Loewestein's rule, which are initially formed probably due to entropic reasons. It is worthwhile to note here that the amount of zeolitic water in this series of **NAT**-type gallosilicate materials decreases with increasing the T-atom ordering, while the opposite holds for the thermal stability. Many important properties of zeolites, most notably their Brønsted acidity and hence catalytic performance, have been speculated to differ according to the distribution of heteroatoms over the available T-sites.[143,144] However, there are few examples where this has been clearly evidenced.

5.7.6.3   Germanium

Ge is directly below Si in the periodic table, and both SiO_2 and GeO_2 may crystallise as quartz, cristobalite and rutile phases. This tetravalent atom can adopt coordination numbers of 4–6 in contrast to Si that is normally tetrahedral in its oxides, therefore, it is not surprising that Ge substitution into several zeolite frameworks including **THO**, **FAU**, **LTA** and **PHI** topologies was successfully attempted five decades ago.[133] Until the late 1990s, however, no attempts to synthesise new microporous germanate frameworks in which all Ge atoms are tetrahedral had been successful.[145,146]

   In 1998, Li and Yaghi reported the synthesis of two pure-germanate zeolites denoted ASU-7 (**ASV**) and ASU-9 (**AST**) in the presence of F⁻ and pyridine, using dimethylamine and 1,4-diazabicyclo[2,2,2]octane (DABCO) as an OSDA, respectively.[147] After a while, Conradsson *et al.* synthesised FOS-5 (**BEC**), the first zeolite beta polymorph C, from a germanate gel with an oxide composition similar to that used for ASU-9 formation.[148] While all three materials contain D4R units, their average Ge-O-Ge angles ($\sim$130°) were determined to be considerably smaller than those (>140°) of typical tetrahedral SiO_2 frameworks. Since Ge has a more ionic character and a longer average T-O distance (1.73 vs 1.61 Å) than Si, this tetravalent atom has been speculated to significantly relax the geometric constraints in the D4R units and hence to stabilise the resulting crystal structures.[149]

Independent of the pure-germanate zeolite syntheses described above, on the other hand, it was in 1995–1997 that Camblor and Valencia found a gradual change in the crystallised product caused by increasing the amount of Ge in the synthesis mixture containing TEA^+ ion as an OSDA for the synthesis of pure-silica Beta zeolite in fluoride media.[150] A series of disordered materials with different proportions of polymorph C intergrown with conventional Beta zeolite was obtained at Ge/(Si+Ge) ratios ≤0.33 and termed ITQ-5.[151] More interestingly, a further increase of the Ge/(Si+Ge) ratio to 0.67 has finally enabled them to crystallise pure polymorph C of zeolite Beta and they hypothesised that Ge may have a profound effect on promoting the formation of an important structural subunit, $i.e$, the D4R or [4⁶] cage.[150] Prompted by this break-through, Corma $et\ al.$ have extensively performed zeolite syntheses through the addition of Ge as a real phase selectivity factor to silicate gel systems containing different types of OSDAs, both in the presence and absence of F^-, and discovered a number of new interesting zeolite structures.[83,112,113,115,152–156] Also, two groups in Mullhouse and Stockholm, respectively, have succeeded in synthesising another series of new germanosilicate zeolites via this approach.[114,157–159] The novel structures synthesised using Ge as a partial substitute of Si thus far, all of which contain D4R units, are listed in Table 5.5.

Among the germanosilicate materials recently reported, of particular interest are ITQ-33 and SU-32 (**STW**).[115,159] As seen in Figure 5.11, ITQ-33 has a three-dimensional pore system consisting of straight, circular 18-ring channels with a crystallographic diameter of 12.2 Å along the c-direction intersected by 10-ring channels.[115] Besides D4R units, the structure of this new extra-large-pore material comprises a previously unseen 14-hedral [$3^24^36^9$] cage with two 3-rings. It has long been hypothesised that 3-rings are essential not only to create more open frameworks, but also to impart higher stability on Ge-substituted zeolites.[160,161] ITQ-33 was prepared at a specific Si/Ge ratio of 2.0 using hexamethonium as an OSDA in either absence or presence of F^-, and the addition of a small amount of Al to the synthesis mixture was required for

Table 5.5 Novel germanosilicate zeolites discovered

Zeolite	Year	IZA code	Pore structure	Si/Ge in the gel	F⁻	(Representative) OSDA	Ref.
ITQ-5	1997	BEC/*BEA	3D, 12 × 12 × 12	0.5–5	Yes	Tetraethylammonium	[150,151]
ITQ-17	2001	BEC	3D, 12 × 12 × 12	0.5–30	Yes	Benzyldiazabicyclo[2.2.2]octane	[152]
ITQ-21	2002		3D, 12 × 12 × 12	2–10	Yes	N-Methylsparteinium	[153]
IM-7/ITQ-13	2003	ITH	3D, 10 × 10 × 9	10–∞	Yes	Hexamethonium	[157,237]
ITQ-22	2003	IWW	3D, 12 × 10 × 8	2	No	1,5-Bis(N-methylpyrollidinium)pentane	[112]
ITQ-24	2003	IWR	3D, 12 × 10 × 10	5–∞	No/yes	Hexamethonium	[113,154]
ITQ-29	2004	LTA	3D, 8 × 8 × 8	2–∞	Yes	4-Methyl-2,3,6,7-tetrahydro-1H,5H-pirido-[3.2.1-ij]quinolinium	[83]
IM-10	2004	UOZ	0D, 6	0–1	Yes	Hexamethonium	[114]
IM-12	2004	UTL	2D, 14 × 12	10/2	No	(6R,10S)-6,10-Dimethyl-5-azoniaspiro[4.5]-decane	[158]
ITQ-33	2006		3D, 18 × 10 × 10	2	Yes/no	Hexamethonium	[115]
ITQ-26	2008	IWS	3D, 12 × 12 × 12	4	Yes	1,3-Bis(triethylphosphoniummethyl) benzene	[155]
ITQ-34	2008		2D, 10 × 9	10	Yes	1,3-Bis(trimethylphosphonium)propane	[156]
SU-15	2008	SOF	3D, 12 × 9 × 9	0.8–2.1	Yes	Diisopropylamine	[159]
SU-32	2008	STW	3D, 10 × 8 × 8 (chiral)	0.8–2.1	Yes	Diisopropylamine	[159]

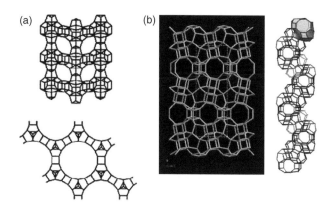

Figure 5.11 Framework structures of (a) ITQ-33 with straight 18-ring channels along the c-axis (bottom) intersected by a two-dimensional 10-ring channel system (top) and of (b) SU-32 viewed along the a-axis (left) and its right-handed helical 10-ring channels (right) for the enantiomorph with the space group $P6_1 22$. All the helical channels in SU-32 are built from the 18-hedral $[4^6 5^8 8^2 10^2]$ cavities connected through the 10-ring windows and intersected at different heights by straight 8-ring channels, in three different directions. (a) Reprinted with permission from A. Corma, M. Diaz-Cabanas, J.L. Jorda, C. Martinez and M. Moliner, *Nature*, **443**, 842. Copyright (2006) Macmillan Publishers Ltd. (b) Reprinted from L. Tang, L. Shi, C. Bonneau, J. Sun, H. Yue, A. Ojuva, B.-L. Lee, M. Kritikos, R.G. Bell, Z. Bacsik, J. Mink and X. Zou, *Nat. Mater.*, **7**, 381. Copyright (2008) Macmillan Publishers Ltd

its crystallisation, like the case of ECR-34,[137] another 18-ring pore material. Further study is necessary to elucidate the role of hexamethonium in the formation of ITQ-33, whose 18-ring pore diameter is much smaller than the OSDA size, and whether this highly flexible OSDA ends up entrapped in a particular range of conformations inside the pore system. This synthesis, as well as that of ECR-34, shows that large OSDAs are not necessarily a requisite for forming extra-large-pores. Rather, the type of structural subunits, substantially influenced by inorganic synthesis factors, appears to be more responsible for the creation of such extra-large-pore zeolites.

In contrast, SU-32 can be synthesised at a wider Si/Ge ratio of 0.8–2.1 using a simpler OSDA, diisopropylamine, with F^- present. This new medium-pore material with intersecting 10- and 8-ring channels is very unique in that it has an intrinsically chiral zeolite structure and each crystal possesses a unique handedness, although both hands are crystallised (as separate crystals) in each synthesis batch. This contrasts with the case of zeolite Beta, where the chiral polymorph A intergrows with other polymorphs and, presumably, both enantiomorphs can be present in the same crystal. One of the enantiomorphs of SU-32, showing only right-handed helical channels, is shown in Figure 5.11. The helical channels in

SU-32 enantiomorphs are built only from the large $[4^6 5^8 8^2 10^2]$ cavity and are intersected at different heights by straight 8-ring channels, which are related by the 6_1 (for right-handed crystals) or 6_5 (for left-handed crystals) screw axes and thus run in three different directions. Under the synthesis conditions described in Tang et al.,[159] which lack any chiral auxiliary that may favour an enantioselective crystallisation, the left-handed and right-handed enantiomorphs are, very likely, equally probable and should, in principle, appear in 50 % proportion in each batch.

While both ITQ-33 and SU-32 were claimed to be thermally stable upon calcination,[115,159] these two zeolites containing expensive Ge cannot have sufficient hydrothermal stability to be used as catalysts and separation media in commercial processes. Thus, one natural progression following the discovery of a number of germanosilicate zeolites including ITQ-33 and SU-32 is to produce the corresponding aluminosilicate analogues with higher surface acidity and hydrothermal stability. More importantly, chirality control in SU-32 so as to produce only one-handed enantiomorphs is necessary before checking its potential usefulness in asymmetric chemistry, i.e. in processes in which its framework chirality is effectively transferred to the products (catalysis or chemical synthesis in the confined chiral space of zeolites) or effectively used to separate enantiomers.

5.7.6.4 Beryllium, Zinc and Lithium

There are also several heteroatoms other than Ge that have proved clearly to drive the synthesis of new zeolite structures with particular building units via their site-specific introduction into silicate frameworks. Lovdarite (LOV) is a natural and synthetic zeolite with a typical composition $[K_4 Na_{12}(H_2O)_{18}]$ $[Be_8 Si_{28} O_{72}]$ and contains a three-dimensional pore system with intersecting 9- and 8-ring channels constructed from the spiro-5 subunit with two fused 3-rings[116] as seen in Figure 5.12. Be has been shown to occupy corners of this rather uncommon building unit, because of its preferential siting into T-sites with narrower T-O-T angles which afford more easily smaller rings and hence new structures together with a high degree of T-atom ordering. It is interesting to note here that another beryllosilicate mineral nabesite (NAB) with two intersecting 9-ring channels possesses the spiro-6 subunit, as well as the spiro-5 subunit.[162] The ability of Be to yield zeolite structures containing 3-rings has been further confirmed by Cheetham et al. who synthesised two new zeolites OSB-1 ($[K_6(H_2O)_9]$ $[Be_3 Si_6 O_{18}]$ – OSO) and OSB-2

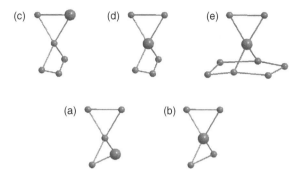

Figure 5.12 The (a,b) spiro-5, (c,d) spiro-6 and (e) spiro-3,5 subunits found in 3-ring-containing silicate-based zeolites. Small circles indicate Si atoms, and large circles represent heteroatoms that have a preference to occupy a position within 3-rings: Be (a–d), Zn (a,c,d), or Li (b,d) atoms. Note that Li is located at the spiro position of these building units, while Zn atoms occupy a peripheral vertex of their 3-rings. By contrast, Be has been found at central (**NAB**) and peripheral (**LOV**) positions of spiro-5 units and at central (**LOV**) or 3-ring peripheral (**LOV, NAB** and **OBW**) positions of spiro-6 units. In **OBW** and **OSO**, 3-rings belong to several spiro-5 units simultaneously and, thus, central and peripheral positions of spiro-5 units are indistinguishable

($[K_{44}(H_2O)_{96}]$ $[Be_{22}Si_{54}O_{150}]$ - **OBW**) using potassium beryllosilicate gels with Si/Be ≤2.6 under wholly inorganic conditions.[163] OSB-1 with intersecting 14- and 8-ring channels is the first zeolite built from 3-rings only. It is also remarkable that a chiral double helix chain composed of 3-rings forms the walls of straight 14-ring channels in this extra-large-pore material. In contrast, 18 out of 19 T-atoms in the asymmetric unit cell of the cage-based, medium-pore zeolite OSB-2 are in the 3-ring position of spiro-5 and spiro-6 subunits. The remaining one T-atom is located in the only one 4-ring of the **OBW** structure in which Be substitution is not energetically favourable. Thus, the Si/B ratio (2.45) of OSB-2 is higher than that (2.0) of OSB-1.

Owing to the existence of 3-rings in several dense zincosilicate phases and its nontoxic nature, Davis and co-workers reasoned that Zn could be an optimum substitute for Be in the formation of microporous structures containing 3-rings. With this in mind, indeed, they were able to synthesise a series of zincosilicate zeolites in the presence of different types of alkali cations alone or together with some OSDA, where new 3-ring-containing materials VPI-7 ($[Na_{26}H_6(H_2O)_{44}]$ $[Zn_{16}Si_{56}O_{144}]$ – **VSV**) and VPI-9 ($[Rb_{44}K_4(H_2O)_{48}]$ $[Zn_{24}Si_{96}O_{240}]$ – **VNI**) are included.[164] A high degree of Si-Zn order in these zincosilicates, as well as a general constancy of their Si/Zn ratios, was found by ^{29}Si MAS NMR spectroscopy, and this has been rationalised by a preference of Zn to occupy T-sites with more acute T-O-Si angles than that shown by Si.[165] Structural studies have further corroborated the high-degree of T-ordering in VPI-7[166] and VPI-9.[167]

The introduction of the monovalent Li atom instead of Be or Zn as a tetrahedral lattice substitute into zeolite frameworks pioneered by Gies and co-workers is another breakthrough in this area. The idea was again based on similarities in their geometrical flexibility and ratio of nonbonded radius to T-O bond length, confirming *a priori* design for zeolites with predetermined structural features. RUB-23 ($|Cs_{10}(Li,H)_{14}(H_2O)_{12}|$ $[Li_8Si_{40}O_{96}]$) and RUB-29 ($[Cs_{14}Li_{24}(H_2O)_{14}]$ $[Li_{18}Si_{72}O_{172}]$) are two successful examples of new microporous lithosilicates containing 3-rings.[168,169] Unlike Be and Zn, however, Li was found to occupy the centre of spiro-5 subunits in these materials. Also, the structure of RUB-29 contains a new spiro-3,5 building unit (Figure 5.12). Since these lithosilicate zeolites have a high ion density resulting from the need for three countercations per Li in the framework as compared with the traditional 1:1 ratio of extraframework cations to Al in aluminosilicate zeolites, they can be useful as an effective solid electrolyte, especially for direct methanol fuel cells (DMFCs) and hydrogen proton exchange membrane fuel cells (PEMFCs) to minimise methanol crossover and loss of hydration and proton conductivity at temperatures above 100 °C, respectively. In this regard, it is remarkable that the partially Na^+ ion-exchanged, dehydrated RUB-29 ($|Cs_6Na_{10}Li_{22}|$ $[Li_{18}Si_{72}O_{172}]$) material exhibits an ionic conductivity of 7×10^{-3} S cm^{-1} at 600 °C,[170] about two orders of magnitude higher than the values ($2–6 \times 10^{-5}$ S cm^{-1}) for the parent material, making it promising for use in high-temperature PEMFCs. With the exception of the aluminosilicate 12-ring material ZSM-18 prepared using tris(2-trimethylamonioethyl)amine as an OSDA,[171] however, none of the 3-ring-containing silicate-based structures known to date shows good thermal stability when converted to the proton form, limiting their potential as new shape-selective catalysts or separation media. Finally, we should note that one natural zeolite containing framework Mg atoms named Mg-BCTT ($[K_{4.56}]$ $[Mg_{2.28}Si_{5.72}O_{16}]$ – BCT) is already known;[172] and a number of Mg-substituted aluminophosphate molecular sieves with different framework topologies have been synthesised as well.[116] Provided that the kinetic limitations due to the low solubility of Mg in alkaline or fluoride solutions is overcome, therefore, we anticipate new zeolite chemistry through the selective substitution of the divalent Mg atom into silicate frameworks.

5.7.7 Zeolite Synthesis from Nonaqueous Solvents

One of the important roles of solvents in zeolite synthesis is to ensure efficient transport and mobility of starting reagents (OSDA, silica, alumina, *etc.*) in synthesis mixtures, but their interactions with any of individual reagents need to be intermediate, helping solubilise both inorganic and organic components, and hence affording a high supersaturation required for zeolite crystallisation. As stated above, the synthesis of high- and pure-silica sodalites from ethylene glycol is arguably the first example where nonaqueous solvents are used in zeolite synthesis.[88] In contrast to the extensive studies on the hydrothermal synthesis of zeolites, however, the body of zeolites being prepared from nonaqueous solvents is still relatively small. The other silicate-based structures that can also be synthesised in essentially nonaqueous systems include ZSM-5, ZSM-35, ZSM-39 (**MTN**), octadecasil (**AST**), P1, Beta and ZSM-48.[173–176] Although small quantities of water from the starting materials are involved in these synthesis and may be of importance, they are certainly at a 'reactive' rather than a 'solvent' level. In addition to ethylene glycol, to date, a number of nonaqueous solvents such as propanol, butanol, hexanol, glycerol, ammonia, sulfonate and pyridine have been used in zeolite synthesis.[173–176]

An interesting feature of the nonaqueous synthesis of zeolites compared with their hydrothermal synthesis is that pure-silica analogues of some known zeolite structures, for instance, of sodalite,[88] ferrierite,[175] and P1,[176] whose aqueous synthesis had been unsuccessful until that date, could be easily obtained in other solvents. It is also remarkable that zeolite Beta can be synthesised as the essentially Al-free form in the presence of TEA^+ and F^- ions in liquid ammonia, despite a high Al content ($SiO_2/Al_2O_3 = 5$) in the starting synthesis mixture, which has been attributed to the very high solubility of aluminium nitride used as an Al source in this nonaqueous solvent system.[176]

On the other hand, many nonaqueous solvents have higher viscosity than water, giving them a lower mass transfer rate by convection, which is beneficial for the formation of large crystals that may be of importance in the development of new optical and electronic materials. Kuperman *et al.* were successful in growing millimetric single crystals of alumino- and pure-silica ferrierite, ZSM-5 and ZSM-39 zeolites with a well defined habit through the combined use of pyridine and/or alkylamine as a solvent and HF as a mineraliser.[175] It has been recognised that the presence of nonaqueous solvents during the crystallisation process of zeolites can modify the distribution and structure of (alumino)silicate species in a significantly different manner than those in aqueous media,[90] due to differences in the physicochemical properties of non-aqueous solvents and water. To date, however, no new silicate-based zeolite structures have been produced as a direct consequence of their use.

Recently, Morris and co-workers have reported a new synthetic strategy in which ionic liquids and eutectic mixtures are used as both solvents and space fillers or OSDAs, leading to a range of novel AlPO$_4$-based molecular sieves.[177,178] It is argued that this strategy removes the competition between OSDA–framework interaction and solvent–framework interaction, although it has thus far resulted in the formation of only one dense silicate phase (SIZ-12).[178] To apply the ionothermal technique as an alternative tool for the discovery of new silicate zeolites, the difficulty in controlling the solubility of silicate species in such organic solvent systems would be the first task to be settled.

5.7.8 The Fluoride Route to Zeolites

Hydroxide is used in zeolite synthesis as a mineraliser, as it helps solubilise and mobilise silica and alumina (and other T-atom sources) affording a high supersaturation and catalysing the breaking and formation of Si-O-T bridges, thus allowing the reorganisation of amorphous gels and other solids to produce zeolites. As far as we know, no other mineraliser (in addition to water itself) acts in the natural crystallisation of zeolite minerals. However, by the end of the 1970s, in another major breakthrough in zeolite synthesis, Flanigen and Patton substituted hydroxide by fluoride to produce an all-silica version of zeolite ZSM-5.[179] The synthesis made use of the classical OSDA for **MFI** zeolites, TPA$^+$, which they also used successfully to produce the same pure-silica zeolite (silicalite) by the hydroxide route at high pH.[180] In the fluoride route, the synthesis proceeded at much lower alkalinities than traditional zeolite syntheses, leaving a pH close to neutral (7.4–10). Calcination of the as-made material (containing TPA$^+$ and F$^-$) afforded a new microporous crystalline polymorph of silica, which was recognised as highly hydrophobic and devoid of any ion-exchange properties.

From that time, the synthesis of zeolites by the so-called 'fluoride route' has developed to a large extent, with great contributions by Guth et al.[181–183] It has been successfully applied to pure-silica, aluminosilicate and heteroatom-substituted silicate zeolites as well as to aluminophosphate and heteroatom-substituted aluminophosphate zeolites, and a very interesting review on the subject has been published.[184] The route has been found particularly well suited for the synthesis of pure-silica zeolites,[182,185] and it has allowed the synthesis of several new zeolitic polymorphs of silica.[186]

Dwelling on the geoinspired interpretation of zeolite synthesis, the hydrothermal fluoride route to pure silica zeolites is a nice example of the concept: substitution of the natural mineraliser by fluoride, together with the use of organic, instead of inorganic, cations, with elimination of Al and framework constituents other than SiO_2, has allowed not only the discovery of new framework zeolite types, but also a dramatic change in some of the most characteristic properties of the obtained zeolite. As opposed to natural zeolites, pure-silica zeolites obtained by the fluoride route are strictly hydrophobic and completely devoid of ion-exchange properties. Thus, pure-silica zeolites have 'lost' the property for which zeolites received their name ('boiling stones' as they 'release bubbles of gas with fire',[187] due to their ability to reversibly adsorb large quantities of water) and also the properties that allowed most of their industrial applications (ion exchange and catalysis). Among the crystalline polymorphs of silica known in nature there is only one rare mineral[188] with a zeolitic structure, melanophlogite (MEP), a material with no pores but with a low FD due to the presence of cavities in its structure. It is interesting to note that albeit for some high- or pure-silica synthetic zeolites (such as NU-87, Beta and ZSM-5) natural analogues have been discovered in minute quantities (gottardiite,[189] tschernichite[190] and mutinaite,[191] respectively); these minerals have always a relatively high Al content (Si/Al ratios of 6.3, 2.7–3.9 and 7.6, respectively).

The chemistry of the fluoride route to zeolites, particularly pure-silica zeolites, is indeed interesting for several reasons. First, in addition to its role as a mineraliser during zeolite synthesis, which allows the crystallisation at pH close to neutral, fluoride ends up occluded into the zeolite, affording charge-balance of the OSDA. Although it was initially believed that fluoride would reside in the main pores of zeolites in the form of a complex or ion pair with the cationic OSDA,[192] the structural characterisation of pure-silica octadecasil located F^- in a small [4^6] cube-like unit within the framework, thus far apart from the cationic charge.[193] A prior work on the crystal structure of the aluminophosphate TPAF-AlPO$_4$-5 (AFI) also located fluoride within interstices of the aluminophosphate wall, far from the cationic OSDA occluded in the channel.[183] Subsequent works[194–211] have shown that this is the general case in silicate zeolites prepared by the fluoride route: fluoride anions are typically located within small cages of the silica framework and around 6 Å far from the atom that formally bears the cationic charge.[186] There is, however, at least one known exception: in pure-silica SSZ-55 (ATS) fluoride occupies two different positions, one of which (F2) may in fact be considered inside a cage ($[4^2 6^4]$), while the other one is close to a small fold in the wall but certainly within the main channel.[212] Even in this case, fluoride does not form an ion pair or complex with the cation, but rather interacts with the silica itself (see below).

Secondly, fluoride in pure silica zeolites is frequently, but not always, covalently bonded to a framework silicon atom that is pentacoordinated, yielding a $[SiO_{4/2}F]^-$ unit in the otherwise tetrahedral silica framework. This was first discovered by single-crystal diffraction in nonasil (NON) synthesised using an organometallic cation, $[Co^{III}(\eta^5\text{-}C_5H_5)_2]^+$, as an SDA.[195] This was the first zeolite synthesised with the aid of an organometallic cation[213] and, in the fluoride-mediated synthesis, its structure contains F^- anions inside a $[4^1 5^8]$ cage and covalently bonded to one of the Si atoms of the cage, which displays a nearly ideal trigonal-bipyramidal geometry. Later on, an investigation by means of ^{29}Si MAS NMR spectroscopy of this material and of as-made silicalite demonstrated that pentacoordinated $[SiO_{4/2}F]^-$ exists in both materials and can be conveniently studied by solid-state NMR techniques.[214] A weak but relatively sharp resonance at a chemical shift around -145 ppm in the room temperature ^{29}Si MAS NMR spectrum of nonasil, whose intensity is selectively enhanced by polarisation transfer from ^{19}F in the ^{29}Si{^{19}F} CPMAS NMR spectrum, was assigned to $[SiO_{4/2}F]^-$ units. For silicalite, a broad resonance centred around -125 ppm in the room temperature spectrum, also enhanced in the ^{29}Si{^{19}F} CPMAS NMR experiment, was substituted by two sharp resonances at -144.1 and -147.0 ppm when the spectrum was recorded at -133 °C. The broad line at -125 ppm was interpreted as an averaged signal due to a dynamic exchange between four- and five-coordinated Si, implying F^- hopping at room temperature to bond different Si atoms. At low temperature F^- hopping is frozen, yielding stable $[SiO_{4/2}F]^-$ units responsible for the sharp lines at -144 and -147 ppm (the presence of two lines is likely due to J coupling).[215] Soon after this report, another study demonstrated by single-crystal microdiffraction that pure silica as-made SSZ-23 (STT) zeolite also contains $[SiO_{4/2}F]^-$ units and ^{29}Si MAS NMR indicated also the existence of a dynamic exchange between four- and five-coordinated Si.[198] Then, a solid-state NMR investigation revealed the existence of pentacoordinated $[SiO_{4/2}F]^-$ units in several as-made pure silica zeolites, either with [silicalite, SSZ-23 and ITQ-4 (IFR)] or without [ZSM-12,

Beta and ITQ-3 (ITE)] fluoride hopping.[215] Both situations can be characterised by ^{29}Si NMR spectroscopy, which reveals either a broad resonance in the $-120/-140$ ppm region or sharper lines around $-140/-150$ ppm, respectively. Fluoride motion was frozen at low temperature (-143 to -133 °C). The number of pure silica zeolites that have been shown to possess five-coordinated silicon has increased since that time: FER,[200,206] STF,[203,205,210] CHA,[206] SFF,[208] EUO,[209] SSZ-73 (SAS)[211] and ATS.[212]

We have to stress, however, that with regard to Si–F interactions in zeolites, in addition to the ('dynamic' or 'fixed') pentacoordinated $[SiO_{4/2}F]^-$ units described above, there is another very common third situation, characterised by the lack of direct Si-F bonding that could be evidenced by experimental techniques, even at low temperature. As far as we know this only occurs when fluoride anions are occluded inside [4^6] silica cages, which is the case in octadecasil,[184,196] ITQ-7 (ISV),[199] ITQ-12 (ITW),[207] and in one of the two types of occluded fluoride in ITQ-13 and IM-7 (ITH).[111,157] We speculate that the available space inside the [4^6] silica cage is too small to allow F$^-$ to approach any of the Si atoms in the vertices, and F$^-$ is constrained in or near the centre of the cage with close contacts with the O atoms in the cage edges. For instance, the distance between O atoms in opposite edges of the [4^6] cage in AST is 5.38 Å, which taking the van der Waals radius of O as 1.35 Å leaves 2.68 Å, only very slightly larger than twice the van der Waals radius of fluoride (1.33 Å). The Si-F distance in AST determined by crystallography[193] and NMR techniques[216] agree well: 2.62 and 2.69 Å, respectively. The ^{19}F NMR chemical shift at around -37 to -40 ppm is a fingerprint of F$^-$ inside [4^6] cages and was indeed used to detect for the first time the real existence of polymorphs different from A and B in zeolites of the Beta family of intergrown materials.[217] Encapsulation in larger cages deshields fluoride to a lesser extent and the ^{19}F chemical shift is typically in the -50 to -80 ppm range. Fluoride in SSZ-55 is even less deshielded, with a chemical shift around -108 ppm.

A third interesting aspect of the fluoride route is that the obtained zeolite is devoid of connectivity defects in its network of siloxane bridges, *i.e.* very few or no Si-O$^-$ or SiOH groups exist.[182,218] This is in sharp contrast to the most common situation when pure-silica or very high-silica zeolites are synthesised at high pH without fluoride anions, since in that case a large concentration of defects, at least four times larger than needed to compensate for cationic charges, are typically observed, forming nests of Si-OH groups hydrogen bonded to SiO$^-$.[219] It may be argued that the lack of defects in F-zeolites is due to a combination of at least two factors: first, while fluoride occlusion affords charge balance of cationic OSDA in the fluoride route, the hydroxide route requires the existence of at least one Si-O$^-$ per cationic charge, because hydroxide itself is not occluded in high silica zeolites; secondly, the high pH of the hydroxide route favours highly charged silicate anions in solution, making difficult the complete condensation of silica.[185,186] The calcination required to remove the organic OSDA also removes fluoride,[182,185,186,193,196] despite theoretical calculations predicting the opposite in the case of F$^-$ occlusion in [4^6] cages.[220] The process leaves a pure microporous SiO$_2$ framework that has been shown by water adsorption experiments to be strictly hydrophobic.[221] When heteroatoms such as Al or Ti are introduced into silica zeolites by the fluoride route, the hydrophobicity of the resulting material lessens to some extent, but they are still more hydrophobic than materials with the same composition prepared by the hydroxide route, a fact that has been shown to be of importance when the zeolites are used in catalytic reactions involving reactants or products of varying polarity.[222–225]

5.7.9 Structure-Direction Issues in the Fluoride Route to Pure-Silica Zeolites

Finally, another very interesting feature of the fluoride route is that it may be argued that the fluoride anion displays significant structure-directing effects.[226] Supporting this, there are several new zeolite structures that, so far, cannot be made in the absence of fluoride, at least as pure-silica materials, particularly structures containing stressed [4^6] cages. Even 4-rings appear to be somehow stressed in pure silica materials, as several studies indicate substitution of some of the Si corners in 4-rings (specially at T-positions shared by several 4-rings)[19] and [4^6] cages[227] by larger and/or more electropositive atoms is energetically favoured, because the increased ionicity should result in an enhanced flexibility. While 4-rings and [4^6] cages are frequent in low-silica aluminosilicates, aluminophosphates, galloaluminophosphates and (silico)germanates, pure-silica zeolites with [4^6] cages need to be stabilised by fluoride. The inclusion of fluoride increases the ionicity of the framework, enhancing its flexibility, as recently concluded from B3LYP periodic calculations in zeolite ITQ-12,[85] a system in which a strong cooperative structure-direction action of fluoride anions and 1,3,4-trimethylimidazolium cations has been claimed.[228]

Considering all the pure silica materials currently known, it appears that fluoride has a strong tendency for occlusion within small cages in pure-silica zeolites (with **ATS** as an exception), and generally these small cages contain at least a 4-ring (with **FER** as an exception), with fluoride located close to this 4-ring (with **STF** as an exception) and interacting with some of the Si atoms in that 4-ring. What is clear is that fluoride has allowed the synthesis of pure-silica materials with $[4^6]$ cages occluding fluoride, and that these structures have so far defeated synthesis for pure-silica systems by the hydroxide route.

However, there are at least two additional important structure-direction factors interplaying with that of fluoride in the synthesis of pure-silica zeolites by the fluoride route: the OSDA and the degree of dilution of the synthesis gel. Villaescusa and Camblor have shown that the general concepts outlined above regarding structure-direction by OSDAs can be successfully used in the search for new pure-silica materials,[186] as shown for instance by the synthesis strategies that finally lead to the three-dimensional large pore zeolite ITQ-7.[229] Nonetheless, we would like to point out at least one peculiarity, in this respect, of the fluoride route: small cations may show by this route a rather specific structure-direction effect (like 1,3,5-trimethylimidazolium, which has a $(C+N)/N^+$ ratio of 8 but shows a rather large specificity towards zeolite ITQ-12 in fluoride aqueous medium).[228]

Camblor *et al.* have also shown that the degree of dilution of the synthesis mixture is of extraordinary importance, with a much limited number of phases crystallising at 'standard' degrees of concentration.[185,186] Very frequently, syntheses that at, say, $H_2O/SiO_2 > 30$, produce only dense phases, yield instead microporous polymorphs of silica at higher concentrations. Frequently, several phases can be obtained under otherwise identical conditions by just varying the concentration. The observation that increasing the concentration tends to produce more and more open frameworks[185] (*i.e.* materials with lower FD) proved to be a very useful empirical trend that we wish to call *Villaescusa's rule*, to honour L.A. Villaescusa's efforts to understand it and apply it to the discovery of new phases.[230] The rule had a precedent in the above-mentioned observation by Gies and Marler that in the amine-mediated synthesis of clathrasils, an increase in concentration tends to yield materials with a larger density of cages.[74,75] Villaescusa's rule has been corroborated by other groups[209,231–234] but still lacks for a convincing rationale. However, it was also observed that the phase sequences that appear upon increasing the degree of dilution parallel the phase sequences occasionally observed when increasing the temperature or the crystallisation time (Figure 5.13), implying that a high concentration favours more metastable phases.[185] This may be due perhaps to an enhanced supersaturation of some of the reactants at low water contents, but since the concentration of reactants at the synthesis temperature is unknown this is at present a mere speculation.

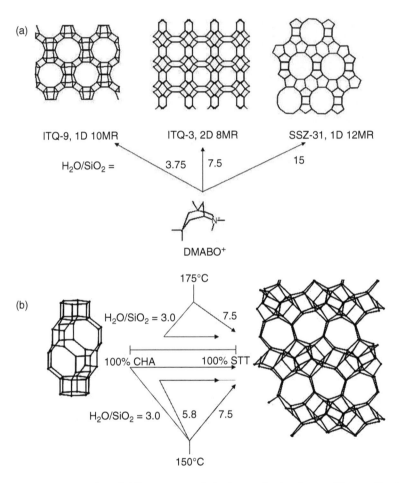

Figure 5.13 Examples of Villaescusa's rule in the synthesis of pure silica zeolites by the fluoride route. (a) At 150 °C using the rigid 1,3,3,6,6-pentamethyl-6-azoniabicyclo[3.2.1]octane cation as an OSDA three zeolites are obtained, their framework density (FD) decreasing by increasing the concentration of the synthesis mixtures. (b) Using *N,N,N*-trimethyl-1-adamantammonium as an OSDA the phases obtained by increasing the concentration are the same as those obtained sequentially by increasing time (direction of the arrows) or by increasing temperature, implying a metastability of the low FD phases obtained at a high concentration. Reproduced with permission from *Top. Catal.*, **9**, 59. Synthesis of all-silica molecular sieves in fluoride media. Copyright (1999) Springer

Be that as it may, Villaescusa's rule allowed Camblor's group to produce several new low density pure-silica materials, including ITQ-3,[235] ITQ-4,[236] ITQ-7,[199] ITQ-12[207] and ITQ-13,[237] as well as some heteroatom substituted materials like the catalytically active Ti-ITQ-7,[238] Ti-Beta,[225] Al-ITQ-4[239] and Al-ITQ-9 (STF),[240] and several known structures that could not be obtained in pure silica form before, always by working with medium to extremely concentrated conditions. This observation and the new structures synthesised prompted a renewed interest in this route.

5.7.10 Topotactic Condensation of Layered Silicates

To end our survey on the synthesis of silicate zeolites, we should mention that it is possible to obtain zeolites also by topotactic condensation of layered silicates. There are several examples in which the as-synthesised materials are layered organic-inorganic hybrids that can be condensed by a subsequent thermal treatment yielding the fully connected zeolites. While the most popular cases are those of MWW,[241] and FER,[242] many other examples of topotactic condensation to zeolites are known. The layered PLS-1, R-RUB-18, Nu-6(1), RUB-39 and Eu-19 materials condense into the pure-silica zeolites CDS-1 (CDO),[243] RUB-24 (RWR),[244] Nu-6(2) (NSI),[245] RUB-41 (RRO)[246] and Eu-20 (CAS-NSI intergrowth),[247] respectively. Thus, this route may be a consistent and convenient way to produce new pure-silica materials without necessarily using fluoride.

Of all these examples of topotactic condensation, the R-RUB-18 to RUB-24 case deserves special comment. The precursor R-RUB-18 is a layered material in which the alkylammonium cation R^+ has been intercalated between the layers of the as-synthesised RUB-18 solid. Although RUB-18[248] is synthesised in the presence of organics (triethanolamine and hexamethylenetetramine), it is a fully inorganic material with a composition [Na$_8$] [Si$_{32}$O$_{64}$(OH)$_8$]·32H$_2$O and, in fact, it has the same structure and composition of the 'ilerite' or 'octosilicate silicate that crystallises from fully inorganic compositions.[249] Na$^+$ in these layered materials can be exchanged by H$^+$, and one can imagine a route to a pure-silica zeolite without intervention of organic additives or fluoride compounds by dehydroxylation and topotactic condensation of a H$^+$-exchanged layered material of adequate structure. While our attempts using ilerite have so far been unsuccessful, possibly due to disorder introduced during the cation-exchange process, we think the basic concept may provide a cheap and safe route to pure-silica materials.

5.8 CONCLUDING REMARKS

The richness of synthetic silicate zeolite chemistry in terms of structures, chemical composition, textural and physicochemical properties and applications largely surpasses that of the natural analogues. All that richness is the result of, chiefly, 60 years of scientific and technical research in academia and in industry, following strategies that, at least in most cases and as a whole, we contend can qualify as *geoinspired*. Synthetic zeolite scientists are learning to some extent to master structure-direction effects of many different kinds in order to annually produce over half a dozen new zeolite topologies that can pass the severe standards of the IZA Structure Commission. We do not foresee any major scientific obstacle that can significantly slow down this pace. New structures, compositions and properties should help tune current zeolite applications and develop new ones.

Major challenges seem today even closer, including the synthesis of homochiral zeolites in the pure and separate form and their evaluation in asymmetric chemical and physicochemical processes, as well as crystalline zeolites with more open spaces, where larger molecules could be processed. Cheaper, faster, environmentally safer synthesis procedures should be sought, in particular with regard to the synthesis of hydrophobic pure-silica materials, where the use of HF should be avoided if a major industrial application is desired. In the coming years we may expect many beautiful new structures and exciting new chemistry within the zeolite field.

ACKNOWLEDGEMENTS

We are grateful for financial support from the Spanish CICYT (project MAT2006-033-56) and the Korea Science and Engineering Foundation through the National Research Lab program (R0A-2007-000-20050-0). Thanks are due to Professor G. Calzaferri and Dr P. Piccione for providing a high resolution image for Figure 5.5, and the original data for Figure 5.6, respectively.

REFERENCES

[1] J.V. Smith, *Chem. Rev.*, **88**, 149 (1988).

[2] J. Rouquérol, D. Avnir, C.W. Fairbridge, D.H. Everett, J.H. Haynes, N. Pernicone, J.D.F. Ramsay, K.S.W. Sing and K.K. Unger, *Pure Appl. Chem.*, **66**, 1739 (1994).

[3] J. Yu and R. Xu, *Acc. Chem. Res.*, **36**, 481 (2003).

[4] Ch. Baerlocher, L.B. McCusker and D.H. Olson, *Atlas of Zeolite Framework Types*, 6th Edn, Elsevier, Amsterdam, 2007.

[5] In fact, the listing of codes also includes seven 'interrupted framework' codes, which are distinguished by a dash before the three-letter code. These are not (4;2)-3D nets and, thus, are not truly 'zeolitic'.

[6] M.A. Camblor, M.-J. Díaz-Cabañas, P.A. Cox, I.J. Shannon, P.A. Wright and R.E. Morris, *Chem. Mater.*, **11**, 2878 (1999).

[7] X.J. Zhang, Y.H. Xing, J. Han, M.F. Ge and S.Y. Niu, *Z. Anorg. Allg. Chem.*, **633**, 2692 (2007).

[8] N. Zheng, X. Bu, B. Wang and P. Feng, *Science*, **298**, 2366 (2002).

[9] S. Correll, O. Oeckler, N. Stock and W. Schnick, *Angew. Chem. Int. Ed.*, **42**, 3549 (2003).

[10] A.J.D. Barnes, T.J. Prior and M.G. Francesconi, *Chem. Commun.*, 4638 (2007).

[11] G. Perego, R. Millini and G. Bellussi, in *Molecular Sieves, Science and Technology*, Vol. **1**, H.G. Karge and J. Weitkamp (Eds), Springer-Verlag, Berlin, 1998, p. 187.

[12] L.B. McCusker, F. Liebau and G. Engelhardt, *Pure Appl. Chem.*, **73**, 381 (2001).

[13] W. Loewenstein, *Am. Mineral.*, **39**, 92 (1954).

[14] K. Sahl, *Z. Kristallogr.*, **152**, 13 (1980).

[15] N. Binsted, S.E. Dann, M.J. Pack and M.T. Weller, *Acta Crystallogr., Sect. B*, **54**, 558 (1998).

[16] S.E. Dann, P.J. Mead and M.T. Weller, *Inorg. Chem.*, **35**, 1427 (1996).

[17] J.A. Tossell, *Am. Mineral.*, **78**, 911 (1993).

[18] J.F. Stebbins, P. Xhao, S.K. Lee and X. Cheng, *Am. Mineral.*, **84**, 1680 (1999) X. Cheng, P. Zhao, and J.F. Stebbins, *Am. Mineral.*, **85**, 1030 (2000).

[19] S.B. Hong, S.H. Lee, C.-H. Shin, A.J. Woo, L.J. Alvarez, C.M. Zicovich-Wilson and M.A. Camblor, *J. Am. Chem. Soc.*, **126**, 13742 (2004).

[20] F.J. Feher, K.J. Weller and J.W. Ziller, *J. Am. Chem. Soc.*, **114**, 9686 (1992).

[21] http://www.zeolites.eu/index_publications.html (accessed on 20 August 2010).

[22] A.M. El-Kamash, M.R. El-Naggar and M.I. El-Dessouky, *J. Hazard. Mater.*, **B136**, 310 (2006).

[23] H. Faghihian, M.G. Marageh and H. Kazemian, *Appl. Radiat. Isot.*, **50**, 655 (1999).

[24] E.M. Flanigen, in *Zeolites Science and Technology*, Vol. 80, F.R. Ribero, A.E. Rodrigues, L.D. Rollmann and C. Naccache (Eds), NATO ASI Series E, Vol. 80, Plenum, New York, 1984, p. 3.

[25] J.D. Sherman, *Proc. Natl. Acad. Sci. USA*, **96**, 3471 (1999).

[26] M. Stöcker, *Microporous Mesoporous Mater.*, **82**, 257 (2005).

[27] P.B. Weisz and V.J. Frilette, *J. Phys. Chem.*, **64**, 382 (1960).

[28] B. Smit and T.L.M. Maesen, *Nature*, **451**, 671 (2008).

[29] A. Pfenninger, in *Molecular Sieves, Science and Technology*, Vol. 2, H.G. Karge and J. Weitkamp (Eds), Springer-Verlag, Berlin, 1999, p. 162.

[30] F.A. Mumpton, *Proc. Natl. Acad. Sci. USA*, **96**, 3463 (1999).

[31] N. Gfeller and G. Calzaferri, *J. Phys. Chem. B*, **101**, 1396 (1997).

[32] S. Megelski and G. Calzaferri, *Adv. Funct. Mater.*, **11**, 277 (2001).

[33] A.S. Ichimura, J.L. Dye, M.A. Camblor and L.A. Villaescusa, *J. Am. Chem. Soc.*, **124**, 1170 (2002).

[34] D.P. Wernette, A.S. Ichimura, S.A. Urbin and J.L. Dye, *Chem. Mater.*, **15**, 1441 (2003).

[35] J.L. Dye, *Science*, **301**, 607 (2003).

[36] M. Comes, M.D. Marcos, R. Martínez-Máñez, M.C. Millán, J.V. Ros-Lis, F. Sancenón, J. Soto and L.A. Villaescusa, *Chem. Eur. J.*, **12**, 2162 (2006).

[37] N.B. Milestone and D.M. Bibby, *J. Chem. Tech. Biotechnol.*, **31**, 732 (1981).

[38] R. Jäger, A.M. Schneider, P. Behrens, B. Henkelmann, K.-W. Schramm and D. Lenoir, *Chem. Eur. J.*, **10**, 247 (2004).

[39] D.H. Olson, M.A. Camblor, L.A. Villaescusa and G.H. Kuehl, *Microporous Mesoporous Mater.*, **67**, 27 (2004).

[40] D.H. Olson, X.B. Yang and M.A. Camblor, *J. Phys. Chem. B*, **108**, 11044 (2004).

[41] P.S. Wheatley, A.R. Butler, M.S. Crane, S. Fox, B. Xiao, A.G. Rossi, I.L. Megson and R.E. Morris, *J. Am. Chem. Soc.*, **128**, 502 (2006).

[42] A.K. Boes, P.S. Wheatley, B. Xiao, I.L. Megson and R.E. Morris, *Chem. Commun.*, 6146 (2008).

[43] *VR Beverage Packaging*, **8**, (2001).

[44] *The Encyclopaedia Britannica*, 11th Edn, Vol. **XII**, Encyclopaedia Britannica, Inc., New York, 1911, p. 696.

[45] H. de St Claire Deville, *C. R. Acad. Sci.*, **54**, 324, (1862), cited by R.M. Barrer, *in Surface Organometallic Chemistry: Molecular Approaches to Surface Catalysis*, Vol. **231**, J.M. Basset, B.C. Gates, J.P. Candy, A. Choplin, M. Leconte, F. Quignard and C. Santini (Eds), NATO ASI Series C, Plenum, New York, 1988, p.221.

[46] L.V.C. Rees, *Biogr. Mems Fell. R. Soc.*, **44**, 37 (1998).

[47] R.M. Barrer, *J. Chem. Soc.*, 127 (1948).

[48] J.A. Rabo and M.W. Schoonover, *Appl. Catal. A*, **222**, 261 (2001).

[49] D.W. Breck, W.G. Eversole and R.M. Milton, *J. Am. Chem. Soc.*, **78**, 2338 (1956).

[50] J.F.V. Vincent, in *Encyclopaedia Britannica Yearbook*, Encyclopaedia Britannica, Inc., New York, 1995, p. 168.

[51] The term geoinspired was proposed by Professor E. Ruiz-Hitzky, Institute of Materials Science of Madrid, to one of the authors; see M.A. Camblor, *Macla*, **6**, 19, (2006).

[52] M.E. Davis and S.I. Zones, in *Synthesis of Porous Materials*, M.L. Occelli and H. Kessler (Eds), Marcel Dekker, New York, 1997, p. 1.

[53] D.W. Lewis, D.J. Willock, C.R.A. Catlow, J.M. Thomas and G.J. Hutchings, *Nature*, **382**, 604 (1996).

[54] M.D. Foster and M.M.J. Treacy, http://www.hypotheticalzeolites.net (accessed on 20 August 2010).

[55] P.M. Piccione, C. Laberty, S. Yang, M.A. Camblor, A. Navrotsky and M.E. Davis, *J. Phys. Chem. B*, **104**, 10001 (2000).

[56] P.M. Piccione, B.F. Woodfield, J. Boerio-Goates, A. Navrotsky and M.E. Davis, *J. Phys. Chem. B*, **105**, 6025 (2001).

[57] P.M. Piccione, S. Yang, A. Navrotsky and M.E. Davis, *J. Phys. Chem. B*, **106**, 3629 (2002).

[58] R.M. Barrer, *Hydrothermal Chemistry of Zeolites*, Academic Press, London, 1982.

[59] H.H. Cho, S.H. Kim, Y.G. Kim, Y.C. Kim, H. Koller, M.A. Camblor and S.B. Hong, *Chem. Mater.*, **12**, 2292 (2000).

[60] R.M. Barrer and P.J. Denny, *J. Chem. Soc.*, 971 (1961).

[61] G.T. Kerr and G.T. Kokotailo, *J. Am. Chem. Soc.*, **83**, 4675 (1961).

[62] G.T. Kerr, *Inorg. Chem.*, **5**, 1537 (1966).

[63] D.W. Breck, W.G. Eversole, R.M. Milton, T.B. Reed and T.L. Thomas, *J. Am. Chem. Soc.*, **78**, 5963 (1956).

[64] R.L. Wadlinger, G.T. Kerr and E.J. Rosinski, US Patent 3,308,069, (1967).

[65] An asterisk preceding a Framework Type Code denotes a framework that has not been obtained pure but only in disordered intergrowths with other frameworks.

[66] J.M. Newsam, M.M.J. Treacy, W.T. Koetsier and C.B. De Gruyter, *Proc. R. Soc. London, Ser. A*, **420**, 375 (1988).

[67] R.J. Argauer and G.R. Landolt, US Patent 3,702,886, (1972).

[68] D.H. Olson, G.T. Kokotailo, S.L. Lawton and W.M. Meier, *J. Phys. Chem.*, **85**, 2238 (1981).

[69] G.T. Kerr, *Science*, **140**, 1412 (1963).

[70] G.T. Kerr, *Inorg. Chem.*, **5**, 1539 (1966).

[71] A. Moini, K.D. Schmitt, E.W. Valyoscik and R.F. Polomski, *Zeolites*, **14**, 504 (1994).

[72] S.L. Lawton and W.J. Rohrbaugh, *Science*, **247**, 1319 (1990).

[73] C.M. Koelmel, Y.S. Li, C.M. Freeman, S.M. Levine, M.J. Hwang, J.R. Maple and J.M. Newsam, *J. Phys. Chem.*, **98**, 12911 (1994).

[74] H. Gies and B. Marler, *Zeolites*, **12**, 42 (1992).

[75] H. Gies, B. Marler and U. Werthmann, in *Molecular Sieves, Science and Technology*, Vol. 1, H.G. Karge and J. Weitkamp (Eds), Springer-Verlag, Berlin, 1998, p. 35.

[76] R.F. Lobo, S.I. Zones and M.E. Davis, *J. Inclus. Phenom. Mol.*, **21**, 47 (1995).

[77] Y. Kubota, M.M. Helmkamp, S.I. Zones and M.E. Davis, *Microporous Mater.*, **6**, 213 (1996).

[78] P. Wagner, Y. Nakagawa, G.S. Lee, M.E. Davis, S. Elomari, R.C. Medrud and S.I. Zones, *J. Am. Chem. Soc.*, **122**, 263 (2000).

[79] S.I. Zones, M.M. Olmstead and D.S. Santilli, *J. Am. Chem. Soc.*, **114**, 4195 (1992).

[80] S.L. Burkett and M.E. Davis, *J. Phys. Chem.*, **98**, 4647 (1994).

[81] A.V. Goretsky, L.W. Beck, S.I. Zones and M.E. Davis, *Microporous Mesoporous Mater.*, **28**, 387 (1999).

[82] S.L. Burkett and M.E. Davis, *Chem. Mater.*, **7**, 1453 (1995).

[83] A. Corma, F. Rey, J. Rius, M.J. Sabater and S. Valencia, *Nature*, **431**, 287 (2004).

[84] P. Behrens, G. van de Goor and C.C. Freyhardt, *Angew. Chem. Int. Ed.*, **34**, 2680 (1995).

[85] C.M. Zicovich-Wilson, M.L. San-Román, M.A. Camblor, F. Pascale and J.S. Durand-Niconoff, *J. Am. Chem. Soc.*, **129**, 11512 (2007).

[86] J.L. Schlenker, J.B. Higgins and E.W. Valyoscik, *Zeolites*, **10**, 293 (1990).

[87] S.-H. Lee, D.-K. Lee, C.-H. Shin, W.C. Paik, W.M. Lee and S.B. Hong, *J. Catal.*, **196**, 158 (2000).

[88] D.M. Bibby and M.P. Dale, *Nature*, **317**, 157 (1985).

[89] W. Hoelderich, L. Marosi, W.D. Mross and M. Schwarzmann, Eur Patent, 051, 741 (1982).

[90] S.B. Hong, M.A. Camblor and M.E. Davis, *J. Am. Chem. Soc.*, **119**, 761 (1997).

[91] D.Y. Han, A.J. Woo, I.-S. Nam and S.B. Hong, *J. Phys. Chem. B*, **106**, 6206 (2002).

[92] P.I. Nagy, W.J. Dunn, G. Alagona and C. Ghio, *J. Am. Chem. Soc.*, **113**, 6719 (1991).

[93] S.-H. Lee, C.-H. Shin, D.-K. Yang, S.-D. Ahn, I.-S. Nam and S.B. Hong, *Microporous Mesoporous Mater.*, **68**, 97 (2004).

[94] S.-H. Lee, C.-H. Shin, G.J. Choi, T.-J. Park, I.-S. Nam, B. Han and S.B. Hong, *Microporous Mesoporous Mater.*, **60**, 237 (2003).

[95] E.W. Valyoscik and L.D. Rollman, *Zeolites*, **5**, 123 (1985).

[96] N.A. Briscoe, D.W. Johnson, M.D. Shannon, G.T. Kokotailo and L.B. McCusker, *Zeolites*, **8**, 74 (1988).

[97] M.D. Shannon, J.L. Casci, P.A. Cox and S.J. Andrews, *Nature*, **353**, 417 (1991).

[98] A.W. Burton, R.J. Accardi, R.L. Lobo, M. Falcioni and M.W. Deem, *Chem. Mater.*, **12**, 2936 (2000).

[99] J. Plevert, Y. Kubota, T. Honda, T. Okubo and Y. Sugi, *Chem. Commun.*, 2363 (2000).

[100] M.J. Díaz-Cabañas, M.A. Camblor, Z. Liu, T. Ohsuna and O. Terasaki, *J. Mater. Chem.*, **12**, 249 (2002).

[101] S.B. Hong, E.G. Lear, P.A. Wright, W. Zhou, P.A. Cox, C.-H. Shin, J.-H. Park and I.-S. Nam, *J. Am. Chem. Soc.*, **126**, 5817 (2004).

[102] B. Han, S.-H. Lee, C.-H. Shin, P.A. Cox and S.B. Hong, *Chem. Mater.*, **17**, 477 (2005).

[103] S.B. Hong, *Catal. Surv. Asia*, **12**, 131 (2008).

[104] S.B. Hong, H.-K. Min, C.-H. Shin, P.A. Cox, S.J. Warrender and P.A.Wright, *J. Am. Chem. Soc.*, **129**, 10870 (2007).

[105] F. Gramm, Ch. Baerlocher, L.B. McCusker, S.J. Warrender, P.A. Wright, B. Han, S.B. Hong, Z. Liu, T. Ohsuna and O. Terasaki, *Nature*, **44**, 79 (2006).

[106] E. Benazzi, J.L. Guth and L. Rouleau, PCT WO 98/17581, (1998).

[107] Ch. Baerlocher, F. Gramm, L. Massüger, L.B. McCusker, Z. He, S. Hovmöller and X. Zou, *Science*, **315**, 1113 (2007).

[108] S.I. Zones, A.W. Burton and K. Wong, PCT WO 2007/079038 (2007).

[109] Ch. Baerlocher, D. Xie, L.B. McCusker, S.-J. Hwang, K. Wong, A.W. Burton and S.I. Zones, *Nat. Mater.*, **7**, 631 (2008).

[110] S.-H. Lee, D.-K. Lee, C.-H. Shin, Y.-K. Park, P.A. Wright, W.M. Lee and S.B. Hong, *J. Catal.*, **215**, 151 (2003).

[111] A. Corma, M. Puche, F. Rey, G. Sankar and S. Teat, *Angew. Chem. Int. Ed.*, **42**, 1156 (2003); see also *corrigenda* in *Angew. Chem. Int. Ed.*, **42**, 2696 (2003).

[112] A. Corma, F. Rey, S. Valencia, J.L. Jorda and J. Rius, *Nat. Mater.*, **2**, 493 (2003).

[113] R. Castaneda, A. Corma, V. Fornes, F. Rey and J. Rius, *J. Am. Chem. Soc.*, **125**, 7820 (2003).

[114] Y. Mathieu, J.-L. Pailaud, P. Caullet and N. Bats, *Microporous Mesoporous Mater.*, **75**, 13 (2004).

[115] A. Corma, M. Díaz-Cabañas, J.L. Jordá, C. Martínez and M. Moliner, *Nature*, **443**, 842 (2006).

[116] Ch. Baerlocher and L.B. McCusker, http://www.iza-structure.org/databases/.

[117] A. Corma, C. Corell and J. Perez-Pariente, *Zeolites*, **15**, 2 (1995).

[118] J. Shin and S.B. Hong, unpublished.

[119] M.A. Camblor, A. Corma and M.-J. Díaz-Cabañas, *J. Phys. Chem. B*, **102**, 44 (1998).

[120] S.I. Zones, S.-J. Hwang and M.E. Davis, *Chem. Eur. J.*, **7**, 1990 (2001).

[121] C.S. Gittleman, A.T. Bell and C.J. Radke, *Catal. Lett.*, **38**, 1 (1996).

[122] S.I. Zones and S.-J. Hwang, *Chem. Mater.*, **14**, 313 (2002).

[123] G.J. Lewis, M.A. Miller, J.G. Moscoso, B.A. Wilson, L.M. Knight and S.T. Wilson, *Stud. Surf. Sci. Catal.*, **154**, 364 (2004).

[124] C.S. Blackwell, R.W. Broach, M.G. Gatter, J.S. Holmgren, D.-Y. Jan, G.J. Lewis, B.J. Mezza, T.M. Mezza, M.A. Miller, J.G. Moscoso, R.L. Patton, L.M. Rohde,

M.W. Schoonover, W. Sinkler, B.A. Wilson and S.T. Wilson, *Angew. Chem. Int. Ed.*, **42**, 1737 (2003).

[125] M.A. Miller, G.J. Lewis, J.G. Moscoso, S.C. Koster, F. Modica, M.G. Gatter and L.T. Nemeth, *Stud. Surf. Sci. Catal.*, **170**, 487 (2007).

[126] M.A. Miller, J.G. Moscoso, S.C. Koster, M.G. Gatter and G.J. Lewis, *Stud. Surf. Sci. Catal.*, **170**, 347 (2007).

[127] M. O'Keeffe and B.G. Hyde, *Crystal Structures: I. Patterns and Symmetry*, Mineralogical Society of America, Washington, DC, 1996.

[128] M. Taramasso, G. Perego and B. Notari, in *Proceedings of the 5th International Zeolite Conference*, L.V.C. Rees (Ed.), Heyden, London, 1980, p. 40.

[129] R. Millini, G. Perego and G. Bellussi, *Top. Catal.*, **9**, 13 (1999).

[130] A.W. Burton, S.I. Zones and S. Elomari, *Curr. Opin. Colloid Interf. Sci.*, **10**, 211 (2005).

[131] A.W. Burton, S. Elomari, R.C. Medrud, I.Y. Chan, C.-Y. Chen, L.M. Bull and E.S. Vittoratos, *J. Am. Chem. Soc.*, **125**, 1633 (2003).

[132] R. Fricke, H. Kosslick, G. Lischke and M. Richter, *Chem. Rev.*, **100**, 2303 (2000).

[133] R.M. Barrer, J.W. Baynham, F.W. Bultitude and W.M. Meier, *J. Chem. Soc.*, 195 (1959).

[134] D.E.W. Vaughan and K.G. Strohmaier, U. Patent 5,096,686, (1992).

[135] S.B. Hong, S.H. Kim, Y.G. Kim, Y.C. Kim, P.A. Barrett and M.A. Camblor, *J. Mater. Chem.*, **9**, 2287 (1999).

[136] Y. Lee, S.J. Kim, G. Wu and J.B. Parise, Chem. Mater., **11**, 879 (1999).

[137] K.G. Strohmaier and D.E.W. Vaughan, *J. Am. Chem. Soc.*, **125**, 16035 (2003).

[138] S.J. Warrender, P.A. Wright, W. Zhou, P. Lightfoot, M.A. Camblor, C.-H. Shin, D.J. Kim and S.B. Hong, *Chem. Mater.*, **17**, 1272 (2005).

[139] B. Han, C.-H. Shin, S.J. Warrender, P. Lightfoot, P.A. Wright, M.A. Camblor and S.B. Hong, *Chem. Mater.*, **18**, 3023 (2006).

[140] M.E. Leonowicz and D.E.W. Vaughan, *Nature*, **329**, 819 (1987).

[141] J. Song, L. Dai, Y. Ji and F.S. Xiao, *Chem. Mater.*, **18**, 2775 (2006).

[142] E. Galli and A.F. Gualtieri, *Am. Mineral.*, **93**, 95 (2008).

[143] W. Souverijns, L. Rombouts, J.A. Martens and P.A. Jacobs, *Microporous Mater.*, **4**, 123 (1995).

[144] O.H. Han, C.-S. Kim and S.B. Hong, *Angew. Chem. Int. Ed.*, **41**, 469 (2002).

[145] J. Cheng and R. Xu, *J. Chem. Soc., Chem. Commun.*, 483 (1991).

[146] C. Cascales, E. Guitierrez-Puebla, M.A. Monge and C. Ruiz-Valero, *Angew. Chem. Int. Ed.*, **37**, 129 (1998).

[147] H. Li and O. Yaghi, *J. Am. Chem. Soc.*, **120**, 10569 (1998).

[148] T. Conradsson, M.S. Dadachov and X.D. Zou, *Microporous Mesoporous Mater.*, **41**, 183 (2000).

[149] M. O'Keeffe and O.M. Yaghi, *Chem. Eur. J.*, **5**, 2796 (1999).

[150] S. Valencia, PhD Thesis, Universidad Politecnia de Valencia, Valencia, 1997.

[151] M.A. Camblor, A. Corma and S. Valencia, Eur. Patent 1,043,274, (1998).

[152] A. Corma, M. Navarro, F. Rey, J. Rius and S. Valencia, *Angew. Chem. Int. Ed.*, **40**, 2277 (2001).

[153] A. Corma, M.J. Díaz-Cabañas, J. Martíney-Triguero, F. Rey, and J. Rius, *Nature*, **418**, 514 (2002).

[154] A. Cantin, A. Corma, M.J. Díaz-Cabañas, J.L. Jorda and M. Moliner, *J. Am. Chem. Soc.*, **128**, 4216 (2006).

[155] D.L. Dorset, K.G. Strohmaier, C.E. Kliewer, A. Corma, M.J. Díaz-Cabañas, F. Rey and C.J. Gilmore, *Chem. Mater.*, **20**, 5325 (2008).

[156] A. Corma, M.J. Díaz-Cabañas, J.L. Jorda, F. Rey, G. Sastre and K.G. Strohmaier, *J. Am. Chem. Soc.*, **130**, 16482 (2008).
[157] N. Bats, L. Rouleau, J.-L. Pailaud, P. Caullet, Y. Mathieu and S. Lacombe, *Stud. Surf. Sci. Catal.*, **154**, 283 (2004).
[158] J.-L. Pailaud, B. Harbuzaru, J. Patarin and N. Bats, *Science*, **304**, 990 (2004).
[159] L. Tang, L. Shi, C. Bonneau, J. Sun, H. Yue, A. Ojuva, B.-L. Lee, M. Kritikos, R.G. Bell, Z. Bacsik, J. Mink and X. Zou, *Nat. Mater.*, **7**, 381 (2008).
[160] G.O. Brunner and W.M. Meier, *Nature*, **337**, 146 (1989).
[161] X. Bu, P. Feng and G.D. Stucky, *J. Am. Chem. Soc.*, **120**, 11204 (1998).
[162] O.V. Petersen, G. Giester, F. Brandstätter and G. Niedermayr, *Can. Mineral.*, **40**, 173 (2002).
[163] A.K. Cheetham, H. Fjellvage, T.E. Gier, K.O. Kongshaug, K.P. Lillerud and G.D. Stucky, *Stud. Surf. Sci. Catal.*, **135**, 788 (2001).
[164] A. Corma and M.E. Davis, *ChemPhysChem*, **5**, 304 (2004).
[165] M.A. Camblor and M.E. Davis, *J. Phys. Chem.*, **98**, 13151 (1994).
[166] C. Rohrig, H. Gies and B. Marler, *Zeolites*, **14**, 498 (1994).
[167] L.B. McCusker, R.W. Grosse-Kunstleve, C. Baerlocher, M. Yoshikawa and M.E. Davis, *Microporous Mater.*, **6**, 295 (1996).
[168] S.H. Park, P. Daniels and H. Gies, *Microporous Mesoporous Mater.*, **37**, 129 (2000).
[169] S.H. Park, J.B. Parise, H. Gies, H. Liu, C.P. Grey and B.H. Toby, *J. Am. Chem. Soc.*, **122**, 11023 (2000).
[170] S.H. Park, A. Senyshyn and C. Paulmann, *J. Solid State Chem.*, **180**, 3366 (2007).
[171] K.D. Schmitt and G.J. Kennedy, *Zeolites*, **14**, 635 (1994).
[172] W.A. Dollase and C.R. Ross, *Am. Mineral.*, **78**, 627 (1993).
[173] Q. Huo, S. Feng and R. Xu, *J. Chem. Soc., Chem. Commun.*, 1486 (1988).
[174] W. Xu, J. Li, W. Li, H. Zhang and B. Liang, *Zeolites*, **9**, 468 (1989).
[175] A. Kuperman, S. Nadimi, S. Oliver, G.A. Ozin, J.M. Garces and M.M. Olken, *Nature*, **365**, 239 (1993).
[176] D.M. Miller and J.M. Garces, in *Proceedings of the 12th International Zeolite Conference*, M.M.J. Treacy, B.K. Marcus, M.E. Bisher and J.B. Higgins (Eds), Materials Research Society, Baltimore, 1998, p. 1535.
[177] E.R. Cooper, C.D. Andrews, P.S. Wheatley, P.B. Webb, P. Wormald and R.E. Morris, *Nature*, **430**, 1012 (2004).
[178] E.R. Parnham and R.E. Morris, *Acc. Chem. Res.*, **40**, 1005 (2007).
[179] E.M. Flanigen and R.L. Patton, U.S. Patent 4,073,865, (1978).
[180] E.M. Flanigen, J.M. Bennett, R.W. Grosse, J.P. Cohen, R.L. Patton, R.M. Kichner and J.V. Smith, *Nature*, **271**, 512 (1978).
[181] J.L. Guth, H. Kessler and R. Wey, *Stud. Surf. Sci. Catal.*, **28**, 121 (1986).
[182] J.L. Guth, H. Kessler, J.M. Higel, J.M. Lamblin, J. Patarin, A. Seive, J.M. Chezeau and R. Wey, *ACS Symp. Ser.*, **398**, 176 (1989).
[183] S. Qiu, W. Pang, H. Kessler and J.L. Guth, *Zeolites*, **9**, 440 (1989).
[184] P. Caullet, J.L. Paillaud, A. Simon-Masseron, M. Soulard and J. Patarin, *C.R. Chim.*, **8**, 245 (2005).
[185] M.A. Camblor, L.A. Villaescusa and M.J. Díaz-Cabañas, *Top. Catal.*, **9**, 59 (1999).
[186] L.A. Villaescusa and M.A. Camblor, *Recent Res. Devel. Chem.*, **1**, 93 (2003).
[187] I.G. Sumelius, in *Molecular Sieves, Vol. 1, Synthesis of Microporous Materials*, M.L. Occelli and H. Robson (Eds), van Nostrand Reinhold, New York, 1992, p. 1.
[188] B.J. Skinner and D.E. Applemann, *Am. Mineral.*, **48**, 854 (1963).

[189] A. Alberti, G. Vezzalini, E. Galli and S. Quartieri, *Eur. J. Mineral.*, **8**, 69 (1996).

[190] R.C. Boggs, D.G. Howard, J.V. Smith and G.L. Klein, *Am. Mineral.*, **78**, 822 (1993).

[191] E. Galli, G. Vezzalini, S. Quartieri, A. Alberti and M. Franzini, *Zeolites*, **19**, 318 (1997).

[192] G.D. Price, J.J. Pluth, J.V. Smith, J.M. Bennett and R.L. Patton, *J. Am. Chem. Soc.*, **104**, 5971 (1982).

[193] P. Caullet, J.L. Guth, J. Hazm, J.M. Lamblin and H. Gies, *Eur. J. Solid State Inorg. Chem.*, **28**, 345 (1991).

[194] B.F. Mentzen, M. Sacerdote-Peronnet, J.L. Guth and H. Kessler, *C.R. Acad. Sci.*, **313**, 177 (1991).

[195] G. Van de Goor, C.C. Freyhardt and P. Behrens, *Z. Anorg. Allg. Chem.*, **621**, 311 (1995).

[196] L.A. Villaescusa, P.A. Barrett and M.A. Camblor, *Chem. Mater.*, **10**, 3966 (1998).

[197] P.A. Barrett, M.A. Camblor, A. Corma, R.H. Jones and L.A. Villaescusa, *J. Phys. Chem. B*, **102**, 4147 (1998).

[198] M.A. Camblor, M.J. Díaz-Cabañas, J. Pérez-Pariente, S.J. Teat, W. Clegg, I.J. Shannon, P. Lightfoot, P.A. Wright and R.E. Morris, *Angew. Chem. Int. Ed.*, **37**, 2122 (1998).

[199] L.A. Villaescusa, P.A. Barrett and M.A. Camblor, *Angew. Chem. Int. Ed.*, **38**, 1997 (1999).

[200] M.P. Attfield, S.F. Weigel, F. Taulelle and A.K. Cheetham, *J. Mater. Chem.*, **10**, 2109 (2000).

[201] I. Bull, L.A. Villaescusa, S.J. Teat, M.A. Camblor, P.A. Wright, P. Lightfoot and R.E. Morris, *J. Am. Chem. Soc.*, **122**, 7128 (2000).

[202] C.A. Fyfe, D.H. Brouwer, A.R. Lewis and J.M. Chézeau, *J. Am. Chem. Soc.*, **123**, 6882 (2001).

[203] L.A. Villaescusa, P.S. Wheatley, I. Bull, P. Lightfoot and R.E. Morris, *J. Am. Chem. Soc.*, **123**, 8797 (2001).

[204] E. Aubert, F. Porcher, M. Souhassou, V. Petříček and C. Lecomte, *J. Phys. Chem. B*, **106**, 1110 (2002).

[205] C.A. Fyfe, D.H. Brouwer, A.R. Lewis, L.A. Villaescusa and R.E. Morris, *J. Am. Chem. Soc.*, **124**, 7770 (2002).

[206] L.A. Villaescusa, I. Bull, P.S. Wheatley, P. Lightfoot and R.E. Morris, *J. Mater. Chem.*, **13**, 1978 (2003).

[207] P.A. Barrett, T. Boix, M. Puche, D.H. Olson, E. Jordan, H. Koller and M.A. Camblor, *Chem. Commun.*, 2114 (2003).

[208] R.J. Darton, D.H. Brouwer, C.A. Fyfe, L.A. Villaescusa and R.E. Morris, *Chem. Mater.*, **16**, 600 (2004).

[209] M. Arranz, J. Pérez-Pariente, P.A. Wright, A.M.Z. Slawin, T. Blasco, L. Gómez-Hortigüela and F. Corà, *Chem. Mater.*, **17**, 4374 (2005).

[210] J.L. Paillaud, B. Harbuzaru and J. Patarin, *Microporous. Mesoporous. Mater.*, **105**, 89 (2007).

[211] D.S. Wragg, R. Morris, A.W. Burton, S.I. Zones, K. Ong and G. Lee, *Chem. Mater.*, **19**, 3924 (2007).

[212] A. Burton, R.J. Darton, M.E. Davis, S.J. Hwang, R.E. Morris, I. Ogino and S.I. Zones, *J. Phys. Chem. B*, **110**, 5273 (2006).

[213] K.J. Balkus Jr and S. Shepelev, *Microporous Mater.*, **1**, 383 (1993).

[214] H. Koller, A. Wölker, H. Eckert, C. Panz and P. Behrens, *Angew. Chem. Int. Ed.*, **36**, 2823 (1997).

[215] H. Koller, A. Wölker, L.A. Villaescusa, M.J. Díaz-Cabañas, S. Valencia and M.A. Camblor, *J. Am. Chem. Soc.*, **121**, 3368 (1999).

[216] C.A. Fyfe, A.R. Lewis, J.M. Chézeau and H. Grondey, *J. Am. Chem. Soc.*, **119**, 12210 (1997).
[217] M.A. Camblor, P.A. Barrett, M.J. Díaz-Cabañas, L.A. Villaescusa, M. Puche, T. Boix, E. Pérez and H. Koller, *Microporous Mesoporous Mater.*, **48**, 11 (2001).
[218] J.-M. Chézeau, L. Delmotte, J.L. Guth and M. Soulard, *Zeolites*, **9**, 78 (1989).
[219] H. Koller, R.F. Lobo, S.L. Burkett and M.E. Davis, *J. Phys. Chem.*, **99**, 12588 (1995).
[220] A.R. George and C.R.A. Catlow, *Zeolites*, **18**, 67 (1997).
[221] T. Blasco, M.A. Camblor, A. Corma, P. Esteve, J.M. Guil, A. Martínez, J.A. Perdigón-Melón and S. Valencia, *J. Phys. Chem. B*, **102**, 75 (1998).
[222] M.A. Camblor, A. Corma., P. Esteve, A. Martínez and S. Valencia:, *Chem. Commun.*, 795 (1997).
[223] M.A. Camblor, A. Corma, S. Iborra, S. Miquel, J. Primo and S. Valencia, *J. Catal.*, **172**, 76 (1997).
[224] M.A. Camblor, A. Corma. H. García, V. Semmer-Herlédan and S. Valencia, *J. Catal.*, **177**, 267 (1998).
[225] T. Blasco, M.A. Camblor, A. Corma, P. Esteve, A. Martínez, C. Prieto and S. Valencia, *Chem. Commun.*, 2367 (1996).
[226] J.L. Guth, H. Kessler, P. Caullet, J. Hazm, A. Merrouche and J. Patarin, in *Proceedings of 9th International Zeolite Conference*, R. von Ballmoos, J.B. Higgins and M.M.J. Treacy (Eds), Butterworth-Heinemann, Boston, 1993, p. 215.
[227] T. Blasco, A. Corma, M.J. Díaz-Cabañas, F. Rey, J.A. Vidal-Moya and C.M. Zicovich-Wilson, *J. Phys. Chem. B*, **106**, 2634 (2002).
[228] X.B. Yang, M.A. Camblor, Y. Lee, H.M. Liu and D.H. Olson, *J. Am. Chem. Soc.*, **126**, 10403 (2004).
[229] L.A. Villaescusa, I. Díaz, P.A. Barrett, S. Nair, J.M. Lloris-Cormano, R. Martínez-Mañez, M. Tsapatsis, Z. Liu, O. Terasaki and M.A. Camblor, *Chem. Mater.*, **19**, 1601 (2007).
[230] L.A. Villaescusa, PhD Thesis, Universidad Politécnica de Valencia, Valencia, 1999.
[231] S.I. Zones, R.J. Darton, R. Morris and S.J. Hwang, *J. Phys. Chem. B*, **109**, 652 (2005).
[232] S.I. Zones, A.W. Burton, G.S. Lee and M.M. Olmstead, *J. Am. Chem. Soc.*, **129**, 9066 (2007).
[233] A.W. Burton, G.S. Lee, S.I. Zones, *Microporous. Mesoporous. Mater.*, **90**, 129 (2006).
[234] S.I. Zones, S.J. Hwang, S. Elomari, I. Ogino, M.E. Davis, A.W. Burton, and C.R. Chimie, **8**, 267 (2005).
[235] M.A. Camblor, A. Corma, P. Lightfoot, L.A. Villaescusa and P.A. Wright, *Angew. Chem. Int. Ed.*, **36**, 2659 (1997).
[236] M.A. Camblor, A. Corma and L.A. Villaescusa, *Chem. Commun.*, 749 (1997).
[237] T. Boix, M. Puche, M.A. Camblor and A. Corma, US Patent 6,471,941, (2002).
[238] M.J. Díaz-Cabañas, L.A. Villaescusa and M.A. Camblor, *Chem. Commun.*, 761 (2000).
[239] L.A. Villaescusa, P.A. Barrett, M. Kalwei, H. Koller and M.A. Camblor, *Chem. Mater.*, **13**, 2332 (2001).
[240] L.A. Villaescusa, P.A. Barrett and M.A. Camblor, *Chem. Commun.*, 2329 (1998).
[241] M.E. Leonowicz, J.A. Lawton, S.L. Lawton and M.K. Rubin, *Science*, **264**, 1910 (1994).

[242] L. Schreyeck, P. Caullet, J.C. Mougenel, J.L. Guth and B. Marler, *Microporous Mater.*, **6**, 259 (1996).

[243] T. Ikeda, Y. Akiyama, Y. Oumi, A. Kawai and F. Mizukami, *Angew. Chem. Int. Ed.*, **43**, 4892 (2004).

[244] B. Marler, N. Ströter and H. Gies, *Microporous Mesoporous Mater.*, **83**, 201 (2005).

[245] S. Zanardi, A. Alberti, G. Cruciani, A. Corma, V. Fornés and M. Brunelli, *Angew. Chem. Int. Ed.*, **43**, 4933 (2004).

[246] Y.X. Wang, H. Gies, B. Marler and U. Müller, *Chem. Mater.*, **17**, 43 (2005).

[247] B. Marler, M.A. Camblor and H. Gies, *Microporous Mesoporous Mater.*, **90**, 87 (2006).

[248] S. Vortmann, J. Rius, S. Siegmann and H. Gies, *J. Phys. Chem. B*, **101**, 1292 (1997).

[249] U. Brenn, H. Ernst, D. Freude, R. Herrmann, R. Jähnig, H.G. Karge, J. Kärger, T. Konig, B. Mädler, U.-T. Pingel, D. Prochnow and W. Schwieger, *Microporous Mesoporous Mater.*, **40**, 43 (2000).

Index

nomenclature   70–1
properties and characterisation
    108–16
silica source   75–9
single crystals   71, 90, 94, 117
surfactant aggregation   72–5
synthetic routes   71–2, 83–108
template removal   79–83
thin films   99–105
Mesoporous transition metal oxide
    materials   153, 157–70
hard template method   155–7
of Group 4 elements   157, 170
of Group 5 elements   170–2
of Group 6 elements   172
of Group 7 elements   172–3
of Groups 8–11 elements   173–4
of lanthanide elements   174
single crystals   157, 172, 173, 174
soft template method   154–5
Metal cyanides
porous magnets   33, 37–40
thermal expansivities   52
Metal formates   37
Metal-ligand charge-transfer
    (MLCT)   49
Metalloligands   29
Metalloporphyrins   30
metal@MOF-5   30
Metal-organic frameworks (MOFs)
    1–3, 31–2, 54–6
as templates to prepare porous
    carbons   230–1
auxetic properties   54
Brønsted acid surface site
    generation   30–1
chirality   6–7, 26–7, 30
compared to zeolites   4, 9–10, 22,
    26, 28
compressibilities   52–4
design principles   3–11
electronic and optical properties   41–51
electron transfer   47–9
enantioselective adsorption   26–7
first generation materials   11
flexible frameworks   12–15, 25–6
framework interconversions   15–18
framework structures and
    properties   3–18
gas adsorption   22–6
guest adsorption and separation   21–7
heterogeneous catalysis   27–31
hydrogen storage   19–21

liquid phase adsorption   26–7
magnetic ordering   32–41
metal site incorporation   28–30
network topologies   5–6
photoluminescence   49–51
post-synthetic modification   8–11
rigid frameworks   11–12, 13, 22–5
second generation materials   11
spin crossover   42–7
storage and release   18–21
structural and mechanical
    properties   51–4
structural response to guest
    exchange   11–18
synthesis   4–8
thermal expansivities   51–2
third generation materials   12
terminology   1n.
Methane storage   20
MF DFT calculations, see Mean-field
    density functional theory
    calculations
Mg-Al layered double hydroxides   230
Mg-BCTT   306
Mg-Ta mixed oxides   171
Micromoulding in capillaries
    104–5, 122
Microporous carbon materials   218
biomass-derived   218–19
clays as hard template   229–31
MOFs as hard template   231
molecular sieves   218
zeolites as hard template   221–9
Microporous materials, IUPAC
    definition   147, 217
see also Microporous carbon
    materials; Microporous transition
    metal oxide materials
Microporous transition metal oxide
    materials   149–53
'Micro true liquid templating'   94
Microwave digestion   82–3
MIL-47   35
MIL-53   14–15
MIL-53(Cr)   35
MIL-101   7, 28, 31
MIL series   7
Mixed metal alkoxides   194
Mixed metal oxides   194–207
MLCT, see Metal-ligand charge-transfer
[$Mn^{II}_3(HCOO)_6$] framework   22–3
β-MnO_2 172–3 see also Manganese
    oxides

Printed and bound in the UK by
CPI Antony Rowe, Eastbourne